Plant Cell Culture

Plant Cell Culture

Essential Methods

Michael R. Davey and **Paul Anthony**

Plant and Crop Sciences Division
School of Biosciences
University of Nottingham
Sutton Bonington Campus
Loughborough, UK

WILEY-BLACKWELL
A John Wiley & Sons, Ltd., Publication

This edition first published 2010

Wiley-Blackwell is an imprint of John Wiley & Sons, formed by the merger of Wiley's global Scientific, Technical and Medical business with Blackwell Publishing.

Registered office: John Wiley & Sons Ltd, The Atrium, Southern Gate, Chichester, West Sussex, PO19 8SQ, UK

Other Editorial Offices:
9600 Garsington Road, Oxford, OX4 2DQ, UK
111 River Street, Hoboken, NJ 07030-5774, USA

For details of our global editorial offices, for customer services and for information about how to apply for permission to reuse the copyright material in this book please see our website at www.wiley.com/wiley-blackwell

Library of Congress Cataloguing-in-Publication Data

Davey, M. R. (Michael Raymond), 1944-
Plant cell culture : essential methods / Michael R. Davey and Paul Anthony.
p. cm.
ISBN 978-0-470-68648-5 (cloth)
1. Plant cell culture. 2. Plant tissue culture. I. Anthony, Paul. II. Title.
QK725.D38 2010
571.6′382 – dc22

2009051020

ISBN: 978-0-470-68648-5

A catalogue record for this book is available from the British Library.

Typeset in 10/12 Times by Laserwords Private Limited, Chennai, India

First impression–2010

Contents

Preface

More than a century has passed since the first attempts were made to culture isolated plant cells in the laboratory, the number of publications confirming the substantial progress achieved in this area of research, especially during the last four decades. In many ways, plant cell culture *per se* has been overshadowed by the recent, phenomenal progress achieved in recombinant DNA technology. Nevertheless, the ability to culture cells and tissues in the laboratory through to the regeneration of fertile plants provides an important base for several technologies. For example, the mass production of elite plants is exploited extensively in present-day commercial enterprises, while techniques such as the generation of haploid plants, *in vitro* fertilization, embryo rescue and somatic hybridization are available to assist the plant breeder in generating hybrid plants. Similarly, the transfer into plants of specific genes by transformation also provides an important underpin to well established techniques of plant breeding, emphasizing the requirement for close liaison between breeders and cell technologists. Many of the approaches associated with the culture of plant cells in the laboratory demand an experienced eye, particularly in the selection of cultures that are most likely to retain and express their totipotency. Consequently, cell culture is, in many respects, as much an art as a science. However, what is remarkable is the ability of individual cells to multiply and to differentiate into intact plants when given the correct environmental conditions in the laboratory. Although cell-to-plant systems have been described for many plants, including some of our most important crops, there are dicotyledons and, in particular, monocotyledons, that are still recalcitrant to regeneration under *in vitro* conditions. These remain a challenge to researchers involved in plant cell culture.

We have had to be selective in the topics that are included in this volume. Consequently, we have focused on aspects of micropropagation, pathways of plant regeneration, mutagenesis, cryopreservation, secondary products, and the technologies associated with hybrid plant production and genetic manipulation. The chapters each provide a general background to the specific areas with appropriate methodology. Whilst the protocols are presented with reference to specific examples, the procedures can be modified accordingly for new material. Our contributors have been asked to provide precise details, however seemingly trivial, of the methods presented, to focus in the 'Troubleshooting' sections on some of the common problems often encountered, and to give detailed advice for the avoidance of such difficulties.

In general, such information is not included in research papers in learned journals. We thank all of the contributors for their patience and understanding during the preparation and extensive editing of the manuscripts. We hope they have also benefited from the experience of providing the detailed protocols that are in routine use in their laboratories.

Michael R. Davey and **Paul Anthony**
University of Nottingham

Contributors

Rownak Afza
Plant Breeding Unit,
International Atomic Energy Agency,
Laboratories Siebersdorf,
Vienna International Centre,
Vienna,
Austria

Fredy Altpeter
Agronomy Department,
Plant Molecular Biology Program,
Genetics Institute,
University of Florida - IFAS,
Gainesville, FL 32611,
USA

Paul Anthony
Plant and Crop Sciences Division,
School of Biosciences,
University of Nottingham,
Sutton Bonington Campus,
Loughborough LE12 5RD,
UK

Francisco J.L. Aragão
Embrapa Recursos Geneticose Biotecnologia,
Parque Estação Biologica – PqEB, Av. W5N,
CP 02372, Brasilia, DF, CEP70770-900,
Brazil

Souleymane Bado
Plant Breeding Unit,
International Atomic Energy Agency,
Laboratories Siebersdorf,
Vienna International Centre,
Vienna,
Austria

Sant S. Bhojwani
Department of Botany,
Dayalbagh Educational Institute (Deemed University),
Dayalbagh,
Agra,
India

Milica Ćalović
University of Florida IFAS,
Citrus Research and Education Center,
Lake Alfred,
FL 33850,
USA

Junfeng Chen
Department of Pharmacy,
Changzheng Hospital,
Second Military Medical University,
Shanghai 200003,
China

Wansheng Chen
Department of Pharmacy,
Changzheng Hospital,
Second Military Medical University,
Shanghai 200003,
China

Ian S. Curtis
Texas A&M AgriLife Research,
2415 East Hwy 83,
Weslaco, TX 78596,
USA

Prem K. Dantu
Department of Botany,
Dayalbagh Educational Institute (Deemed University),
Dayalbagh,
Agra,
India

Michael R. Davey
Plant and Crop Sciences Division,
School of Biosciences,
University of Nottingham,
Sutton Bonington Campus,
Loughborough LE12 5RD,
UK

Antje Doil
University of Applied Sciences and Research Institute for Horticulture, Weihenstephan,
Am Staudengarten 8,
D-85354 Freising,
Germany

Aisling Dunne
Institute of Bioengineering and Agroecology,
National University of Ireland,
Maynooth,
Ireland

Jim M. Dunwell
School of Biological Sciences,
University of Reading,
Whiteknights,
Reading RG6 6AS,
UK

Aloma Ewald
Institute of Vegetable and Ornamental Crops,
Kuehnhaeuser Str. 101,
D-99189 Kuehnhausen,
Germany

Alena Gajdošová
Institute of Plant Genetics and Biotechnology SAS,
Akademicka 2,
95007 Nitra,
Slovakia

Jude Grosser
University of Florida IFAS,
Citrus Research and Education Center,
Lake Alfred,
FL 33850,
USA

Bridget V. Hogg
Institute of Bioengineering and Agroecology,
National University of Ireland,
Maynooth,
Ireland

Ivan Iliev
University of Forestry,
Faculty of Ecology and Landscape Architecture,
10 Kliment Ohridski blvd.,
1756 Sofia,
Bulgaria

Shri Mohan Jain
Plant Breeding Unit,
International Atomic Energy Agency,
Laboratories Siebersdorf,
Vienna International Centre,
Vienna,
Austria
*Current address – Department of Applied Biology,
University of Helsinki,
PL-27 Helsinki,
Finland

E.R. Joachim Keller
Genebank Department,
Leibniz Institute of Plant Genetics and Crop Plant Research (IPK),
Corrensstrasse 3,
D-06466 Gatersleben,
Germany

Spiridon Kintzios
Agricultural University of Athens,
75 Iera Odos,
EL-11855 Athens,
Greece

Cristiano Lacorte
Embrapa Recursos Geneticos e Biotecnologia,
Parque Estação Biologica – PqEB, Av. W5N,
CP 02372, Brasilia, DF, CEP70770-900,
Brazil

Cilia L.C. Lelivelt
Rijk Zwaan Breeding B.V.,
1e Kruisweg 9,
4793 RS Fijnaart,
The Netherlands

Gabriela Libiaková
Institute of Plant Genetics and Biotechnology
SAS,
Akademicka 2,
95007 Nitra,
Slovakia

Eliezer S. Louzada
Texas A&M University–Kingsville,
Citrus Center,
Weslaco,
TX 78599,
USA

Chikelu Mba
Plant Breeding Unit,
International Atomic Energy Agency,
Laboratories Siebersdorf,
Vienna International Centre,
Vienna,
Austria

Kim-Hong Nguyen
Institute of Bioengineering and Agroecology,
National University of Ireland,
Maynooth,
Ireland

Kim E. Nolan
School of Environmental and Life Sciences,
The University of Newcastle,
NSW 2308,
Australia

Jacqueline M. Nugent
Institute of Bioengineering and Agroecology,
National University of Ireland,
Maynooth,
Ireland

Sergio J. Ochatt
Laboratoire de Physiologie Cellulaire,
Morphogenèse et Validation (PCMV),
Centre de Recherches INRA de Dijon,
B.P. 86510,
21065 Dijon,
France

Deval Patel
Plant and Crop Sciences Division,
School of Biosciences,
University of Nottingham,
Sutton Bonington Campus,
Loughborough LE12 5RD,
UK

J. Brian Power
Plant and Crop Sciences Division,
School of Biosciences,
University of Nottingham,
Sutton Bonington Campus,
Loughborough LE12 5RD,
UK

Elíbio L. Rech
Embrapa Recursos Geneticos e
Biotecnologia,
Parque Estação Biologica – PqEB, Av. W5N,
CP 02372, Brasilia, DF, CEP70770-900,
Brazil

Sandra Reinhardt
Institute of Vegetable and Ornamental Crops,
Department of Plant Propagation,
Kuehnhaeuser Str. 101,
D-99189 Kuehnhausen,
Germany

Ray J. Rose
School of Environmental and Life Sciences,
The University of Newcastle,
NSW 2308,
Australia

Sukhpreet Sandhu
Agronomy Department,
Plant Molecular Biology Program,
Genetics Institute,
University of Florida – IFAS,
Gainesville, FL 32611,
USA

Rajbir S. Sangwan
Laboratoire AEB,
Universite de Picardie Jules Verne,
33, Rue Saint Luc,
80039 Amiens,
France

Angelika Senula
Genebank Department,
Leibniz Institute of Plant Genetics and Crop Plant Research (IPK),
Corrensstrasse 3,
D-06466 Gatersleben,
Germany

Xiaofen Sun
State Key Laboratory of Genetic Engineering,
School of Life Sciences,
Fudan University,
Shanghai 200433,
China

Michio Tanaka
Faculty of Agriculture and Graduate School of Agriculture,
Kagawa University,
Miki-cho,
Ikenobe 2393,
Kagawa-ken, 761–0795,
Japan

Kexuan Tang
Plant Biotechnology Research Center,
Fudan-SJTU-Nottingham Plant Biotechnology R&D Center,
School of Agriculture and Biology,
Shanghai Jiao Tong University,
Shanghai 200240,
China

Jaime A. Teixeira da Silva
Faculty of Agriculture and Graduate School of Agriculture,
Kagawa University,
Miki-cho,
Ikenobe 2393,
Kagawa-ken, 761–0795,
Japan

Giovanni Vianna
Embrapa Recursos Geneticos e Biotecnologia,
Parque Estação Biologica – PqEB, Av. W5N,
CP 02372, Brasilia, DF, CEP70770-900,
Brazil

Traud Winkelmann
Institute of Floriculture and Woody Plant Science,
Leibniz University Hannover,
Herrenhaeuser Str. 2,
D-30419 Hannover,
Germany

Ying Xiao
Department of Pharmacy,
Changzheng Hospital,
Second Military Medical University,
Shanghai 200003,
China

Lei Zhang
Department of Pharmacognosy,
School of Pharmacy,
Second Military Medical University,
Shanghai 200433,
China

1

Plant Micropropagation

Ivan Iliev[1], Alena Gajdošová[2], Gabriela Libiaková[2] and Shri Mohan Jain[3]*

[1]*Faculty of Ecology and Landscape Architecture, University of Forestry, Sofia, Bulgaria*
[2]*Institute of Plant Genetics and Biotechnology SAS, Nitra, Slovakia*
[3]*Plant Breeding Unit, International Atomic Energy Agency, Laboratories Siebersdorf, Vienna, Austria*
**Current address – Department of Applied Biology, University of Helsinki, Helsinki, Finland*

1.1 Introduction

The technique of plant tissue culture is used for growing isolated plant cells, tissues and organs under axenic conditions (*in vitro*) to regenerate and propagate entire plants. 'Tissue culture' is commonly used as a blanket term to describe all types of plant cultures, namely callus, cell, protoplast, anther, meristem, embryo and organ cultures [1]. It relies on the phenomenon of cell totipotency, the latter being the ability of single cells to divide, to produce all the differentiated cells characteristic of organs, and to regenerate into a whole plant. The different techniques of culturing plant tissues may offer certain advantages over traditional methods of propagation. Growing plants *in vitro* in a controlled environment, with in-depth knowledge of the culture conditions and the nature of the plant material, ensures effective clonal propagation of genetically superior genotypes of economically important plants. Tissue cultures represent the major experimental systems used for plant genetic engineering, as well as for studying the regulation of growth and organized development through examination of structural, physiological, biochemical and molecular bases underlying developmental processes. Micropropagation has become an important part of the commercial propagation of many plants [2–6] because of its advantages as a multiplication system [7–9]. Several techniques for *in vitro* plant propagation have been devised, including the induction of axillary and adventitious shoots,

Plant Cell Culture Edited by Michael R. Davey and Paul Anthony

the culture of isolated meristems and plant regeneration by organogenesis and/or somatic embryogenesis [10–12].

Fertile plants can be regenerated either by the growth and proliferation of existing axillary and apical meristems, or by the regeneration of adventitious shoots. Adventitious buds and shoots are formed *de novo*; meristems are initiated from explants, such as those of leaves, petioles, hypocotyls, floral organs and roots.

This chapter summarizes the application of the most commonly used *in vitro* propagation techniques for trees, shrubs and herbaceous species that can be implemented on a continuous basis throughout the year.

1.2 Methods and approaches

1.2.1 Explants and their surface disinfection

Small pieces of plants (explants) are used as source material to establish cells and tissues *in vitro*. All operations involving the handling of explants and their culture are carried out in an axenic (aseptic; sterile) environment under defined conditions, including a basal culture medium of known composition with specific types and concentrations of plant growth regulators, controlled light, temperature and relative humidity, in culture room(s) or growth cabinet(s). The disinfection of explants before culture is essential to remove surface contaminants such as bacteria and fungal spores. Surface disinfection must be efficient to remove contaminants, with minimal damage to plant cells. This chapter focuses on the general procedures for developing *in vitro* cultures, illustrated by protocols for specific plants and explants.

PROTOCOL 1.1 Surface Disinfection of Explants

Equipment and Reagents

- Autoclave
- Laminar flow cabinet
- Ultraviolet lamp
- Scalpels, forceps, scissors, rest for supporting axenic instruments (Duchefa), glass beakers (100 ml), glass Petri dishes (100 × 15 mm), white cotton gauze[a] (15 × 15 cm), magnetic mini-stirrer (ScienceLab) and stirring bars, filter paper (Whatman, Standard Grade; 10 mm diameter circles), aluminium foil, funnel and suction flask, glass beakers (100 ml–1 l in volume).
- Unifire Gasburner (Uniequip), glass bead sterilizer (Duchefa) or alcohol lamp
- Distilled water: 350 ml aliquots in 500 ml bottles
- Tween 20 (Sigma)
- Ethanol: 95 and 70% (v/v)
- NaClO or $Ca(ClO)_2$: 0.5–5% or 3–7% (w/v) aqueous solutions, respectively (Chemos GmbH)

- $HgCl_2$ (Sigma): 0.1–0.2% (w/v) aqueous solution[b]
- H_2SO_4: 96% (v/v) solution[c]
- Bacteriocidal soap
- Culture vessels with sterile culture medium (See Protocol 1.2 for preparation of culture medium).

Method

1. Place several filter papers into each of the glass Petri dishes. Wrap the Petri dishes, glass beakers, scissors, scalpels, forceps, funnel, white gauze and suction flask in aluminium foil.
2. Disinfect the material from Step 1 and bottles of distilled water in an autoclave at 120 °C, 118 kPa (1.18 bar) steam pressure for 20 min.
3. Disinfect the laminar flow cabinet by exposing the work bench to ultraviolet illumination for 3 h. Spray the work surface of the cabinet with 95% (v/v) ethanol; allow to dry.
4. Remove the epidermis from stem segments and scale leaves from buds of woody species[d].
5. Wash the explants under running tap water for 5 min.
6. Wash hands thoroughly with bacteriocidal soap before commencing work.
7. Disinfect the explants in the laminar flow cabinet. Place the explants in a beaker (autoclaved). Wash the explants (by stirring on magnetic mini-stirrer) in 70% (v/v) ethanol (2 min) and 5% (w/v) NaClO, containing 20 drops per litre of Tween 20 (15–30 min). After immersion in each solution, wash the explants 3 times with sterile distilled water for 3, 5 and 10 min; discard the washings
8. After surface disinfection, keep the plant material in distilled water in Petri dishes in the laminar flow cabinet to prevent drying.
9. Before preparing the explants, disinfect the forceps and scalpels using a glass bead sterilizer, Unifire Gasburner, or by flaming using the alcohol lamp for 10–15 s.
10. Remove the cut ends of the explants[e] (e.g. apical or axillary buds, leaves, petioles, flowers, seedling segments) with a sterile scalpel before placing the explants on the culture medium.

Notes

[a]Place small plant parts, such as tiny seeds or buds, into gauze bags to facilitate manipulation during disinfection.

[b]Mercuric chloride ($HgCl_2$) is a highly effective surface sterilant but is extremely toxic. Local regulations must be enforced with its use. The duration of surface disinfection in 0.1% (w/v) aqueous solution is 1–3 min for leaves and stems of herbaceous plants, 8–10 min for nodal and apical segments of woody plants, and 10–20 min for seeds.

[c]Use for 4–5 min to disinfect seeds with a hard testa.

[d]Removal of the epidermis from the stem segments and scale leaves from buds may increase the disinfection efficiency in woody species.

[e]Cut the ends of the explants in the laminar flow cabinet on sterile filter papers or on a sterile white tile.

1.2.2 Culture media and their preparation

Culture media contain macroelements, microelements, vitamins, other organic components (e.g. amino acids), plant growth regulators, gelling agents (if semisolid) and sucrose. Gelling agents are omitted for liquid media. The composition of the culture medium depends upon the plant species, the explants, and the aim of the experiments. In general, certain standard media are used for most plants, but some modifications may be required to achieve genotype-specific and stage-dependent optimizations, by manipulating the concentrations of growth regulators, or by the addition of specific components to the culture medium. Commercially available ready-made powdered medium or stock solutions can be used for the preparation of culture media. A range of culture media of different formulations, and plant growth regulators are supplied by companies such as Duchefa and Sigma-Aldrich. Murashige and Skoog medium (MS) is used most extensively [13]. A procedure for the preparation of MS medium supplemented with plant growth regulators for raspberry micropropagation [14] is given in Protocol 1.2.

PROTOCOL 1.2 Preparation of Culture Medium

Equipment and Reagents

- Culture vessels: 25 × 150 mm sterile plastic disposable culture tubes with screw-caps (Sigma-Aldrich), Full-Gas Microbox culture jars (jar and lid OS60 + ODS60; Combiness), Erlenmeyer 'Pyrex' flasks 125 ml capacity (Sigma-Aldrich) or Petri dishes (60 × 15 mm or 100 × 15 mm; Greiner Bio-One). Glass Petri dishes, if used, must be disinfected by autoclaving or dry heat treatment
- Autoclave
- Laminar flow cabinet
- Refrigerator/freezer
- Distilled water (water purification system)
- Electronic heated stirrer
- Analytical balances
- pH meter
- Microwave oven
- Pipettes and measuring cylinders

- Beakers, 100 ml and 1–2 l, 100 ml flasks, funnels, aluminium foil
- PP/PE syringes without needles, capacity 50 ml (Sigma-Aldrich)
- Acrodisc syringe membrane filters (25 mm, 0.2 μm pore size; Sigma-Aldrich)
- 1 M HCl and KOH
- MS packaged powdered medium, including macro and microelements and vitamins (Duchefa)
- Plant growth regulators for raspberry micropropagation: benzylaminopurine (BAP) and β-indolebutyric acid (IBA; Duchefa)
- Other plant growth regulators: auxins – naphthaleneacetic acid (NAA), indole-3-acetic acid (IAA), 2,4-dichlorophenoxyacetic acid (2,4-D); cytokinins – kinetin, zeatin, 6-γ-γ-(dimethylallylamino)-purine (2-iP), thidiazuron (TDZ); gibberellins – gibberellic acid (GA_3); abscisic acid (ABA); organic components – sucrose, plant agar, citric acid, ascorbic acid (Duchefa)
- Plant preservative mixture – PPM (Plant Cell Technology, Inc.).

Method

1. To prepare 1 l MS medium, dissolve 4.406 g powdered medium in 500 ml of double distilled water in a 2 l beaker.
2. Prepare separate stock solutions of each plant growth regulator.
3. Add heat stable supplements to the medium before autoclaving, such as 30 g sucrose, 8 g agar, the desired plant growth regulators in a specific volume of stock solution (e.g. 5 ml BAP and 5 ml IBA) to reach the required final concentrations (1 mg/l BAP and 0.1 mg/l IBA for raspberry micropropagation). Adjust the medium to the final volume (1 l) by adding double distilled water[a].
4. Adjust the pH of the medium to 5.6–5.8 with 1 M HCl or KOH[b] and heat in microwave oven until the gelling agent is dissolved.
5. Autoclave the medium at 1 kg/cm (15 psi) at 121 °C for 20 min[c].
6. Dispense the medium into the culture vessels (15 ml per culture tube, 50 ml per Erlenmeyer bank, 50 ml per Full-Gas Microbox culture jar, 30 ml per 9 cm Petri dish) in the laminar flow cabinet. Close the vessels.

Preparation of Stock Solutions

1. Prepare separate stock solution for each plant growth regulator. Weigh the plant growth regulators to obtain a quantity 20 times the quantity given in the formulation for the medium (e.g. 20 mg BAP and 2 mg IBA), and dissolve in 100 ml distilled water[d].
2. Dissolve auxins (NAA, IAA, IBA and 2,4-D) in 1 ml ethanol and make up to 100 ml with distilled water.
3. Dissolve cytokinins (kinetin, zeatin, BAP, 2-iP) and ABA in 1 ml 1 M NaOH or 1 M KOH; make up to 100 ml with distilled water.

4 Store the stock solutions in 100 ml flasks in a refrigerator (not frozen) for not more than 2 months[e].

Filter Sterilization of Heat Sensitive Compounds

1 Wrap a funnel and 100 ml flask in aluminum foil and autoclave.

2 Fill the PP/PE syringe with the solution of heat labile constituents (e.g. zeatin, 2-iP, IAA, GA_3, citric acid, ascorbic acid). Mount an Acrodisc syringe membrane filter on the syringe and filter the solution into the funnel and into a sterile flask. Dispense the filter sterilized solution into convenient aliquots (e.g. 10–20 ml) in sterile, screw-capped vessels. Perform this operation in a laminar flow cabinet. Store the filter sterilized solutions at −20 °C.

Notes

[a]Heat labile constituents, such as some growth regulators and organic compounds (e.g. zeatin, 2-iP, IAA, GA_3, citric acid, ascorbic acid), should not be autoclaved but filter sterilized before adding to the autoclaved culture medium after the medium has cooled to 40–50 °C in the laminar flow cabinet.

[b]The pH of the culture medium is usually adjusted to 5.6–5.8. For acid-loving species, a lower pH is required (4.5 or less).

[c]To minimize contamination by micro-organisms, a broad-spectrum biocide/fungicide for plant tissue culture [Plant Preservative Mixture (PPM); Plant Cell Technology, Inc.] may be added to the medium at a concentration of 2–20 ml/l, which effectively prevents or reduces microbial contamination. Some plant species are more sensitive to PPM than others. Rooting in less tolerant plant species may be partially inhibited. In this case, the explants should be exposed to PPM for only a limited time.

[d]Cytokinins (BAP, kinetin, 2-iP, zeatin) are added to the culture medium to induce axillary or adventitious shoots. Auxins (2,4-D, NAA, IAA) induce callus formation. IBA is generally used to induce adventitious roots. GA_3 or polyamines added to the medium will promote shoot elongation.

[e]Culture media should be used within 2 to 4 weeks of preparation and may be kept for 6 weeks before use, if refrigerated.

1.2.3 Stages of micropropagation

The following distinct stages are recognized for the micropropagation of most plants:

Stage I: Establishment of axenic cultures – introduction of the surface disinfected explants into culture, followed by initiation of shoot growth. The objective of this stage is to place selected explants into culture, avoiding contamination and providing an environment that promotes shoot production [15]. Depending on the type of explant, shoot formation may be initiated from apical and axillary buds (pre-existing meristems), from adventitious meristems that originate on excised shoots, leaves, bulb scales, flower stems or cotyledons (direct organogenesis), or from callus that develops at the cut surfaces of explants (indirect organogenesis). Usually 4–6 weeks are required to complete this stage and to generate explants that are ready to be moved to Stage II [16]. Some woody plants may take up to 12 months to complete Stage I [15], termed 'stabilization'. A culture is stabilized when explants produce a constant number of normal shoots after subculture [16].

Stage II: Multiplication – shoot proliferation and multiple shoot production. At this stage, each explant has expanded into a cluster of small shoots. Multiple shoots are separated and transplanted to new culture medium [16]. Shoots are subcultured every 2–8 weeks. Material may be subcultured several times to new medium to maximise the quantity of shoots produced.

Stage III: Root formation – shoot elongation and rooting. The rooting stage prepares the regenerated plants for transplanting from *in vitro* to *ex vitro* conditions in controlled environment rooms, in the glasshouse and, later, to their ultimate location. This stage may involve not only rooting of shoots, but also conditioning of the plants to increase their potential for acclimatization and survival during transplanting. The induction of adventitious roots may be achieved either *in vitro* or *ex vitro* in the presence of auxins [17–19]. The main advantage of *ex vitro* compared to *in vitro* rooting is that root damage during transfer to soil is less likely to occur. The rates of root production are often greater and root quality is optimized when rooting occurs *ex vitro* [20–23].

Stage IV: Acclimatization – transfer of regenerated plants to soil under natural environmental conditions [16]. Transplantation of *in vitro*-derived plants to soil is often characterized by lower survival rates. Before transfer of soil-rooted plants to their final environment, they must be acclimatized in a controlled environment room or in the glasshouse [24, 25]. Plants transferred from *in vitro* to *ex vitro* conditions, undergo gradual modification of leaf anatomy and morphology, and their stomata begin to function (the stomata are usually open when the plants are in culture). Plants also form a protective epicuticular wax layer over the surface of their leaves. Regenerated plants gradually become adapted to survival in their new environment [26].

1.2.4 Techniques of micropropagation

Cultures of apical and axillary buds

Currently, the most frequently used micropropagation method for commercial mass production of plants utilizes axillary shoot proliferation from isolated apical or axillary buds under the influence of a relatively high concentration of cytokinin. In this procedure, the shoot apical or axillary buds contain several developing leaf primordia. Typically, the explants are 3–4 mm in diameter and 2 cm in length. Development *in vitro* is regulated to support the growth of shoots, without adventitious regeneration.

PROTOCOL 1.3 Propagation by Culture of Apical and Axillary Buds

Equipment and Reagents

- Culture facilities – culture room or plant growth cabinet with controlled temperature, light and humidity; culture vessels
- Laminar flow cabinet, ultraviolet lamp
- Scalpels, forceps, scissors, a rest for holding sterile tools (Duchefa), 50 ml beakers
- Unifire Gasburner (Uniequip), glass bead sterilizer (Duchefa) or glass alcohol lamp
- Ethanol 70% and 95% (v/v); Tween 20 (Sigma); NaClO (Chemos GmbH); $HgCl_2$ (Sigma)
- Bacteriocidal soap
- Murashige and Skoog medium (MS-Duchefa)
- Anderson's Rhododendron medium (AN-Duchefa)
- Plant growth regulators and organic components: BAP, 2-iP, zeatin, TDZ, adenine sulfate, NAA, IAA, IBA, sucrose, agar
- Distilled water
- Activated charcoal (Duchefa)
- Commercial plastic multi-pot containers (pot diam. 40 mm) with covers
- Peat, perlite, vermiculite

Method

Explant selection and disinfection:

1. Select the explants as single-node segments, preferentially from juvenile[a], rejuvenated plants[b,c], or *in vitro*-derived plants.
2. For commercial large-scale micropropagation, it is preferable to use pathogen-indexed stock plants as a source of explants.
3. See Protocol 1.1 for surface disinfection of explants.

Establishment of cultures:

1. Place isolated disinfection apical and axillary buds, from which the upper scale leaves have been removed, on culture medium (MS-based medium for *Lavandula dentata* L. and AN medium for *Vaccinium corymbosum* L.). See Protocol 1.2 for preparation of culture media. Carry out these operations in a laminar flow cabinet after UV and ethanol disinfection (See Protocol 1.1).
2. Add cytokinins to the medium to induce axillary shoots: BAP (0.01–5 mg/l), 2-iP (0.01–10 mg/l), zeatin (2–15 mg/l), TDZ (0.01–10 mg/l), adenine sulfate (40–120 mg/l). Add auxins (NAA, IAA, IBA) in low concentrations (0.01–0.1 mg/l) to

the medium to support shoot growth[d]. Optimize experimentally the cytokinin and auxin types and concentrations for each species[e].

3 Culture the explants for 4 weeks on cytokinin-containing medium in the growth cabinet at 23 ± 2 °C with a 16 h photoperiod (50 μmol/m^2/s; white fluorescent lamps).

Shoot multiplication:

1 Separate *in vitro* regenerated axillary shoots and transfer the shoots onto the appropriate culture medium (MS medium for *L. dentata* and AN medium for *V. corymbosum*) supplemented with the same or a reduced cytokinin concentration.

2 Cut the regenerated shoots into one-node segments and culture on cytokinin-supplemented medium to stimulate shoot proliferation.

3 Repeat the procedure depending on the number of shoots required. Some of the regenerated shoots *in vitro* can be retained for use to provide an axenic stock of explants for further multiplication.

Rooting of regenerated shoots:

Root the regenerated shoots by two approaches:

1 *Ex vitro* rooting by 'pulse treatment' – immerse the stem bases of 15–20 mm long regenerated shoots into an auxin solution (e.g. IBA at 1–10 mg/l) in 50 ml beakers for 3–7 days, followed by planting in commercial plastic multi-pot containers with soil or a mixture of peat, perlite and vermiculite (equal volumes). Cover the containers and shoots to maintain soil and air humidity.

2 *In vitro* rooting on culture medium supplemented with IBA at a concentration of 1 mg/l and activated charcoal at 1–10 g/l[f]. Reduction of the components of the culture medium to half strength, darkness during culture[g] and inoculation with mycorrhizal fungi[h], may stimulate rooting.

Examples

Micropropagation of *Lavandula dentata* by culture of apical and axillary buds (27).

1 Excise stem segments (each 2–3 cm in length) bearing apical or lateral axillary buds from 5-year-old plants between September and December.

2 Disinfect the stem segments by immersion in 70% (v/v) ethanol for 30 s, and sodium hypochlorite (NaClO) solution (1 g/l) containing 0.01% (v/v) Tween-20 for 20 min; rinse thoroughly with sterile distilled water.

3 Culture the dissected apical and lateral buds vertically on MS culture medium supplemented with sucrose (30 g/l), agar (6 g/l; Merck), cytokinin (BAP; 0.5 mg/l) and auxin (IBA; 0.5 mg/l) at pH 5.6–5.8.

4 Maintain the cultures in the growth cabinet at 25 ± 2 °C under a 16 h photoperiod (50 μmol/m^2/s; white fluorescent illumination).

5 Root the isolated shoots on MS medium supplemented with 0.5 mg/l NAA.

Micropropagation of *Vaccinium corymbosum* by culture of apical and axillary buds [17].

1 Harvest branches with dormant buds from mature donor plants during February and at the beginning of March; cut the branches into single-node segments.

2 Disinfect the segments with apical and axillary buds by washing under running tap water for 1 h, followed by immersion in 70% (v/v) ethanol for 2 min. Transfer the cuttings into 300 ml 0.1% (w/v) mercuric chloride with three drops of Tween for 6 min. Wash the explants thoroughly with sterile distilled water (three changes, each 15 min). Retain all the washings and discard according to local regulations for toxic chemicals.

3 Culture the isolated dormant apical and axillary buds, from which the upper scales are removed after disinfection, on AN medium supplemented with sucrose (30 g/l), Phytoagar (8 g/l) and zeatin (2 mg/l), at pH 4.5–5.0.

4 Maintain the cultures in the growth cabinet at $23 \pm 2\,^{\circ}C$ with a 16 h photoperiod (50 $\mu mol/m^2/s$, white fluorescent illumination).

5 For further proliferation of *in vitro* regenerated axillary shoots, culture the shoots on the same medium with zeatin (0.5 mg/l) with subculture every 5 weeks.

6 Root the regenerated shoots (each 15–20 mm in height) *ex vitro* by dipping (2–3 min) into IBA solution (0.8 mg/l), followed by planting in commercial plastic multi-pot containers (pot diam. 40 mm) filled with peat-based compost, or *in vitro* on AN medium with IBA (0.8 mg/l) and activated charcoal (0.8 g/l).

Notes

[a]The branches from the basal part of the crown, near to the trunk and highest order of branching, are more juvenile than others in the crown of the plant. More juvenile are epicormics, shoots originating from spheroblasts, severely pruned trees, stump and root sprouts [28].

[b]Rejuvenation may be initiated by grafting scions from mature trees onto juvenile rootstocks. Use explants for culture from trees 1–3 years after grafting [29].

[c]Keeping the cut branches in the sterile liquid medium without growth regulators or in water, in a growth cabinet for 4–5 days, may force the plant material into growth.

[d]Synthetic auxins are more stable and most effective. They include IBA and NAA at 0.1–10 mg/l, 2,4-D at 0.05–0.5 mg/l and the natural auxin IAA (1–50 mg/l). IBA is the most effective auxin for adventitious root induction.

[e]Prepare the MS culture medium with several combinations of growth regulators and grow the same type of explant (dormant bud) for 5 weeks. During testing for the optimal culture medium, change only one factor at a time in the composition of the medium. In order to determine appropriate cytokinin type and concentration for shoot induction, combine different concentrations (0.5, 1, 2, 3 and 5 mg/l) of cytokinins with 0.05 mg/l auxin. Evaluate the number of regenerated shoots and select the most efficient cytokinin concentration. Use the most efficient cytokinin concentration in combination with different auxin concentrations (0.05, 0.1, 0.2, 0.5) to determine the optimal auxin concentration.

[f]For some plants, such as *Sequoiadendron giganteum* and *Fraxinus excelsior*, rooting is optimal by maintaining the shoots in auxin-supplemented medium (induction medium) for 1–5 days, followed by transfer to an auxin-free medium for root formation.

[g]Some plants form roots more rapidly in the dark during auxin treatment.

[h]Mycorrhizae are a close relationship between specialized soil fungi (mycorrhizal fungi) and plant roots. Mycorrhizae may stimulate the rooting of some species [30–34].

Meristem and single- or multiple-node cultures (shoot cultures)

Meristems are groups of undifferentiated cells that are established during plant embryogenesis [35]. Meristems continuously produce new cells which undergo differentiation into tissues and the initiation of new organs, providing the basic structure of the plant body [36]. Shoot meristem culture is a technique in which a dome-shaped portion of the meristematic region of the stem tip is dissected from a selected donor plant and incubated on culture medium [37]. Each dissected meristem comprises the apical dome with a limited number of the youngest leaf primordia[a], and excludes any differentiated provascular or vascular tissues. A major advantage of working with meristems is the high probability of excluding pathogenic organisms, present in the donor plant, from cultures[b]. The culture conditions are controlled to allow only organized outgrowth of the apex directly into a shoot, without the formation of any adventitious organs, ensuring the genetic stability of the regenerated plants.

The single-or multiple-node technique involves production of shoots from cultured stem segments, bearing one or more lateral buds, positioned horizontally or vertically on the culture medium[c]. Axillary shoot proliferation from the buds in the leaf axils is initiated by a relatively high cytokinin concentration[d]. Meristem and node cultures are the most reliable for micropropagation to produce true-to-type plants[e].

PROTOCOL 1.4 Propagation by Meristem and Nodal Cultures

Equipment and Reagents

- Culture facilities (culture room or plant growth cabinet) with automatically controlled temperature, light, and air humidity; sterile disposable Petri dishes (60 and 100 mm; Greiner Bio-One), Full-Gas Microbox culture jars (jar and lid OS60 + ODS60; Combiness)
- Laminar flow cabinet, ultraviolet lamp
- Stereomicroscope
- Unifire Gasburner (Uniequip), glass bead sterilizer (Duchefa) or glass alcohol lamp
- Scalpel, needles, fine tweezers, rest for holding sterile tools (Duchefa)
- Detergent Mistol (Henkel Ibérica, SA), ethanol 70% and 95% (v/v); Tween 20 (Sigma); NaClO (Chemos GmbH); $HgCl_2$ (Sigma)
- 'Keep Kleen' disposable vinyl gloves (Superior Glove Works Ltd.)

- Bacteriocidal soap
- Plant growth regulators and organic components: BAP, GA_3, IBA, myoinositol, sorbitol, thiamine, nicotinic acid, glycine, phloroglucinol, agar, sucrose, ribavirin (Duchefa)
- Double distilled water
- Activated charcoal (Duchefa)
- Quoirin and Lepoivre medium (QL; Duchefa)
- Driver and Kuniyuki medium (DKW; Duchefa)
- Filter paper bridges made from Whatman filter paper[f]

Method

Explant selection and disinfection:

1. Select the explants, single-or multiple-node segments, preferentially from juvenile, rejuvenated plants, *in vitro* derived plants, or branches with dormant buds in the case of woody species.
2. Disinfect the explants according to Protocol 1.1. *In vitro*-derived plants should already be axenic.

Meristem cultures:

1. Isolate the meristems under the stereomicroscope in the laminar hood. Remove the upper leaves from each bud. Hold shoot segments with each bud and carefully remove the remaining leaves and leaf primordia one by one using dissection instruments. Disinfect the equipment (needle, scalpel and tweezers) regularly during this procedure using the gasburner. Excise each meristem (0.1 mm in diam.; 0.2–0.5 mm high) with one to two leaf primordia and transfer to the surface of semi-solid QL culture medium [38].
2. Culture the isolated meristems on semi-solid QL medium, or in the same liquid medium by placing the meristems on semisubmerged filter paper bridges. Use a similar composition of growth regulators as for bud cultures. Determine the optimal types and concentrations of growth regulators for each species.

Nodal cultures:

1. Culture the nodal explants in a vertical or horizontal position on cytokinin-enriched medium (see Protocol 1.3).
2. Avoid inserting the explants too deeply into the medium and submerging the nodes.
3. Culture for 4 weeks on cytokinin-containing medium.

See Protocol 1.3 for shoot multiplication and rooting.

Examples

Micropropagation of *Prunus armeniaca* from cultured meristems [38].

1 Collect branches from adult apricot field-grown trees between January and March, when buds are starting to swell.

2 Cut the shoots into two- or three-nodal sections; wash with water and detergent (e.g. Mistol; Henkel Ibérica, SA), shake for 5 min in 70% (v/v) ethanol and 20 min in a 20% (v/v) solution of sodium hypochlorite (Chemos GmbH; 0.8% final concentration). Wash three times with sterile distilled water.

3 Dissect out buds and meristems from lateral and apical buds perform in a laminar flow cabinet using sterile disposable Petri dishes and steriler instruments. Wearing sterile 'Keep Kleen' disposable vinyl gloves, hold the basal end of the stem; disinfect the instruments frequently. Remove the bark surrounding each bud followed by the outer bud scales; continue until the meristematic dome and a few leaf primordia are exposed. Remove the meristem by cutting its base leaving an explant approx. 0.5–1 mm long with a wood portion that allows further manipulations and culture.

4 Prepare culture medium consisting of QL macro-and micronutrients and vitamins (38), supplemented with myoinositol (50 mg/l), 2% (w/v) sorbitol and semi-solidified with 0.6% (w/v) agar (Hispanlab); adjust the pH to 5.7. In order to induce development of the rosette of leaves, add 0.5–2.0 mg/l BAP. For elongation, add 2.0–4.0 mg/l GA and 0.5–1.0 mg/l BAP.

5 Subculture the meristems to new culture medium every 2 weeks and maintain the cultures in the growth chamber at $23 \pm 1\,^{\circ}C$ under a 16 h photoperiod ($55\,\mu mol/m^2/s$, white fluorescent lamps).

6 For proliferation of elongated shoots, transfer the shoots to Full-Gas Microbox culture jars (jar and lid OS60 + ODS60) each containing 50 ml of proliferation medium with QL macronutrients, DKW (38) micronutrients (DKW; Duchefa), sucrose (30 g/l), thiamine (2 mg/l), nicotinic acid (1 mg/l), myoinositol (100 mg/l), glycine (2 mg/l) and the growth regulators 0.04 mg/l IBA and 0.40–0.70 mg/l BAP.

7 Root isolated shoots on medium containing half strength QL macronutrients, DKW micronutrients, sucrose (20 g/l), thiamine (2 mg/l), nicotinic acid (1 mg/l), myoinositol (100 mg/l), glycine (2 mg/l), plus 40 mg/l phloroglucinol and 0.20–0.60 mg/l IBA.

Micropropagation of *Prunus armeniaca* from cultured nodes [38].

1 Excise shoots from rapidly growing branches during spring; remove the expanded leaves.

2 Follow the procedure as described for meristem culture to surface disinfect the explants.

3 Cut nodal explants, each 2 cm long, and culture the explants vertically with the basal end of each node embedded a few mm into the culture medium.

4 For culture establishment, use the same proliferation medium as described for meristem cultures supplemented with BAP (0.4 mg/l) and IBA (0.04 mg/l).

5 Transfer sprouted and elongated shoots to Full-Gas Microbox culture jars (OS60 + ODS60) each containing 50 ml of the same proliferation medium but with 0.04 mg/l IBA and 0.40–0.70 mg/l BAP.

6 Root the isolated shoots in the same way as described for meristem cultures. The original protocols are described by Pérez-Tornero and Burgos [38].

Notes

[a]The size of the isolated explant (meristem only or meristem with leaf promordia) is crucial for survival and regeneration. Meristems alone have less chance of survival. However, obtaining virus-free plants is more probable with only meristems.

[b]To generate virus-free plants, thermotherapy (cultivation for 6 weeks at 35–38 °C) or chemotherapy (treatment with 40 mg/l ribavirin for several weeks) can be used during meristem culture.

[c]Sometimes one dormant bud develops and inhibits elongation of other shoots. In this case, the shoot may be excised and the base recultured. GA_3 at 0.1–10.0 mg/l [39] and activated charcoal at 1–10 g/l [40] is sometimes used to promote shoot elongation [16].

[d]High concentration of cytokinins may induce vitrification (pale and glassy appearance of cultures followed by growth reduction). Vitrification can be prevented by replacing BAP with 2-iP, by reducing chloride, ammonium and/or growth regulator concentrations in the culture medium [42]. Gelrite (Duchefa) should be avoided, but may be used in combination with agar at 3 : 1 (w : w). Vitrification can be prevented by subculture of the shoots from a semi-solid to a liquid medium, by incubating at low temperature (8–10 °C) for 1–2 months, or by increasing the concentration of agar to 0.8–1.0% (w/v) (if the concentration of agar increases, growth may be depressed because of increased osmotic pressure).

[e]During multiplication, off-type propagules sometimes appear, depending on the plant and method of regeneration. Restricting the multiplication phase to three subcultures is recommended to avoid development of off-type shoots in some plants, such as Boston fern (*Nephrolepis exaltata* 'Bostoniensis'). Exploiting procedures that decrease the potential for variability (e.g. reduce the growth regulator concentrations and avoiding callus formation that may result in adventitious shoots) [43]. Sometimes regenerated shoots deteriorate with time, lose their leaves and the potential to grow [44].

[f]Cut the Whatman filter paper into 1.5–2 cm strips and fold over.

Adventitious shoot formation

Adventitious shoot formation is one of the plant regeneration pathways *in vitro*, and is employed extensively in plant biotechnology for micropropagation and genetic transformation, as well as for studying plant development [45]. Adventitious meristems develop *de novo* and *in vitro* they may arise directly on stems, roots or leaf explants, often after wounding or under the influence of exogenous growth regulators (direct organogenesis). Cytokinins are often applied to stem, shoot or leaf cuttings to promote adventitious bud and shoot formation [46]. Adventitious buds and shoots usually develop near existing vascular tissues enabling the connection with vascular tissue to be observed. Adventitious organs sometimes also originate in callus that forms at the cut surface of explants (indirect organogenesis). Somaclonal variation, which may be useful or detrimental, may occur during adventitious shoot regeneration.

PROTOCOL 1.5 Induction of Adventitious Buds and Shoots

Equipment and Reagents

- Culture facilities (culture room or plant growth cabinet) with automatically controlled temperature, light, and air humidity; sterile disposable Petri dishes (60 and 100 mm, Greiner Bio-One), Full-Gas Microbox culture jars (jar and lid OS60 + ODS60, Combiness)
- Laminar flow cabinet, ultraviolet lamp
- Unifire Gasburner (Uniequip), glass bead sterilizer (Duchefa) or glass alcohol lamp
- Scalpel, fine tweezers, rest for holding sterile tools (Duchefa)
- Plant growth regulators and organic components: zeatin, Plant agar, sucrose, (Duchefa)
- Anderson's Rhododendron medium (AN; Duchefa)

Method

Selection of explants:

1. Excise cotyledons, hypocotyls, petioles, segments of laminae, flower stems of immature inflorescences, or bulb scales, preferentially from *in vitro*-growing plants[a].
2. Disinfect explants according to Protocol 1.1.

Establishment of cultures:

1. Place explants on the AN medium for adventitious shoot regeneration in *Vaccinium corymbosum*. Wounding of the explants using a scalpel may improve adventitious bud regeneration.
2. For the induction of adventitious buds in many plant species, a high cytokinin concentration and low auxin concentration are required in the medium, as in the case for axillary bud induction (see Protocol 1.3). Cytokinins and their concentrations need to be optimized experimentally for each species.
3. Culture for 4 weeks on a cytokinin-rich medium; transfer to medium with a low cytokinin concentration to promote further shoot growth and elongation.

See Protocol 1.3 for shoot multiplication and rooting.

Example

Micropropagation of *Vaccinium corymbosum* by adventitious shoot regeneration [17]:

1. Excise the upper three to four leaves from *in vitro*-grown plants of *V. corymbosum* cv. Berkeley and wound each explant on the midrib using a scalpel held vertically. Place leaf explants with their adaxial surfaces on the culture medium in 60 or 100 mm diam. Petri dishes.
2. Use AN medium with sucrose (30 g/l), plant agar (8 g/l) and zeatin (0.5 mg/l), at pH 4.5–5.0, to induce adventitious buds.
3. After 5 weeks, transfer the explants to AN medium in Full-Gas Microbox culture jars. The medium should be of the same composition and cytokinin concentration as used for shoot regeneration and multiplication.
4. For long-term proliferation of *in vitro* regenerated shoots, maintain material on the same medium containing 0.5 mg/l zeatin and subculture every 4–5 weeks.
5. Increase shoot proliferation by excising regenerated shoots and cutting the shoots into segments, each with one node. Culture the explants on medium with 0.5 mg/l zeatin.
6. Maintain the cultures in the growth cabinet at $24 \pm 2\,^{\circ}C$ under a 16 h photoperiod ($50\,\mu mol/m^2/s$; white fluorescent illumination)[b,c].
7. Use the procedure described in Protocol 1.3 for *ex vitro* or *in vitro* rooting of isolated shoots.

Notes

[a]Juvenile or rejuvenated explants regenerate adventitious shoots more easily than older material.

[b]Light intensity and quality play important roles in adventitious shoot regeneration, mainly during the initiation phase. Keep the cultures in the light during the first 3–5 days to initiate adventitious buds.

[c]A higher temperature (24–25 °C) is favourable for adventitious shoot regeneration in many species.

[d]Rich culture medium (such as MS-based medium) with vitamins, has a stimulatory effect on adventitious shoot regeneration.

Somatic embryogenesis

Somatic embryogenesis was defined by Emons [47] as the development from somatic cells of structures that follow a histodifferentiation pattern which leads to a body pattern resembling that of zygotic embryos. This process occurs naturally in some plant species and can be also induced *in vitro* in others species. There is considerable information available on *in vitro* plant regeneration from somatic cells by somatic embryogenesis. Somatic embryogenesis may occur directly from cells or organized tissues in explants or indirectly through an intermediate callus stage [48, 49, 50].

It has been confirmed in many species that the auxins 2,4-D and NAA, in the correct concentrations, play a key role in the induction of somatic embryogenesis. Application of the cytokinins, BAP or kinetin, may enhance plant regeneration from somatic embryos after the callus or somatic embryos have been induced by auxin treatment. However, in some species (such as *Abies alba*) cytokinins on their own induce somatic embryogenesis [51].

PROTOCOL 1.6 Induction of Somatic Embryogenesis

Equipment and Reagents

- Culture facilities (culture room or plant growth cabinet) with automatically controlled temperature, light, and air humidity; sterile disposable Petri dishes (60 and 100 mm, Greiner Bio-One), six-well Falcon Multiwell dishes, culture jars such as Full-Gas Microboxes (jar and lid OS60 + ODS60; Combiness)
- Laminar flow cabinet, ultraviolet lamp
- Gasburner Unifire (Uniequip), glass bead sterilizer (Duchefa) or glass alcohol lamp
- Stereomicroscope
- Glasshouse
- Scalpel, needles, fine tweezers, rest for holding sterile tools (Duchefa)
- Bacteriocidal soap
- 10% (v/v) H_2O_2 containing one drop of Silwet (Union Chemicals)
- Plant growth regulators and organic components: 2,4-D, NAA, BAP, ABA, Plant agar, Gelrite (Duchefa), sucrose, maltose, activated charcoal (Duchefa)
- PEG-4000
- Distilled water
- Initiation and maintenance medium (EDM6); embryo maturation media (EMM1 and EMM2); germination medium (BMG-2)
- Nylon cloth (30 μm pore size; Spectrum Laboratory Products, Inc.)
- Plastic food wrap; aluminium foil
- Peat and pumice
- Hyco V50 trays with plastic lids

Method

Selection of explants:

1. Cotyledons, hypocotyls, petioles and leaf segments, flower stems of immature inflorescences, bulb scales, mature and immature zygotic embryos (excise embryos from disinfected seeds under sterile conditions using the stereomicroscope), preferentially from juvenile *in vitro*-growing plants.
2. Disinfect the explants, if not from *in vitro*-grown plants, according to Protocol 1.1.

Induction of somatic embryogenesis and embryo development:

1 For many plant species (e.g. *Arachis hypogaea, Brassica napus*), culture the explants in sterile Petri dishes on medium supplemented with a high auxin concentration (2,4-D, NAA at 2–6 mg/l), but for some species (e.g. *Abies alba, Dendrobium* sp., *Corydalis yanhusuo*) on cytokinin-containing medium[a].

2 Proliferate embryogenic tissues by culture on new culture medium of the same composition.

3 Transfer the cultures to growth regulator-free medium for further somatic embryo development (pre-maturation). Embryos in globular, heart, torpedo and cotyledonary stages, the latter coinciding with the initiation of root primordia, should be visible on the surface of explants or in any induced callus [52]. In conifers, embryonal-suspensor masses are formed composed of small dense meristematic cells with long transparent suspensor cells.

4 In order to induce the maturation of somatic embryos (initiation of embryo growth and accumulation of storage products), transfer the embryogenic calli to medium supplemented with ABA (abscisic acid) with a decreased osmotic potential achieved by application of PEG-4000[b], or by increasing the carbohydrate content (maltose) for 8 weeks [53, 54].

5 Apply a desiccation treatment for embryo germination and conversion to plants. Isolate well-formed somatic embryos and transfer to unsealed 90 mm Petri dishes (six-well Falcon Multiwell dishes) placed in a sterile desiccator containing sterile distilled water for 2 weeks. Germinate the somatic embryos on hormone-free medium containing 1% (w/v) activated charcoal [55].

Example

Micropropagation of *Pinus radiata* by somatic embryogenesis [56]:

1 Collect cones approx. 8–10 weeks after fertilization. Remove the seeds from the cones, surface disinfect the seeds in 10% (v/v) H_2O_2 containing one drop of Silwet (Union Chemicals) for 10 min. Rinse two to three times in sterile water. Remove aseptically the seed coats.

2 Place whole megagametophytes containing immature embryos, at the torpedo to precotyledonary stages, onto initiation medium (EDM6) with sucrose (30 g/l), Gelrite (3 g/l), BAP (0.6 mg/l), auxin 2,4-D (1 mg/l), at pH 5.7 [56].

3 Maintain the cultures in the growth chamber at $24 \pm 1\,^{\circ}$C under low illumination (5 µmol/m^2/s).

4 After 2–6 weeks when the embryos are expelled from the megagametophytes onto the medium and embryogenic tissue reaches 10 mm in diameter, separate the tissue from the original explant and transfer to maintenance medium of the same composition as the initiation medium (EDM6). Maintain the cultures by serial transfer to new medium every 14 days.

5 To induce embryo maturation, take five portions (each 10 mm in diam.) of embryogenic tissue after 7 days of culture on EDM6 medium and place the tissues onto

Embryo Maturation Medium (EMM1) [56] supplemented with sucrose (30 g/l), Gelrite (6 g/l) and abscisic acid (15 mg/l). After 14 days, transfer onto the second maturation medium (EMM2) which has the same composition as EMM1 except for a lower concentration of Gelrite (4.5 g/l). Transfer to new EMM2 medium every 14 days until mature somatic embryos develop (6–8 weeks). Maintain the cultures at 24 ± 1 °C under low intensity illumination (5 µmol/m^2/s).

6 To germinate the somatic embryos, harvest the white somatic embryos with well formed cotyledons and place them on nylon cloth contained in each of three wells of six-well Falcon Multiwell dishes (several embryos per week). Half fill the remaining three wells with sterile water. Seal the dishes with plastic food wrap. Wrap each dish in aluminium foil and store at 5 °C for at least 7 days. Transfer the nylon cloth containing the embryos to germination medium (BMG-2) and incubate for 7 days at 24 °C in the light and 20 °C in the dark (16 h photoperiod with 90 µmol/m^2/s, cool white fluorescent illumination). Remove the embryos from the nylon cloth and place the embryos horizontally on the germination medium. After 6–8 weeks, transplant germinating embryos into Hyco V50 trays containing a mixture of peat : pumice (2 : 1, v : v), and cover the trays with plastic lids. Gradually acclimatize the plants to glasshouse conditions by removing the lids for increasing periods.

7 Media formulations and additional procedure details are given in the original protocol.

Notes

[a]Stress-related stimuli, such as osmotic shock, the presence of heavy metals and auxin starvation induce somatic embryogenesis [57].

[b]PEG stimulates embryo maturation but induces alterations in somatic embryo morphology and anatomy that may lead to reduced germination and survival.

1.3 Troubleshooting

- Some explants placed on culture medium exude dark coloured compounds into the culture medium (phenols, pigments) that are released from the cut ends of the explants. This can cause browning of tissue and the medium, which is often connected with poor culture establishment and reduced regeneration ability. Minimize the wounding of explants during isolation and surface disinfection to reduce this response. Other approaches to prevent tissue browning include removal of these compounds by washing of explants in sterile water for 2–3 h, frequent subculture of explants to new medium with the excision of brown tissues, initial culture in liquid medium with subsequent transfer to semi-solid medium, culture on a porous substrate (paper bridges) and adsorption with activated charcoal (AC) or PVP (polyvinylpyrrolidone) by addition of these compounds to the culture medium. However, AC can also adsorb growth regulators or be toxic to some tissues. The use of antioxidants, such as ascorbic acid, citric acid, L-cysteine or mercaptoethanol, can also prevent browning of tissues in culture [58]. Excessive browning may cause serious problems in the different stages of shoot regeneration.

- Hyperhydricity (vitrification), i.e. the appearance of transparent and watery structures, is a physiological disorder occuring in plant tissue cultures [36, 59, 60]. Major problems are not encountered up to the weaning stage when it is limited in extent. Hyperhydricity can be caused by a high cytokinin concentration, high water retention capacity when the container is too tightly closed, or by a low concentration of gelling agent.

- Sometimes decline of vigour in culture with stagnacy in shoot growth and proliferation is observed which may be caused by several factors. These include unsuitable composition of the culture medium, lack of some nutrients, calcium deficiency in the apices, which causes necrosis, the presence of latent persistent microbial contaminants, cytokinin habituation (extensive proliferation of short shoots on cytokinin-free medium without elongation and rooting ability), loss of regeneration ability in long-term cultures (due to epigenetic variation) and culture aging, including transition from the juvenile to a mature stage.

- Somaclonal variation may arise during *in vitro* regeneration [61]. Chromosomal rearrangements are an important source of this variation [62]. Somaclonal variation is not restricted to, but is common in plants regenerated from callus. Variation can be genotypic or phenotypic which, in the later case, can be either genetic or epigenetic in origin [41]. Cytological, biochemical and molecular analyses are required to confirm clonal fidelity of vegetatively propagated plant material. Such analyses enable efficient and rapid testing of undesired genetic variability in comparison with traditional methods based on morphological and physiological assays.

- Detailed information on *in vitro* propagation techniques for a broad spectrum of plant species are available in Jain and Gupta [63], Rout *et al.* [64] and Jain and Häggman [65].

References

***1. George EF (1993) *Plant Propagation by Tissue Culture: The Technology*. Exegetics Ltd., Edington, UK.

Fundamental information on tissue culture methods.

2. Dirr MA, Heuser Jr CW (1987) *The Reference Manual of Woody Plant Propagation: From Seed to Tissue Culture*. Varsity Press, Athens, GA, USA.

3. George EF, Sherrington PD (1984) *Plant Propagation by Tissue Culture:Handbook and Directory of Commercial Laboratories*. Exegetics Ltd., Eversley, UK.

4. Zimmerman RH, Greisbach FA, Hammerschlag FA, Lawson RH (1986) (eds) *Tissue Culture as a Plant Production System for Horticultural Crops*. Martinus Nijhoff Publishers, Dordrecht, The Netherlands.

5. Stimart DP (1986) Commercial micropropagation of florist flower crops. In: *Tissue Culture as a Plant Production System for Horticultural Crops*. Edited by RH Zimmerman, FA

Greisbach, FA Hammerschlag and RH Lawson. Martinus Nijhoff Publishers, Dordrecht, The Netherlands, pp. 301–315.

6. Fiorino P, Loreti F (1987) *HortScience* **22**, 353–358.

7. Debergh PC (1987) In: *Plant Tissue and Cell Culture*. Edited by CE Green, DA Somers, WP Hacket and DD Biesboer. Alan R. Liss, New York, pp. 383–393.

8. Pierik RLM (1997) *In Vitro Culture of Higher Plants*, 4th edn. Kluwer Academic Publishers, Dordrecht, The Netherlands.

***9. Razdan MK (2003) *Introduction to Plant Tissue Culture*. Science Publishers Inc., Enfield, NH, USA.

Clearly written, well-documented introductory information on plant tissue culture methods.

10. Williams EG, Maheswaran G (1986) *Ann. Bot.* **57**, 443–462.

11. Gautheret RJ (1983) *Bot. Mag.* **96**, 393–410.

12. Gautheret RJ (1985) In: *Cell Culture and Somatic Cell Genetics of Plants*. Edited by IK Vasil. Academic Press, New York, USA. Vol. 2, pp. 1–59.

13. Murashige T, Skoog F (1962) *Physiol. Plant.* **15**, 473–497.

14. Gajdošová A, Ostrolucká MG, Libiaková G, Ondrušková E, Šimala D (2006) *J. Fruit Ornamental Plant Res.* **14**, 61–76.

15. McCown BH (1986) In: *Tissue Culture as a Plant Production System for Horticultural Crops*. Edited by RH Zimmeman, RJ Griesbach, FA Hammerschlag and RH Lawson. Martinus Nijhoff Publishers, Dordrecht, The Netherlands, pp. 333–342.

***16. Hartmann HT, Kester DE, Davies FT Jr., Geneve RL (2002) *Hartmann and Kester's Plant Propagation: Principles and Practices*, 7th edn. Prentice-Hall, Inc., Englewood Cliffs, NJ, USA

Volume covering all aspects of plant propagation.

17. Ostrolucká MG, Gajdošová A, Libiaková G, Hrubíková K, Bežo M (2007) In: *Protocols for Micropropagation of Woody Trees and Fruits*. Edited by SM Jain and H Häggman. Springer-Verlag, Berlin, Heidelberg, pp. 445–455.

18. Ostrolucká MG, Gajdošová A, Libiaková G (2007) In: *Protocols for Micropropagation of Woody Trees and Fruits*. Edited by SM Jain and H Häggman. Springer-Verlag, Berlin, Heidelberg, pp. 85–91.

19. Gajdošová A, Ostrolucká MG, Libiaková G, Ondrušková E (2007) In: *Protocols for Micropropagation of Woody Trees and Fruits*. Edited by SM Jain & H Häggman. Springer, Dordrecht, The Netherlands, pp. 447–464.

20. Bonga J, von Aderkas P (1992) *In vitro Culture of Trees*. Kluwer Academic Publishers, Dordrecht, The Netherlands.

21. De Klerk GJ, Van der Krieken W, De Jong JC (1999) *In Vitro Cell. Dev. Biol-Plant.* **35**, 189–199.

22. De Klerk GJ (2001) In: *Plant Roots: the Hidden Half*. Edited by Y Waisel, A Eshel and U Kafkafi. Marcel Dekker Publishers, New York, Basel, pp. 349–357.

23. De Klerk GJ., Arnold-Schmitt B, Lieberei R, Neumann KH (1997) *Bio. Plant.* **39**, 53–66.

24. Preece JE, Sutter EG (1991) In: *Micropropagation Technology and Application*. Edited by PC Debergh and RH Zimmerman. Martinus Nijhoff Publishers, Dordrecht, The Netherlands, pp. 71–93.

25. Rohr R, Iliev I, Scaltsoyiannes A, Tsoulpha P (2003) *Acta Hort.* **616**, 59–69.

26. Donelly D & Tisdall L (1993) In: *Micropropagation of Woody Plants*. Edited by MR Ahuja. Kluwer Academic Publishers, Dordrecht, Boston, London, pp. 153–166.

27. Echeverrigaray S, Basso R, Andrade LB (2005) *Biol. Plant.* **49**, 439–442.

*28. Bonga J (1982) In: *Tissue Culture in Forestry*. Edited by J Bonga and D Durzan. Martinus Nijhoff/Dr.W. Junk Publishers, Amsterdam, The Netherlands, pp. 387–412.

Application of tissue culture methods to woody plants.

29. Franklet A, Boulay M, Bekkaoui F, Fouret Y, Verschoore-Martouze B, Walker N (1987) In: *Cell and Tissue Culture in Forestry*. Edited by JM Bonga and DJ Durzan. Martinus Nijhoff Publishers, Dordrecht, The Netherlands, Vol. 1. pp. 232–248.

30. David A, Faye M, Rancillac M (1983) *Plant Soil* **71**, 501–505.

31. Stein A, Fortin JA (1990) *Can. J. Bot.* **68**, 492–498.

32. Piola F, Rohr R, von Aderkas P (1995) *Physiol. Plant.* **95**, 575–580.

33. Douds Jr DD, Bécard G, Pfeffer PE, Doner LW (1995) *HortScience* **30**, 133–134.

34. Grange O, Bärtschi H, Gay G (1997) *Trees*, **12**, 49–56.

35. Hay A, Tsiantis M (2005) *Development*, **132**, 2679–2684.

36. Castellano MM, Sablowski R (2005) *Curr. Opin. Plant Biol.* **8**, 26–31.

***37. Wang PJ, Charles A (1991) In: *Biotechnology in Agriculture and Forestry*. Vol. 17. *High-Tech and Micropropagation*. Edited by YPS Bajaj. Springer-Verlag, Heidelberg, pp. 32–53.

Basic information on micropropagation through meristem culture.

38. Pérez-Tornero O, Burgos L (2007) In: *Protocols for Micropropagation of Woody Trees and Fruits*. Edited by SM Jain and H Häggman. Springer-Verlag, Berlin, Heidelberg, pp. 267–278.

39. Aitken-Christie J, Jones C (1985) *Acta Hort.* **166**, 93–100.

40. Gaspar Th, Kevers C, DeBergh P, Maene L, Paques M, Boxus Ph. (1987) In: *Cell and Tissue Culture in Forestry*. Edited by JM Bonga and DJ Durzan. Martinus Nijhoff Publishers, Dordrecht, The Netherlands, Vol. 1, pp. 152–166.

41. Kitin P, Iliev I, Skaltsoyiannes A, Nellas Ch, Rubos A, Funada R (2005) *Plant Cell, Tiss. Org. Cult.* **82**, 141–150.

42. Dumas E, Monteuuis O (1995) *Plant Cell Tissue Org. Cult.* **40**, 231–235.

43. Anderson WC (1980) In: *Proceedings of the conference on nursery production of fruit plants through tissue culture. Applications and feasibility*. pp. 1–10. Edited by RH Zimmermann. US Department of Agricultural Science and Education Administration ARR-NE-11.

44. Geneve RL (1989) *Proc. Internatl. Plant Propagtors Soc.* **39**, 458–462.

45. Zhang S, Lemaux P (2004) *Critical Rev. Plant Sci.* **23**, 325–335.

46. Vooková B, Gajdošová A (1992) *Biol. Plant.* **34**, 23–29.

47. Emons AMC (1994) *Acta Bot. Neerl.* **43**, 1–14.

48. Williams EG, Maheswaran G (1986) *Ann. Bot.* **57**, 443–462.

49. Fehér A, Pasternak TP, Dudits D (2003) *Plant Cell Tissue Org. Cult.* **74**, 201–228.

50. Castellanos M, Power B, Davey M, (2008) *Propag. Ornam. Plants.* **8**, 173–185.

51. Vooková B, Gajdošová A, & Matúšová R (1998) *Biol. Plant.* **40**, 523–530.

*52. Dodeman VL, Ducreux G, Kreis M (1997) *J. Exp. Bot.* **313**, 1493–1509.

Identification of the mechanism underlying the developmental stages of somatic and zygotic embryogenesis.

***53. von Arnold S, Sabala I, Bozkhov P, Daychok J, Filonova L (2002) *Plant Cell Tiss. Organ Cult.* **69**, 233–249.

Factors regulating somatic embryogenesis.

54. Salaj T, Matúšová R, Salaj J (2004) *Biol. Cracoviensia Series Botanica*, **46**, 159–167.

55. Salajova T, Salaj J, Kormut'ák A (1999) *Plant Sci.* **145, 33–40.

Efficient protocol for conifer somatic embryogenesis.

56. Walter Ch, Find JI, Grace LJ (2005) In: *Protocols for Somatic Embryogenesis in Woody Plants*. Edited by SM Jain SM and PK Gupta. Springer-Verlag, Berlin, Heidelberg, pp. 11–24.

57. Quiroz-Figueroa FR, Rojas-Herrera R, Galas-Avalos RM, Loyola-Vargas VM (2006) *Plant Cell Tiss. Organ Cult.* **86**, 285–301.

***58. George EF (1996) *Plant Propagation by Tissue Culture: In Practice*. Exegetics Ltd., Edington, UK.

Fundamental information on tissue culture methods.

59. Ziv M (1991) In: *Micropropagation: Technology and Application*. Edited by PC Debergh and RH Zimmerman. Kluwer Academic Publishers, Dordrecht, The Netherlands, pp. 45–70.

60. Debergh P, Aitken-Christie J, Cohen D *et al.* (1992) *Plant Cell Tissue Org. Cult.* **30**, 135–140.

61. Jain SM (2001) *Euphytica* **118**, 153–166.

**62. Jain, SM, Brar DS, Ahloowalia BS (1998) *Somaclonal Variation and Induced Mutations in Crop Improvement*. Kluwer Academic Publishers, Dordrecht, The Netherlands.

Information on genetic variability arising during *in vitro* culture and its use in plant breeding.

**63. Jain SM, Gupta PK (2005) *Protocols for Somatic Embryogenesis in Woody Plants*. Springer-Verlag, Berlin, Heidelberg.

Detailed information on somatic embryogenesis in a broad spectrum of plant species.

64. Rout GR, Mohapatra A, Jain SM (2006) *Biotechnol. Adv.* **24**, 531–560.

**65. Jain SM, Häggman H (2007) *Protocols for Micropropagation of Woody Trees and Fruits*. Springer-Verlag, Berlin, Heidelberg.

Detailed information on *in vitro* propagation techniques for a broad spectrum of plant species.

2

Thin Cell Layers: The Technique

Jaime A. Teixeira da Silva and Michio Tanaka
Faculty of Agriculture and Graduate School of Agriculture, Kagawa University, Kagawa-ken, Japan

2.1 Introduction

The advent of the concept of thin cell layers (TCLs) began about 35 years ago with the ground-breaking work by Khiem Tranh Than Van in which she demonstrated that by excising thin, transverse slices of tissue from pedicels of flowering *Nicotiana tabacum* it was possible to induce flowers, vegetative buds and roots *in vitro* [1]. At that time, much work had already been focused on the tissue culture of tobacco, including the fundamental study by Murashige and Skoog [2] that eventually led to the establishment of a basal medium. The latter proved to be the most commonly used medium in plant tissue culture. Certainly, it was neither the ability to culture tobacco tissue under axenic (sterile) conditions, nor it the possibility to culture plant cells *in vitro* to create a complete plant (the original concept of totipotentiality or totipotency which Haberlandt proposed almost 75 years earlier), that was revolutionary about TCLs. Rather, it was the capacity to control more strictly the outcome of an organogenic 'programme', not so much by the contents and additives of the medium or the surrounding environment, but rather by the size of the explant itself, that captivated the attention of plant tissue culture scientists since 1973. In the 35 years or so that have elapsed, TCLs have been shown to be veritable tools in the controlled organogenic potential of almost every group of plants, with hundreds of examples having been put successfully to the test [3, 4].

This methods chapter focuses on what was once considered to be a particularly difficult-to-propagate plant, namely *Cymbidium* hybrids. However, TCL technology, now allows for easy and reproducible tissue culture of this valuable ornamental and cut-flower pot plant and the possibility for micropropagation, including the use of bioreactors, without the need for expensive labour and technology. Such

Plant Cell Culture Edited by Michael R. Davey and Paul Anthony

an in-depth method cannot be found anywhere in the literature, mainstream or otherwise, despite several decades of orchid tissue culture research.

2.2 Methods and approaches

2.2.1 TCL

Plant tissue culture has always had a basic, fundamental and common vision; how to perfect a protocol such that a desired organ or plant of interest can be generated, inexpensively, reproducibly and in large numbers. Just over a century after Haberlandt postulated that any living plant cell could generate a complete clonal product, his concept was put into practice to produce an endless list of successful protocols for an ever-increasing range of plants.

Initially, the concept of TCL was applied to thin sections of *N. tabacum* pedicels [1]. At that time, it was suggested that a 1 mm-thick layer of cells as epidermal peels (of variable dimensions) should be defined as a longitudinal TCL or lTCL, while a transverse slice, a few millimetres thick, should be termed a transverse TCL or tTCL. In a recent paper, the first author contested the entire premise behind the terminology originally used and now widely adapted, and suggested that the term should be adjusted to thin tissue layer or TTL [5]. This author hopes that the present chapter may provide some consistent ground-rules and guidelines for plant tissue culture scientists and explains in some detail a protocol that facilitate the concept of a TCL to be more easily understood and applied.[a]

Disclaimer

[a]The claims and successes/cautions explained herein are only applicable for *Cymbidium* hybrid Twilight Moon 'Day Light'. This chapter does not in any way insinuate or imply the success of the technique to any other *Cymbidium* or orchid species, or any other plant.

2.2.2 Choice of material: *Cymbidium* hybrid

The focus on *Cymbidium* hybrid orchids has been selected for three main reasons. Until recently, only terrestrial cymbidiums had been propagated *in vitro* [6], mainly through the culture of shoot tips [7], whereas *Cymbidium* hybrids were much more difficult to propagate. Because it is a difficult plant to propagate efficiently, being able to manipulate organogenesis precisely *in vitro* makes it a suitable model plant. By showing that the TCL technique is applicable to an expensive ornamental market commodity, hope is created to exploit the technique in both developing and developed countries for the mass propagation of conventional cash crops, as well as other difficult-to-propagate species.

Cymbidium tissue culture has been reviewed elsewhere [8]. Consequently, only the most important and fundamental concepts are defined here. The first of these is that of a protocorm-like body (PLB), which is an organ that resembles a protocorm, but is not such a structure, since a protocorm must derive from a seed. A

PLB does not derive from a seed, although a PLB may derive from a protocorm. That said, from where does the original PLB derive? This basic, important fact is always overlooked in almost every single tissue culture and micropropagation protocol available for almost every orchid, but one whose record must be clarified. In the case of hybrid *Cymbidium*, where plantlets, originally derived from the tissue culture of sterilized shoot tips, are cultured on a highly organic substrate (e.g. one supplemented with banana), from a flask of about 100 rooted shoots, about 1% of plants *spontaneously* form a PLB at the base of the leaf sheath. This primary (1°) PLB, once cultured on appropriate medium, can then form secondary (2°) PLBs [9], albeit at a low multiplication rate. Every time a PLB is used (whole or in part) for subculture, it is considered a 1° PLB and any PLB that is derived from a 1° PLB is a 2° PLB. A tertiary (3°) PLB is essentially the same as a 2° PLB (in terms of its origin), although it is strictly clonal, that is of the same size, shape and dimensions, and would be used in commercial micropropagation. The capacity for PLBs to be suitable explants for callus formation and 'somatic embryogenesis' was demonstrated later [10]. Note how the term somatic embryogenesis has been placed in inverted commas. The term somatic embryogenesis is often incorrectly and loosely used by many plant tissue culture scientists, often without histological evidence. Orchid tissue culture scientists often classify a highly compact and dense cluster of immature PLBs as somatic embryos, which is incorrect. Histological, cytometric and genetic analyses have showed that a PLB is a somatic embryo (i.e. they are not separate entities), a revolutionary finding in the field of orchid tissue culture [11], although the consequences of this finding appear not yet to be fully appreciated.

The methodology below does not include any process of the tissue culture protocol that goes beyond the plantlet stage *in vitro*, as that is beyond the scope of the TCL technology, and is unrelated to the technique of focus in this chapter.

PROTOCOL 2.1–3

The following equipment and reagents are required for Protocols 2.1–2.3.

Equipment and Reagents

- Glasshouse facilities for stock plants
- Laminar flow cabinet for aseptic procedures
- Constant temperature facilities
- Binocular dissection microscope
- Sterile double distilled water (SDDW)
- 1.5% (w/v) sodium hypochlorite solution
- Glass beakers (250 ml)
- Erlenmeyer flasks (100 ml, 250 ml)
- Forceps
- Blades (Hi stainless platinum or carbon steel; Feather Safety Razor Co., Ltd.)

- Petri dishes (100 mm diam., 15 mm deep; Falcon)
- Half strength Murashige and Skoog (MS)-based medium [12] lacking growth regulators
- Medium of Vacin and Went [13] with Nitsch microelements [14], 2.0 mg/l tryptone, 0.1 mg/l α-naphthaleneacetic acid (NAA), 0.1 mg/l kinetin, 2% (w/v) sucrose, 8.0 g/l Bacto agar
- Kinetin (Tissue Culture [TC] grade; Sigma)
- NAA (TC grade; Sigma)
- Benzyladenine (BA) (TC grade; Sigma)
- Tryptone (TC grade; Sigma)
- Activated charcoal (AC) (acid washed; Sigma)
- Bacto agar (Difco Laboratories)
- Gelrite gellan gum (TC grade; Sigma)
- Coconut water (obtained from fresh, green coconuts, free of flesh, frozen immediately)
- Whatman No. 1 filter paper (9 cm diam.)

PROTOCOL 2.1 Induction of Primary Protocorm-like Bodies (1° PLBs) from Shoots of Mature Plants

Method

1. Excise young shoots of *Cymbidium* hybrid Twilight Moon 'Day Light' from 3-year-old mature plants growing in a glasshouse lacking any visible symptoms of bacterial, fungal or viral infection.
2. Place shoots under running tap water in a suitable container (e.g. 250 ml beaker) for 30 min. Working in a laminar flow cabinet, surface sterilize the explants in 1.5% (w/v) sodium hypochlorite solution in a 250 ml flask for 15 min. Transfer shoots to new sterilizing solution for another 15 min. Rinse shoots three times with sterile distilled water (~5 min each wash) and place in a sterile Petri dish.
3. Isolate apical meristems (~5–10 mm terminal tips). Culture the apical explants on plant growth regulator-free half-strength MS-based salts medium [12] to induce 1° PLBs[a,b].
4. After ~6 months, 1° PLB(s) should appear at the base of rooted shoots. Excise these PLBs and subject them to Protocols 2.1–2.3.[c]
5. A 'Universal' medium for 2° PLB formation is 10 1° PLBs on 40 ml/100 ml flask of PLB-induction medium, based on the formulation of Vacin and Went [13] supplemented with Nitsch microelements [14], 2 mg/l tryptone, with NAA and kinetin each at 0.1 mg/l. To this medium, add 2% (w/v) sucrose. Adjust the pH to 5.8 ± 0.1 and add 8 g/l Bacto agar; autoclave at 835 kPa (121 psi) for 21 min.

6 Culture 1° and 2° PLBs at 25 ± 0.5°C under a 16 h photoperiod provided by fluorescent tubes (FL 20 SS-BRN/18, Cool White, Plant Lux, 18 W, Toshiba) with a low photon flux density of 30–40 μmol/m^2/s.[d]

Notes

[a]1° PLBs are not guaranteed to form on this medium. Ideally a banana-based medium (half strength MS-based medium, 0.2% (w/v) activated charcoal, 8% (w/v) banana homogenate; modified from [15] will yield more 1° PLBs. Once the initial shoots begin to elongate (before roots elongate, or cut off roots), transfer to 0.5% (w/v) Gelrite supplemented with 2% (w/v) ripe banana and 10% (v/v) coconut water; this results in strong growth of the shoots and roots of plantlets.

[b]This process/medium combination usually yields 100% survival with the cultivar Twilight Moon 'Day Light'.

[c]Use at least 40 replicates, and to repeat the experiment at least three times for statistical treatment. Wherever possible, use more than one cultivar for comparison.

[d]The authors' experience is that a high level of irradiation (>80 μmol/m^2/s) may inhibit 2° PLB formation, sometimes completely. Conversely, darkness is not so effective, and it is better to substitute 0.1 mg/l kinetin with 1.0 mg/l BA. In this case, 2° PLBs form, but these are white and not as numerous. Moreover, they regain their photosynthetic capacity once transferred to light. Another alternative is to supplement the kinetin/BA medium with AC at 1% (w/v), and place the cultures in the light; it is possible that the AC mirrors a darkened natural environment of *Cymbidium* in its tree-top habitat.

PROTOCOL 2.2 Conventional (i.e. 2°) PLB Formation from Complete 1° PLBs

Method

1 As the 1° PLB grows, 2° PLBs form on the 1° PLB. These may simply be separated and placed on the same medium to induce tertiary (3° PLBs)[a–e] (see Figure 2.1).

Notes

[a]Most protocols in the literature on '*Cymbidium*' are mainly on terrestrial cymbidiums, which, like *Dendrobium* spp., are much easier to propagate *in vitro*. Most, if not all, of these protocols, use this method or small variations thereof.

[b]This results in very few (average = 1.68, $n = 40$) 2° PLBs per 1° PLB. Hypothetically, subculture to subculture would yield a 13.4 × multiplication rate after five consecutive subcultures (2 months each); i.e. with a single initial 1° PLB, a total of 5353° PLBs can be obtained after a 10-month period, assuming that every 1° and 2° PLB is used, that every 1° and 2° PLB survives, and that every 1° and 2° PLB is able to differentiate.

[c]A typical subculture should be made once every 2 months before the apical meristems have time to develop into shoots, and before roots can emerge from the base of the 1° PLBs.

[d]Oxidation and browning of 1° PLBs can take place rapidly (within 10 min). Consequently, 1° PLBs should be placed on culture medium immediately following dissection.

[e]The size of the 1° PLB will differ, depending on the cultivar. However, for the cultivar Twilight Moon 'Day Light', 1° PLBs of standard diameter (4–6 mm) should be used. Larger 1° PLBs may be too advanced developmentally, and may have started to form a shoot and adventitious roots. This tends to reduce the PLB-inducing potential of 1° PLBs. Too small a 1° PLB will result in poor 2° PLB formation because of too much tissue damage and reduced surface area.

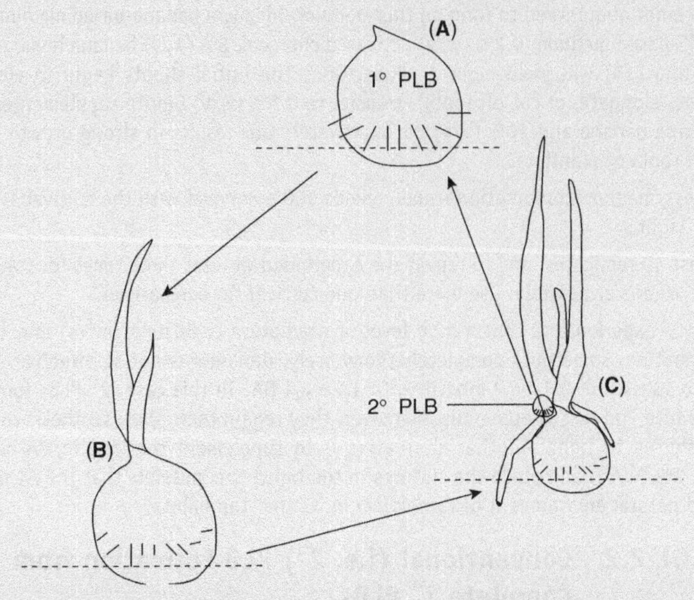

Figure 2.1 Protocol 2.2. Culture of a complete 1° PLB (**A**) results in the formation of a plantlet (shoot and adventitious root formation. (**B**) 2° PLBs, whose formation is erratic after 30–45 days (**C**), and whose rate of formation is low, can be harvested and employed as 1° PLBs (**A**) in a second round of 2° PLB formation. This method is not recommended for micropropagation (i.e. 3° PLB formation) due to differences in size, shape and developmental stage. Dashed line indicates the culture medium surface. Figure not to scale.

PROTOCOL 2.3 Improved (i.e. 2°) PLB Formation from Half 1° PLBs [10, 16, 17]

Method

1. When each 1° PLB grows, 2° PLBs form on the 1° PLB, usually at the base. Separate the 2° PLBs and place in an autoclaved glass Petri dish with a double sheet of Whatman No. 1 filter paper laid on the base.[a]
2. Using a feather blade, cut the top 1 mm of the 1° PLB, which contains the apical meristem. Slice away the bottom, brown part of the 1° PLB, if applicable[a–e] (see Figure 2.2).

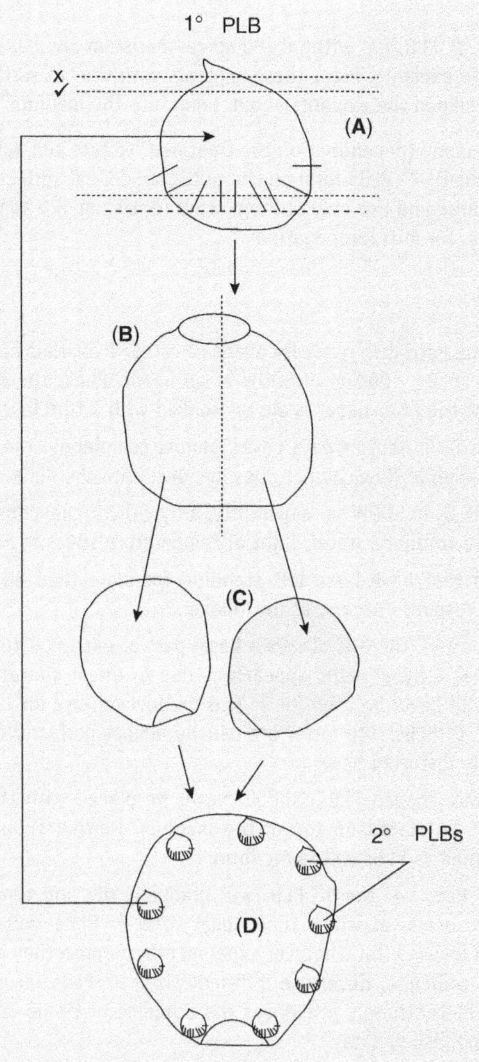

Figure 2.2 Protocol 2.3. A complete 1° PLB (**A**) is NOT cultured (unlike Protocol 2.2); instead, the apical meristem and basal part of the PLB that is in contact with the culture medium are dissected/removed, yielding a 'trimmed' 1° PLB (**B**). Note: trimming should take place before the shoot apical meristem begins to elongate into a shoot. This 'trimmed' 1° PLB is cut symmetrically length-wise to yield two half-moon-shaped explants (**C**). When each half-moon-shaped PLB explant is re-placed on the same culture medium, several 2° PLBs form near, or at, cut surfaces (primarily) and on the surface after 30–45 d (**D**); the rate of formation is greater than in Protocol 2.2, and can be harvested and employed as 1° PLBs (**A**) in a second round of 2° PLB formation. This method is recommended for micropropagation (i.e. 3° PLB formation) because of high rates of PLB formation, each of more-or-less uniform size, shape and developmental stage. Dashed line indicates the culture medium surface. Dotted lines indicate lines of sectioning. ✓ = correct level (removes the shoot apical meristem); ✗ = incorrect level (does not remove the shoot apical meristem). Figure not to scale.

3 Slice the 'trimmed' 1° PLB, i.e. without the apical meristem and base, symmetrically to yield two half-moon explants. Place these explants with the cut surface in contact with the medium. Embed the explants about 1 mm into the medium[f].

4 After 45–60 days, many (depending on the treatment to test and apply, i.e. the actual experimental protocol), 2° PLBs form on the outer, epidermal surface of each PLB. Allow these to enlarge and use only uniform sized (optimum = 4–6mm) 2° PLBs for PLB production, i.e. for micropropagation[g,h].

Notes

[a]It is useful to have one Petri dish ready for every 10–20 1° PLBs that need to be prepared. For a total of 1000 1° PLBs, 1000 ml of SDDW is sufficient. Place 10–20 ml of SDDW into each Petri dish so that the filter paper is always soaked with a thin layer of SDDW.

[b]Never allow the PLBs to dry-out; always cover, almost completely, the Petri dish so that the air flow from the laminar flow cabinet does not desiccate the PLBs.

[c]Never submerge the PLBs in SDDW as, apparently, a hyperhydric response occurs and PLBs are extremely sensitive to injury, water, light or temperature stress in SDDW.

[d]Discard any 1° PLBs that have been left standing for more than 30 min in SDDW. An apparent hyperhydric response occurs, as in [c] above.

[e]In 1° to 2° PLB formation, there is always a basal part of each PLB that is callus-like in appearance, or that has a hyperhydric appearance due to direct contact with the culture medium. 1° PLBs should never be used for 3° PLB production; use only 2° PLBs that form on the outer layer of 1° PLBs. The latter are usually almost perfectly round, and do not have a morphologically distorted base.

[f]Explants (1° half-moon shaped PLBs) should never be placed with their intact surface down on the medium, or placed on top of the medium. Neither should they be totally embedded in the medium as PLBs will rarely form.

[g]Usually the 'mother' PLB, i.e. the 1° PLB, will gradually die and turn brown. This will take about 60 days to occur, at which time, ideal sized 2° PLBs, will have formed. The latter can, and should be used, for whatever experimental purpose they are required, or for micropropagation. In principle, never use different sized 2° PLBs for experiments, since the initial size of 2° PLBs strongly influences the outcome of tissue culture experiments (Teixeira da Silva, unpublished data).

[h]The sharpness of the blade is one of the most important factors that determines the success of Protocols 2.3 and 2.4, in particular Protocol 2.4, which requires thin explants. Feather blades, made in Japan by the Feathers Safety Razor Co., Ltd., should be sterilized by autoclaving for at least 17 min., boiling, then immersing in 98% ethanol (no need to flame). They will remain sharp for explant preparation. Several other makes of blade from other suppliers around the world do not give the same perfect 'slice'.

[i]This results in a large number (mean = 8.21, $n = 40$) of 2° PLBs per 1° PLB half-moon. Hypothetically, subculture to subculture should yield a 4000× multiplication rate after 4 consecutive subcultures (3 months each). Thus, with two initial 1° PLB half-moon explants, a total of ~36 350 3° PLBs can be obtained after a 12-month period, assuming that every 1° and 2° PLB is used, that every 1° and 2° PLB survives and that every 1° and 2° PLB is able to differentiate.

PROTOCOL 2.4 TCL-induced (2°) PLB Formation from 1° PLB tTCLs [11, 18, 19]

Method

1. When the 1° PLB grows, 2° PLBs form on the 1° PLB, usually at the base. Following the general guidelines for Protocol 2.3, select only ideal sized and shaped 2° PLBs.
2. Using a new feather blade for every six to eight PLBs, make a 0.5–1.0 mm deep incision in the shape of a square, 3–5 × 3–5mm in area. Slice this area to separate the epidermal 0.5–1.0 mm in one continuous movement, thus creating an lTCL[a,b] (see Figure 2.3b–d).
3. Using a new feather blade for every six to eight PLBs, and only using the central 5 mm girth of the 1° PLB, make a 0.5–1.0 mm transverse slice throughout the whole PLB, thus creating a tTCL[a,b] (Figure 2.3e–h).

Notes

[a]It is important to prepare the lTCL in a single stroke (e.g. as one would when opening an envelope with a new letter opener). If the explant is prepared in several strokes (e.g. as in slicing an object with a bread knife), the explant itself tends to become damaged on both upper- and under-surfaces.

[b]Although the inner tissue (subepidermal layers and below) of a PLB never, in any treatment tested [11], forms 2° PLBs, any damage to this tissue results in rapid browning (within 1 week) and eventual necrosis (within 1–2 weeks) of the TCL. It is thus imperative to change the feather blade regularly and to water the cut lTCLs/tTCLs with SSDW.

[c]This results in a very large number (average = 14.48, $n = 40$) of 2° PLBs per 1° PLB lTCL, but in much fewer (average = 6.08, $n = 40$) 2° PLBs per 1° PLB tTCL (the reason is related to the total surface area of a tTCL being much less than that of an lTCL). Note that two lTCLs can be prepared from an ideal-sized 1° PLB, while five tTCLs can be prepared from the same mother explant. Hypothetically, subculture to subculture should yield a 24 280× multiplication rate after three consecutive subcultures (3 months each) for lTCLs. Thus, with two initial 1° PLB lTCLs, a total of ~351 700 3° PLBs can be obtained after a 9-month period, assuming that every 1° and 2° PLB is used, that every 1° and 2° PLB survives, and that every 1° and 2° PLB is able to differentiate. For tTCLs, these values are lower, but still significant, if considering a commercial micropropagation facility. Hypothetically, subculture to subculture would yield a 4620× multiplication rate after three consecutive subcultures (3 months each) for tTCLs. Thus, from five initial 1° PLB tTCLs, a total of ~28 100 3° PLBs can be obtained after a 9-month period, assuming that every 1° and 2° PLB is used, that every 1° and 2° PLB survives and that every 1° and 2° PLB is able to differentiate.

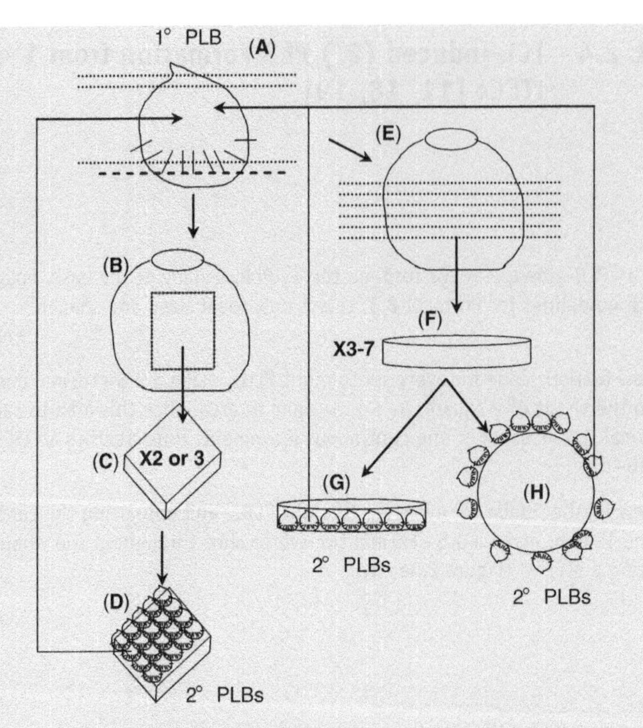

Figure 2.3 Protocol 2.4. A whole 1° PLB (**A**) is *not* cultured (as in Protocol 2.3); rather, the apical meristem and basal part of the PLB that is in contact with the medium are dissected/removed, yielding a 'trimmed' 1° PLB (**B**). Note: trimming should take place before the shoot apical meristem begins to elongate into a shoot. This 'trimmed' 1° PLB now enters the lTCL (**B–D**) or the tTCL (**E–H**) pathways. In the lTCL pathway, two to three lTCLs (0.5 mm thick, 3 × 3 mm) can be prepared from a single 'trimmed' 1° PLB (**C**). When each lTCL is re-plated on the same medium, numerous 2° PLBs form over the entire surface after 20–25 days, and can be harvested at 30–45 days (**D**). The rate of formation is higher than in Protocols 2.2 and 2.3, and can be harvested and employed as 1° PLBs (**A**) in a second round of 2° PLB formation. This method is recommended for micropropagation (i.e. 3° PLB formation) because of high rates of PLB formation, each of more uniform size, shape and developmental stage than those harvestable from Protocol 2.3. In the tTCL route, a single 'trimmed' 1° PLB can yield three to seven (best is five) 'slices' or tTCLs (**E, F**). When each tTCL is re-placed on the same medium, numerous 2° PLBs form only on the surface containing PLB surface (internal tissue never forms PLBs; **G** = side view, **H** = top view) after 20–25 days, and can be harvested at 30–45 d (**G, H**); the rate of formation is greater than in Protocol 2.2, but never more than Protocol 2.3 or the lTCL method, and can be harvested and employed as 1° PLBs (**A**) in a second round of 2° PLB formation. This method is not recommended for micropropagation (i.e. 3° PLB formation) because of low rates of PLB formation, even though each is uniform in size, shape and developmental stage (as for the lTCL route). Dashed line indicates the culture medium surface. Dotted lines indicate lines of sectioning. ✓ = correct level (removes the shoot apical meristem); ✗ = incorrect level (does not remove the shoot apical meristem). Figure not to scale.

2.3 Troubleshooting

- It is always useful to run any experiment using Protocol 2.2 as the 'positive' control since Protocol 2.2 is used most commonly for many orchids, and *Cymbidium*, in particular. This is especially useful if the objective of a particular experiment is to quantify the number of PLBs formed as the result of an experimental procedure.

- The choice of a suitable, uniform sized PLB for Protocols 2.2, 2.3 or 2.4 is essential. The developmental stage of the PLB is also vital for successful experimental design. If these two factors are not considered carefully, then spurious results are likely to be obtained, independent of the number of replicates. To avoid protocol error, the authors recommend that at least two to three PLB subcultures be performed using Protocol 2.2 to select uniform sized and shaped PLBs. PLBs (Twilight Moon 'Day Light') greater than 5–6 mm are usually too advanced in their developmental programme and are likely to lead to shoot formation, which interferes with the regeneration potential of the explant. Similarly, PLBs < 3 mm in diameter are difficult to handle, even using a dissecting microscope, are prone to injury, yield few TCLs and are developmentally immature.

- The sharpness of the blade used to prepare PLB explants cannot be over-emphasized. Poorly or roughly prepared PLBs, ones that have suffered excessive damage, will die. In a single explant preparation session, in which it is estimated that 50–75 tTCLs can be prepared in 1 h, the blade must be changed for every 10–20 TCLs. Similarly, PLB explants that have been left, after preparation, for more than 0.5 h, should be discarded since their PLB-generating potential is low.

2.3.1 General comments

- TCL technology is an *in vitro* technique, based on the same principles that apply to any general plant tissue culture protocol, as far as experimental design and execution are concerned. One exception is Protocol 2.4 pertaining exclusively to TCLs, which needs particular attention to size, technique and care of the explants.

- TCL technology does not involve any high-technology histological, biochemical, or genetic techniques. However, TCLs have incredible potential when used in conjunction with any of these approaches, for assessing cellular and ultrastructural processes, controlling developmental events, assisting genetic transformation protocols, and improving regeneration and micropropagation of difficult-to-propagate species [3].

- An extremely useful technique, namely flow cytometry, can be used to assess the 'purity' of an explant (Teixeira da Silva, unpublished data). By understanding the ploidy level of explants through a rapid (<30 min) assay, their origin can be determined and, hence, the appropriateness for the proposed study.

- *Cymbidium* hybrid has been selected as an example for this methods chapter because it represents a fascinating and complex model system for plant development *in vitro*. Protocols 2.2–2.4 are linked and any person wishing to maximize the culture of orchids *in vitro* should pass sequentially through Protocols 2.2, 2.3 and 2.4, in order to harvest standard sized and shaped 2° PLBs. Credit should be given to Professor Michio Tanaka, Japan for the initial perfection of the technique underlying Protocol 2.2, which was initially applied to *Phalaenopsis* and *Vanda* [20].

- Protocol 2.2 results in mixed organogenesis, including PLBs, adventitious roots, shoots and callus. Protocol 2.3 results primarily in PLBs, some callus and, occasionally, shoots. Protocol 2.4 results exclusively in PLB production. Due to the multiple organogenic pathways that would result from the use of different protocols, in particular from Protocol 2.2, the estimated output (total number of 3° PLBs) would be extremely skewed, slightly skewed or almost not skewed when referring to Protocols 2.2, 2.3 and 2.4, respectively. To give the reader a more realistic perspective, 3° PLBs, i.e. of uniform size, shape and developmental stage, would/could be the material used to generate clonal hybrid *Cymbidiums* in a commercial orchid micropropagation unit, since shoots would all emerge very much synchronously and root and shoot development would result in very little variation. Protocols that have been established from Protocol 2.2, as is found in (>95% of all papers published for any orchid species *in vitro*, result in an organogenic outcome, but the programme is not 'pure', and is thus not very useful for commercial exploitation. Protocol 2.4 strengthens the importance of TCLs as tools for controlling organogenesis in an academic and a business setting.

- Essentially, one of the strong positive points of TCLs is the inherent capacity to strictly control an organogenic programme more than with a conventional explant, which has many advantages and applications in plant tissue culture. This was demonstrated for tobacco florigenesis (1), chrysanthemum rhizogenesis [21], *Lilium* somatic embryogenesis [22], and several other examples (Teixeira da Silva, in preparation).

- In relation to genetic stability and somaclonal variation, a resulting plant that is derived from an *in vitro* event, should be subjected to one or more rigorous tests for variation. Flow cytometry is a simple, but informative technique for testing culture 'purity'. Genetic fidelity can also be tested using molecular markers such as RAPDs [23].

References

*1. Tran Thanh Van M (1973) *Nature* **246**, 44–45.

The original TCL technique is described.

2. Murashige T, Skoog F (1962) *Physiol. Plant*. **15**, 495–497.

**3. Nhut DT, Van Le B, Tran Thanh Van K, Thorpe T (eds) (2003) *Thin Cell Layer Culture System: Regeneration and Transformation Applications*, Kluwer Academic Publishers, Dordrecht, The Netherlands, pp. 1–16.

An in-depth assessment of the origin of TCL technology with applications to almost every plant group.

4. Teixeira da Silva JA, Tran Thanh Van K, Biondi S, Nhut DT, Altamura MM (2007) *Floriculture Ornamental Biotech*. **1, 1–13.

The most recent review available on the application of TCL technology with a focus on ornamentals.

5. Teixeira da Silva JA (2008) *Int. J. Plant Dev. Biol*. **2**, 79–81.

6. Hasegawa A (1987) *Mem. Fac. Agric. Kagawa Univ*. **50**, 1–108.

7. Morel GM (1960) *Am. Orchid Soc. Bull*. **29**, 495–497.

8. Nayak NR, Tanaka M, Teixeira da Silva JA (2006) In: *Floriculture, Ornamental and Plant Biotechnology: Advances and Topical Issues*, edited by JA Teixeira da Silva. Global Science Books, Isleworth, UK, Vol IV, pp. 558–562.

9. Begum AA, Tamaki M, Kako S (1994) *J. Jpn. Soc. Hortic. Sci*. **63**, 663–673.

10. Huan LVT, Tanaka M (2004) *J. Hort. Sci. Biotech*. **79**, 406–410.

***11. Teixeira da Silva JA, Tanaka M (2007) *J. Plant Growth Regul*. **25**, 203–210.

A ground-breaking publication that challenges the conventional definition of a somatic embryo in orchid tissue culture and which shows clearly that a PLB is a somatic embryo.

12. Murashige T, Skoog F (1962) *Physiol. Plant*. **15**, 473–497.

13. Vacin E, Went FW (1949) *Bot. Gaz*. **110**, 605–613.

14. Nitsch C, Nitsch JP (1967) *Planta* **72**, 371–384.

15. Shiau Y-J, Sagare AP, Chen U-C, Yang S-R, Tsay H-S (2002) *Bot. Bull. Acad. Sin*. **43**, 123–130.

16. Teixeira da Silva JA, Chan M-T, Sanjaya, Chai M-L, Tanaka M (2006) *Sci. Hort*. **109**, 368–378.

17. Teixeira da Silva JA, Singh N, Tanaka M (2005) *Plant Cell, Tiss. Organ Cult*. **84**, 119–128.

18. Teixeira da Silva JA, Giang DTT, Chan M-T, *et al*. (2007) *Orchid Sci. Biotech*. **1**, 15–23.

19. Teixeira da Silva JA, Norikane A, Tanaka M (2007) *Acta Hort*. **748**, 207–214.

20. Tanaka M, Hasegawa A, Goi M (1975) *J. Jpn. Soc. Hort. Sci*. **44**, 47–58.

21. Teixeira da Silva JA (2003) *Plant Growth Regul*. **39**, 67–76.

22. Nhut DT, Huong NTD, Bui VL, Teixeira da Silva JA, Fukai S, Tanaka M (2002) *J. Hort. Sci. Biotech*. **77**, 79–82.

23. Teixeira da Silva JA, Tanaka M (2006) *Acta Hort*. **725**, 203–209.

3

Plant Regeneration – Somatic Embryogenesis

Kim E. Nolan and Ray J. Rose

School of Environmental and Life Sciences, The University of Newcastle, NSW, Australia

3.1 Introduction

In somatic embryogenesis (SE), embryos form asexually from somatic cells. SE is most commonly associated with the *in vitro* culture of excised tissues in a nutrient medium containing exogenously supplied plant growth regulators. However, SE can occur naturally as on the succulent leaves of *Kalanchoë* [1], and a type of SE can also occur naturally *in vivo* through the process of apomixis. Plants which undergo apomixis develop embryos in the ovule without fertilization [2] and fertile seed is produced with the same genotype as the parent. The methods in this chapter are concerned with SE *in vitro* and the use of the term 'SE' will be in the context of the *in vitro* form. SE is used in transformation procedures for many species.

For SE to occur, the differentiated plant cell needs to dedifferentiate (unless the cell is already meristematic) and form a stem cell, which develops through characteristic embryological stages to produce every cell type of the new plant. Therefore, the progenitor cell of a somatic embryo is a totipotent stem cell. Adventive shoots arising from culture can resemble somatic embryos. The main feature that defines a somatic embryo in comparison to an adventive bud is an anatomically discrete radicular end with no vascular connection to the maternal tissue [3].

The development of plant somatic embryos *in vitro* was first demonstrated in 1958 by Reinert [4] and Steward [5]. SE is classified into two types:

Plant Cell Culture Edited by Michael R. Davey and Paul Anthony

1 Indirect SE, where the explant tissue initially undergoes rapid cell division to form a relatively disorganized mass of cells called 'callus.' Somatic embryos then arise from the callus tissue.

2 Direct SE where embryos form directly from the explant without an intervening callus phase [6].

In both types of SE, the embryos resemble zygotic embryos and, for example, in dicotyledonous plants, go through the globular, heart, torpedo and cotyledonary stages, as do zygotic embryos. The embryos may then germinate and produce fertile plants. One major difference between somatic and zygotic embryogenesis is that somatic embryos do not go through the desiccation and dormancy observed in zygotic embryos, but rather tend to continue development into the germination phase as soon as they are fully formed [7].

There is considerable variation in the methods used to induce SE in different plants. Initially, a great deal was learnt about the importance of the type of explant and the role of exogenously supplied auxins and cytokinins, as well as other culture conditions. In some species, unspecified genotypic differences between plants were found to affect embryogenic competence (for a short review, see reference [8]). In more recent years, the roles have been discovered of other factors, such as stress and secreted proteins in the culture medium. These factors, along with the exploding field of gene discovery, have provided a wealth of new knowledge as reviewed in [9]. Of special note in this respect is the ability of the over-expression of certain transcription factors to induce SE, independent of exogenous growth regulators [10]. However, even after the publication of many papers on SE, we still do not understand how a cell is reprogrammed to become competent to form a somatic embryo.

The aim of this chapter is to describe generic methods that will enable SE to be initiated in any laboratory, but should be used with the caveat that species-specific adjustments will be required.

3.2 Methods and approaches

The procedures described here focus on methods of producing somatic embryos from explants that pass through a callus phase (indirect SE).

3.2.1 Selection of the cultivar and type of explant

For a given species SE is genotype dependent with significant variation in response between cultivars. Examples of this can be seen in cotton [11], soybean [12, 13], safflower [14], barley [15] and wheat [16]. In the genus *Medicago*, special genotypes have been produced by selection and breeding in the case of *M. sativa* [17] and in *M. truncatula* by an initial cycle of tissue culture to produce a regenerated plant which was used as a source of seed for selection over four generations [18, 19]. The enhancement of SE in plants that have been regenerated from tissue culture has also been observed in sunflower [20], carrot [21] and wheat [22]. If a species has not been regenerated previously by SE, then it is important to test a number of different cultivars. If the plant is recalcitrant but some regeneration occurs, then test the tissue from the regenerated plants for increased SE. The seed from the regenerated plants may be a source of a more embryogenic genotype, but as in *M. truncatula*, this trait may segregate [19].

The next question that arises is what tissue should be used as an explant since a wide range of source tissues have been used. The first piece of information that is helpful is to determine what explants are used in a closely related species, genus or family. Zygotic embryos (or other tissues in a meristematic state) are a popular source of tissue, as somatic embryos will form more readily from cells that are already in an embryonic state. However, embryos can be tedious to isolate. Seedling tissue is easier with which to work and is still in a juvenile state. In the more regenerable species more developed tissues such as leaves, roots, petioles or stems can be used. From broad considerations, it is known that the Solanaceae is more amenable to regeneration than the Fabaceae or the Gramineae, while the monocotyledons often require less differentiated tissue, and utilizing the embryo at the appropriate stage can be important [23]. Tissue selected as an explant source should always be young and healthy. A good reference for a summary of different explant sources used for different plants, as well as information on culture conditions, can be found in Thorpe [24].

3.2.2 Culture media

The culture media employed must supply all the essential nutrients for plant growth, a source of carbon and appropriate growth regulators for explant growth and the induction of somatic embryos. Although a myriad of different types of media are used, many culture media are based on a few original formulations, such as those of Gamborg (B5 medium; [25]), Murashige and Skoog (MS medium; [26]), Nitsch and Nitsch [27] or Schenk and Hildebrandt (SH medium; [28]). Table 3.1 gives the composition of these media. Some of these media and variation in their components may be purchased commercially. Plant growth regulators should be considered separately to the basal nutrient medium.

Basal media: nutrient components

In addition to requiring adequate nutrition for cells to grow and divide, SE can be enhanced by regulating the type and concentration of the nutrients of the culture medium. The most important nutrient in this respect is nitrogen [29, 30], the type of nitrogen supplied having a strong influence on the induction of SE. Often the presence is required of ammonium or some other source of reduced nitrogen, such as glycine, glutamate or casein hydrolysate. The ratio of ammonium to nitrate has also been shown to affect SE. Optimization of the carbon source, potassium, calcium or phosphorus has also been shown to positively affect SE.

A source of carbon, generally sucrose, is needed as the plant tissue is no longer able to supply its own through photosynthesis. Thiamine and *myo*-inositol appear to be the most important vitamins in culture media [31]. Presumably, they are

Table 3.1 Composition of 1 l of culture media.

	MS	B5	SH	Nitsch	P4*
Major salts	(mg)	(mg)	(mg)	(mg)	(mg)
KNO_3	1900	2500	2500	950	1875
$MgSO_4.7H_2O$	370	250	400	185	225
KCl	–	–	–	–	225
$NH_4H_2PO_4$	–	–	300	–	–
$(NH_4)_2SO_4$	–	134	–	–	–
NH_4NO_3	1650	–	–	720	600
KH_2PO_4	170	–	–	68	–
$NaH_2PO_4.H_2O$	–	150	–	–	–
$CaCl_2$	–	–	–	166	–
$CaCl_2.2H_2O$	440	150	200	–	300
Minor salts	(mg)	(mg)	(mg)	(mg)	(mg)
$MnSO_4.4H_2O$	22.3	–	–	25	–
$MnSO_4.H_2O$	–	10	10	–	10
H_3BO_3	6.2	3	5	10	3
$ZnSO_4.7H_2O$	8.6	2	1	10	2
KI	0.83	0.75	1	–	0.75
$Na_2MoO_4.2H_2O$	0.25	0.25	0.1	0.25	0.25
$CuSO_4.5H_2O$	0.025	0.025	0.2	0.025	0.025
$CoCl_2.6H_2O$	0.025	0.025	0.1	–	0.025
$FeSO_4.7H_2O$	27.8	–	15	27.85	9.267
$Na_2EDTA.2H_2O$	37.3	–	20	37.25	37.25
Sequestrene 330 Fe	–	28	–	–	–
Vitamins	(mg)	(mg)	(mg)	(mg)	(mg)
Myoinositol	100	100	1000	100	100
Thiamine HCl	0.1	10	5	0.5	10
Nicotinic acid	0.5	1	5	5	1
Pyridoxine HCl	0.5	1	0.5	0.5	1
Folic acid	–	–	–	0.5	–
Biotin	–	–	–	0.05	–
Others					
Glycine	2 mg	–	–	2 mg	–
Casein hydrolysate	1 g	–	–	–	250 mg

Table 3.1 (*continued*).

	MS	B5	SH	Nitsch	P4*
Sucrose	30 g	20 g	30 g	20 g	30 g
Agar	10 g	6–8 g	6 g	8 g	8 g
pH	5.7–5.8	5.5	5.8–5.9	5.5	5.8

Note: growth regulators are not included
*P4 medium from Thomas *et al.* [53].

necessary because of an inability for them to be synthesized by the cultured tissue. The addition of casamino acids (casein hydrolysate) provides essential amino acids that may not be readily synthesized by cultured tissues. A review on media nutrients is provided by Ramage and Williams [32].

When working with a previously uncultured species, it is important to check the culture media that have been used for other closely related species and use that information as a starting point. Additionally, it may be worthwhile to assess two or three types of media (Table 3.1) to assess which one is the best. Our standard medium has been P4 (Table 3.1).

Naturally occurring plant hormones and commercially available growth regulators

Although the basal medium can influence the hormone response, it is the plant growth regulators that drive somatic embryogenesis. The most important hormone in the induction of SE is auxin. The synthetic auxin 2,4-dichlorophenoxyacetic acid (2,4-D) is the auxin most often used to induce SE, although 1-naphthaleneacetic acid (NAA), indole-3-butyric acid (IBA), picloram (4-amino-3,5,6-trichloropicolinic acid) and indole-3-acetic acid (IAA) are also commonly used (7, 30, 33, 34). IAA, a naturally occurring auxin, tends to be weaker and more readily broken down than synthetic auxins such as 2,4-D and NAA (35). Auxin stimulates the formation of proembryogenic masses (PEMs), which are cell clusters within the cell population that are competent to form somatic embryos. Once PEMs have formed, they may develop to the globular stage of embryogenesis, but then their further development is blocked by auxin. The removal or reduction of auxin in the culture medium allows the PEMs to develop into somatic embryos [7, 36]. Some plant species are able to form somatic embryos using auxin as the sole growth regulator, but others also require cytokinin.

There are reports of somatic embryo induction and formation on media with cytokinin as the sole growth regulator. These reports are very few relative to those reporting induction by auxin alone, or auxin plus cytokinin. SE induced by cytokinin alone tends not to have a callus phase – that is via direct SE. In most cases, the cytokinin used was thidiazuron [1-phenyl-3-(1,2,3-thidiazol-5-yl)urea], a herbicide, which is classed as a cytokinin, but which mimics both auxin and cytokinin effects on growth and differentiation [37]. There is a trend for other cytokinins to be employed only when zygotic embryos are used as the explant source. The most common cytokinins employed in embryogenic cultures are 6-benzylaminopurine (BAP), kinetin, zeatin, 6-(γ, γ-dimethylallylamino)purine (2iP) and thidiazuron (TDZ).

Rarely, SE can be induced by stress alone or the stress-related hormone, abscisic acid (ABA) [38–40]. Although these approaches illustrate the role of stress in SE, they would not be the treatments to use initially when attempting to regenerate a previously uncultured species via SE. ABA in conjunction with auxin, either with or without cytokinin, can have a positive effect on SE. Another hormone that has been reported to influence SE is gibberellic acid (GA_3). GA_3 can be inhibitory to SE formation, although some stimulation has also been reported [41]. GA_3 is more likely to be beneficial after SE formation to promote their germination.

3.2.3 Preparation of culture media

Culture medium stock powders can be purchased commercially (see Sigma web site http://www.sigmaaldrich.com/Area_of_Interest/Life_Science/Plant_Biotechnology/Tissue_Culture_Protocols.html). These are convenient, but the scope for altering the composition of the medium to suit a specific culture situation is limited. In order to simplify the preparation of media, stock solutions are made and stored. Preparation of media involves the mixing of aliquots of several stock solutions. Culture media contain major salts, minor salts, vitamins, sucrose and hormones. There may also be other organic additives such as glycine, yeast, casamino acids or coconut milk. Major salts are generally required at millimolar (mM) concentrations and provide the major inorganic nutrients, while minor salts are provided at micromolar (μM) concentrations. Major salt stock solutions, except for calcium, are made up at a 10 × normal concentration. Calcium tends to precipitate when present with the other salts and is made up as a separate 100 × stock. Major salts contain sources of nitrogen, phosphorus, potassium, calcium, magnesium and chloride. Minor salts and vitamins can be made up as a 1000 × stock solutions. Minor salts are prepared without iron, again for precipitation reasons. Iron solutions need to be made separately and chelated with EDTA (ethylenediaminetetraacetic acid) to prevent precipitation and to increase availability (see Protocol 3.1). Stock solutions of major salts, calcium, minor salts and vitamins are all stored at $-20\,^{\circ}C$ and need to be thawed prior to the preparation of medium. Such solutions should be frozen in small volumes to prevent repeated freeze/thaw cycles. Iron and some hormone stock solutions are stored at $4\,^{\circ}C$.

Protocol 3.1 outlines how to prepare a 200 × $FeNa_2$ EDTA stock solution for tissue culture. This is based on the iron in MS medium [26], but the iron concentration is at one third the concentration described in the formulation for MS medium. The reduction in iron is based on the work of Dalton *et al.*, who drew attention to problems with precipitation of iron in MS medium and showed that iron concentrations could be reduced without affecting plant growth [42].

Culture media can be either liquid or semi-solid. Cell suspension cultures are grown in liquid medium or cultures can be grown on filter paper soaked with liquid medium. Semi-solid medium cultures, solidified with agar, have explants cultured in vessels such as Petri dishes. The type of agar used can affect tissue growth. Difco Bacto agar at 0.8% (w/v) works well for SE.

Growth regulator stock solutions are made at 1000 μM concentrations. Some papers give concentrations in g/l values, but molarity values should be used for accurate comparison of hormone concentrations. The Sigma web site gives a useful table of plant hormone storage conditions, notes on how to dissolve the hormones in stock solutions and information on whether the compounds are suitable for autoclaving. It is vital that hormones are made up correctly and stay in solution once prepared. We routinely dissolve auxins (2,4-D, NAA and IAA) by heating and stirring until dissolved. Cytokinins can be dissolved in the same way, but this is facilitated by the addition of a small amount of 1 N HCl. The acidic solution prevents the cytokinin from precipitating, which has been known to cause problems in culture experiments. A number of hormones are co-autoclavable, enabling them to be added to the culture medium before sterilization by autoclaving. Others lose activity through autoclaving and must be filter sterilized (through a sterile 0.22 μm filter) and added to the medium under aseptic (axenic) conditions after the medium is autoclaved. Most of the commonly used hormones are co-autoclavable. There may be slight loss of activity for some of them, but that may be compensated by the addition of a slightly higher concentration. For example, we routinely autoclave BAP, but our culture protocol was optimized using autoclaved BAP in the medium. Therefore, any loss of activity through autoclaving would have been compensated in the optimization process. If a hormone does need to be added after autoclaving, adjust the pH of the hormone stock solution to that of the medium, so addition of the hormone after adjustment of the pH of the medium does not cause a change in overall pH. The two most commonly used auxins, 2,4-D and NAA and the commonly used cytokinin, BAP are co-autoclavable and can be stored at 4 °C.

PROTOCOL 3.1 Preparing a Chelated $FeNa_2$ EDTA Stock Solution

To make 1000 ml of a 200 × stock solution.

Equipment and Reagents

- $FeSO_4.7H_2O$ (analytical grade)
- $Na_2EDTA.2H_2O$ (analytical grade)
- MilliQ (MQ) water
- 1 l beaker
- 1 l volumetric flask
- Funnel
- Balance
- Magnetic stirring block with heater
- Thermometer

Method

1. Dissolve 7.44 g of $Na_2EDTA.2H_2O$ in approximately 900 ml MQ water.
2. While stirring, bring the solution to 98–99 °C and slowly add 1.853 g of $FeSO_4.7H_2O$.
3. Keep stirring while solution is allowed to cool in a beaker open to the air.
4. Adjust the volume to 1000 ml with MQ water[a][b].

Notes

[a]The colour of the solution should be straw yellow. If the EDTA is not heated sufficiently, the chelation reaction does not go to completion and the pH is more acidic ($pH = 1 - 2$). The H_2EDTA will precipitate out of the medium. Usually, a small amount of precipitate does form in this solution after storage. If this happens, do not stir the solution; take the solution from the bottle but avoid any precipitate. No adverse effects have been found on cultures from using solutions with some precipitate.

[b]The solution is light sensitive; store in an amber-coloured bottle at 4°C.

PROTOCOL 3.2 Preparing Agar Medium from Stock Solutions

This protocol is to prepare culture medium containing 3% (w/v) sucrose, casamino acids at 250 mg/l, 0.8% (w/v) agar, 10 μM 2,4-D and 5 μM BAP. The type of culture medium and hormones will vary according to the situation, but this protocol describes how to generically prepare culture medium.

Equipment and Reagents

- Medium components
- Balance
- Pipettors/measuring cylinders for measuring stock solutions
- Magnetic stirrer and stirring bar
- pH meter
- 1 l beaker
- Funnel
- 1 l volumetric flask
- 2 l conical flask

Method

The recipe below outlines the quantity of each component required.

Stock	Stock concentration	Amount (ml) to add for 1 l of medium
Major salts	10 ×	100
Calcium	100 ×	10
Casamino acids	100 ×	(25 g/l) 10
Iron	200 ×	5
Minor salts	1000 ×	1
Vitamins	1000 ×	1
2,4-D	1000 µM	10
BAP	1000 µM	5
Sucrose		30 g
Agar		8 g

1. Thaw frozen stock solutions either by placing the containers in warm water or in a microwave oven. Stock solutions stored at −20 °C are major salts, calcium, casamino acids, minor salts and vitamins.
2. Weigh out sucrose and place in beaker.
3. Weigh out agar and put into the conical flask (to be ready for autoclaving).
4. Add stock solutions to beaker according to the recipe.
5. Add about 800 ml of MQ water.
6. Stir until the sucrose has dissolved.
7. Pour through the funnel into a volumetric flask and make up to just below the 1 l mark. Mix and pour back into the beaker.
8. Adjust the pH using 1 M KOH, 0.1 M KOH or 0.5 N HCl (if necessary).
9. Make to correct volume in the volumetric flask.
10. Pour the medium into the conical flask (in 3). Plug the opening with a cotton wool plug wrapped in cheesecloth, and cover with aluminum foil.
11. Sterilize the medium by autoclaving at 121 °C, 105 kPa for 15–20 min.
12. After sterilization, allow the medium to cool to about 55 °C. At this stage, any filter sterilized ingredients can be added under aseptic conditions (i.e. using sterilized plugged pipette tips on a clean pipettor and working in a laminar air flow cabinet or biohazard hood).
13. Swirl gently to mix. Pour into Petri dishes in a laminar flow cabinet or biohazard hood, which has been presterilized using UV light for 20 min. A 9 cm Petri dish holds approx. 25 ml of medium.
14. Allow the agar medium to set for a minimum of 20 min with the lids off the dishes.
15. Replace lids, pack Petri dishes of medium back into the original Petri dish bag and seal with tape to maintain sterility. Leave the medium for 2 days at room temperature to check for growth of any contaminants and store at 4 °C until required. Tissue culture medium (without antibiotics) can be stored for up to 3 months. Medium containing antibiotics should not be stored for more than 1 month.

Calculating the Volume of Stock Solution to Add to Medium

Different recipes will contain different concentrations of components, especially hormones. Use the basic equation below to calculate how much stock to add. Remember to keep units the same, for example, volume in ml on one side of the equation must be in ml on the other side.

$$V_I C_I = V_F C_F$$

V_I is initial volume of solution required

C_I is the concentration of the initial (stock) solution

V_F is the final volume (of the medium)

C_F is the final concentration

A rearrangement of this equation gives:

$$V_I = \frac{V_F C_F}{C_I}$$

If 500 ml of medium is required with a 10 μM concentration of 2,4-D and the stock solution concentration is 1000 μM, $V_F = 500$ ml, $C_F = 10$ μM and $C_I = 1000$ μM.

$$V_I = \frac{500 \times 10}{1000}$$
$$V_I = 5 \text{ ml}$$

Five ml of 1000 μM 2,4-D stock solution needs to be added to 500 ml of medium to give a final concentration of 10 μM 2,4-D.

3.2.4 Sterilization of tissues and sterile technique

Healthy plant tissue is generally aseptic internally, but will harbor microorganisms on its surface. These microorganisms must be destroyed to prevent their overgrowth under culture conditions. Once the tissue is sterilized, all manipulations must be performed in a sterile environment to prevent contamination of the culture. Sources of contamination include the air, instruments, the work area, the researcher, culture vessels and water for rinsing tissue. Tissue culture should be conducted in a laminar flow cabinet or biohazard hood, fitted with a UV light to enable sterilization of the cabinet prior to use.

Sterilization of instruments

Clean metal instruments, such as forceps and scalpels can be sterilized by dipping the working end in a container (e.g. a Coplin jar) of 95% ethanol, draining and evaporating excess ethanol, and flaming over a Bunsen burner or spirit lamp. As this technique is potentially hazardous, a high degree of caution should be used to prevent accidental fire. Never put a hot instrument back into the ethanol container and keep a fireproof cover for the ethanol container close by in case of accident. After flaming, instruments can be set to cool with their base supported by a stand and the working ends of the instrument suspended in the air. An alternative is the use of glass bead sterilizers, where instruments are inserted into heated glass beads for sterilization and then cooled. Generally, this method is less effective than the flame sterilization method. Work with two sets of instruments, so that instruments can be re-sterilized frequently throughout the culture procedure. One set can be cooling on the stand while the second set is in use.

Sterilization of tissue

Explant tissue is sterilized using one or more sterilizing solutions, followed by rinsing in sterile distilled water. Calcium or sodium hypochlorite and 70% (v/v) ethanol are efficient sterilizing solutions. A detergent or wetting agent can be used to allow better contact of the solution with the tissue surface. The sterilization process can be preceded by washing tissue under running tap water for 20–30 min to physically remove most microorganisms. Household bleach (sodium hypochlorite) can be used to disinfect tissue. Bleach as purchased contains about 4–5% (v/v) available chlorine. This can be diluted for tissue sterilization. A protocol that allows sterilization without damaging the tissue may need to be determined empirically. A sterilization process that entails a short pretreatment with 70% (v/v) ethanol followed by bleach treatment is generally effective. Distilled water for rinsing tissue can be autoclaved in individual polycarbonate containers with screw lids. Protocol 3.3 outlines a sterilization procedure we routinely use for sterilizing leaf tissue and seeds of *Medicago truncatula*.

PROTOCOL 3.3 Sterilization of *Medicago truncatula* Leaf Tissue for Tissue Culture

Equipment and Reagents

- Laminar air flow cabinet or biohazard hood
- 70% (v/v) ethanol
- White King bleach (Sara Lee Household & Body Care Pty., Ltd.) diluted 1 in 8 (v : v) with distilled water (0.5% available chlorine)
- Tea infuser (from supermarket – autoclaved)
- Autoclaved containers for holding sterilization solutions
- Autoclaved containers of distilled water with screw top lids
- Sterile forceps (sterilization as described in text)

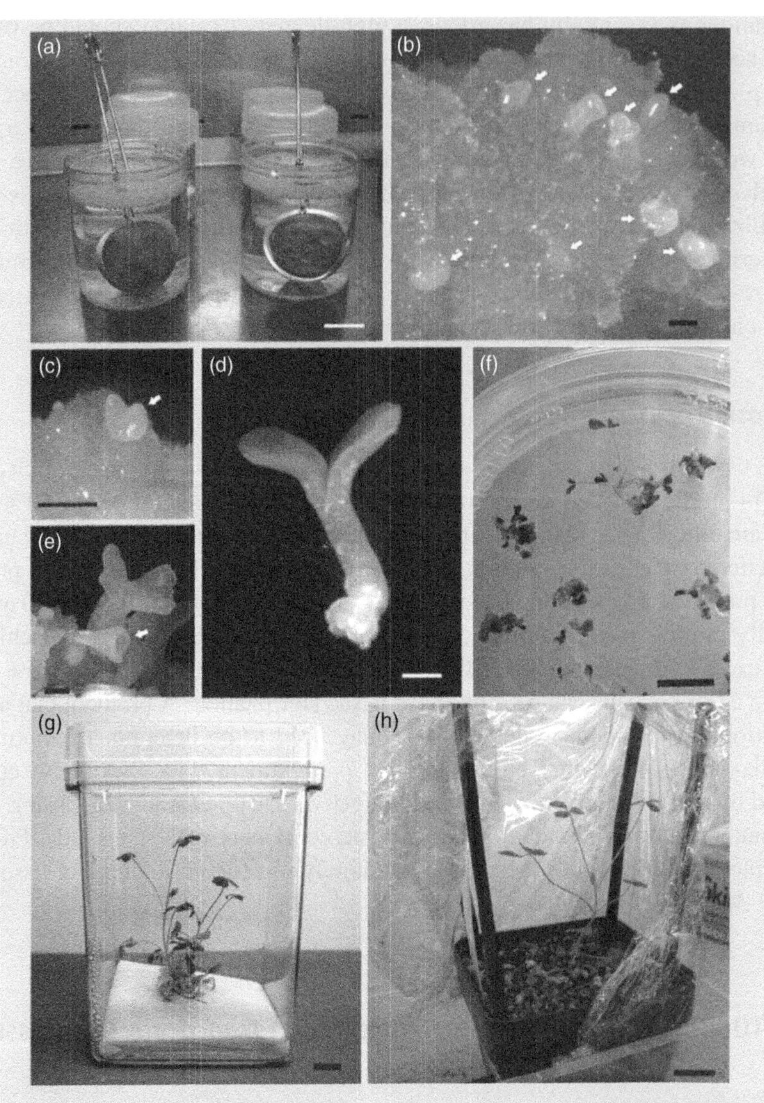

Figure 3.1 Plant regeneration by somatic embryogenesis in *Medicago truncatula*. (a) Sterilization of tissue. Tissue placed in a tea infuser and immersed in sterilization solutions and sterile water in autoclaved polycarbonate containers. Bar = 2 cm. (b) A 5 week-old culture of *M. truncatula*. SEs (arrows) appear as smooth protuberances on the rough surface of the callus tissue. Bar = 1 mm. (c) A heart stage SE (arrow) emerging from callus. Bar = 1 mm. (d) A cotyledonary stage SE. Bar = 1 mm. (e) An example of abnormal embryo development. A SE with fused cotyledons (arrow). Bar = 1 mm. (f) Germinated SEs on hormone-free agar medium in a 2 cm high Petri dish. Bar = 1 cm. (g) A regenerated plant growing on a filter paper bridge, soaked in liquid medium in a Magenta vessel. Bar = 1 cm. (h) A regenerated plant after transfer to soil. The plastic cling wrap tent, supported by stakes has been opened to allow the regenerated plant to 'harden' to the environment. Bar = 2 cm.

Method

1 Place leaves into the ball of the tea infuser (Figure 3.1a). Avoid placing too many leaves into the infuser or tissue damage will occur and sterilization will be impeded[a].

2 Immerse tea infuser containing leaves in 70% (v/v) ethanol solution for 30 sec then remove, draining excess ethanol from the tea infuser (Figure 3.1a).

3 Immerse in bleach solution and leave for 10 min[b]. Gently swirl the tea infuser several times during the sterilization process. Remove from solution and drain excess solution[c].

4 Immerse in sterile distilled water and swirl gently. Remove and drain excess water[d].

5 Open the tea infuser and, using sterile forceps, transfer the leaf tissue to another container of sterile distilled water. Close the lid and gently invert and swirl to rinse. Leave tissue in the rinse water until ready to cut up.

Notes

[a]The tea infuser facilitates the transfer from one solution to another, while ensuring that the tissue is fully immersed in the solution. Otherwise, gentle shaking will be needed to maintain surface contact with the solution and sterile forceps used to transfer tissue from one solution to another.

[b]White King bleach contains a detergent which enhances surface contact.

[c]The sterilization process can be calibrated to suit tissue by changing the time in 70% (v/v) ethanol [70% (v/v) is the best concentration for sterilization] and/or by changing the dilution of the bleach solution or time in the sterilant. If tissue is damaged by bleach, a longer time at a lower concentration may be more suitable. Conversely, if tissue is more robust, a shorter time at a stronger concentration may be possible.

[d]An extra rinse/s can be added as appropriate.

3.2.5 Culture and growth of tissue

The sterilized tissue needs to be cut into explants and placed onto culture medium. Sterile technique must be maintained at all times. Explant size should be small, $<1\,cm^2$. It is beneficial to have cut surfaces at the edges of the tissue, as stress is important in the induction of SE [9]. If the tissue was cut into pieces prior to sterilization, the edges should be trimmed to provide a newly cut surface and to remove any cells damaged by the sterilant. Tissue is cut using a scalpel, scissors or a hole punch. See Section 3.2.4 for sterilization of instruments. A sterile flat surface for the manipulation of the tissue is also required. Disposable sterile Petri dishes, or the less expensive lids from disposable Chinese food containers can be autoclaved and used.

Explants should be placed on the medium, making sure there is good surface contact of the tissue with the medium, but without pushing the explants beneath the surface. This allows efficient uptake of nutrients without inhibiting air supply. Petri dishes are sealed with a strip of Parafilm (Pechiney Plastic Packaging) stretched around the edge of the dish. Liquid cultures should be shallow to allow air exchange or, as in the case of cell suspension cultures, rotated vigorously to enhance aeration.

Cultures are maintained in a controlled temperature room at about 27 °C. Whether to incubate cultures in the light or in darkness will depend on the plant material. Callus formation is usually enhanced in darkness; initiating cultures in darkness before transfer to light can be useful. Light-grown cultures are grown under Cool White fluorescent illumination ($5 - 50\,\mu mol/m^2/s$) with day length varying between 12 and 16 h. In some plants, culture in the light will increase the number of somatic embryos initiated (43,44), whereas in others more embryos form in darkness (45). Independent of the number of embryos that form, conversion of embryos to plants is best carried out in the light [44, 46, 47].

3.2.6 Culture and induction of somatic embryos

Cultures should be transferred or subcultured to new medium every 1–4 weeks, to maintain the supply of nutrients and growth regulators. A 'typical' culture medium for induction of SE contains an auxin, commonly 2,4-D and perhaps a cytokinin. In such a medium, cell division is initiated and callus tissue develops. Callus is a 'relatively disorganized' mass of dividing cells. The term 'relatively disorganized' is used as although the original tissue structure of the explant is lost, vascular tissue growth still occurs through the callus. In response to the auxin in the medium, the formation of PEMs occurs at this stage, but under most circumstances further development is blocked by auxin. After the induction period, the auxin concentrations are reduced or auxin is removed from the medium to allow the PEMs to develop into somatic embryos. However, there are some examples where reduction of the auxin concentration in the medium is not necessary for SE development. Sometimes an auxin pulse is used to induce SE. This consists of an increased concentration, or a more potent auxin, applied for a short period of time, generally only a few days.

When somatic embryos form, they first appear as smooth protuberances on the surface of the callus. The main visual distinction between an embryo and the callus is the smoother surface of the embryo compared with the roughness of the callus (Figure 3.1b). They are generally lighter in color or they may be green if the tissue is cultured in the light. Somatic embryos undergo the same developmental stages as zygotic embryos (Figure 3.1c, d), but there is also a higher incidence of abnormal types of morphology than would occur *in vivo* (Figure 3.1e). A somatic embryo has a closed end with no vascular connection with the callus. In contrast, a shoot maintains vascular connection with the callus (3). It may be necessary to prepare samples for histology to clearly demonstrate whether structures present on the callus are embryos or adventive shoots.

3.2.7 Embryo development

Somatic embryos usually require a different culture medium for development than that used for induction. Often this medium is devoid of growth regulators, or has reduced auxin concentrations and/or a weaker auxin. Ideally a somatic embryo would develop on medium lacking growth regulators, since, by this stage, its development should be auto-regulated as in a normal germinating seed. However, exogeneous hormones or other treatments may assist with development.

Some species, particularly conifers, require a separate embryo maturation treatment. During zygotic embryogenesis, the process of embryo maturation is regulated by ABA and involves the accumulation of seed storage proteins and late embryogenesis abundant (LEA) proteins, the development of desiccation tolerance and inhibition of precocious germination. Similarly, somatic embryo maturation treatments often involve treatment with ABA (10–50 μM) for a period of several weeks and/or a treatment that will simulate desiccation through the provision of osmotic agents. High molecular weight (>4000) polyethylene glycol (PEG) is particularly good for this purpose [34]. Focus on somatic embryo maturation may be of benefit if there is a problem with germination of somatic embryos. During embryo maturation, the accumulation of seed storage proteins supports the growth of the embryo after germination. Lack of accumulation of these proteins may influence germination, or survival after germination.

A common problem in germinated somatic embryos is their lack of root development. Low concentrations of auxin in the medium assist rooting. IBA, IAA and NAA are auxins frequently used for this purpose. If shoot development is impeded, some cytokinins in the medium may be beneficial, or medium containing a low concentration of auxin with respect to cytokinin. GA_3 is sometimes added to the medium at this stage to assist embryo germination and development. Germinating somatic embryos should be transferred to the light if they had been cultured previously in the dark.

As SE-derived plants develop, the culture container may need to be changed. There are a number of culture containers of varying sizes on the market. Initially, taller than normal Petri dishes can be used (Figure 3.1f). Magenta pots (BioWorld, Dublin, OH, USA) are popular (Figure 3.1g). Developing plants can be grown in agar-solidified medium or on filter paper bridges soaked in liquid medium (Figure 3.1g). Plants with a strong shoot and root system are ready for transfer to soil.

PROTOCOL 3.4 Plating of Explants and Regeneration of Plants via Somatic Embryogenesis from Cultured Leaf Tissue of *Medicago truncatula*

This regeneration protocol has been developed using the 2HA seed line of *M. truncatula*, which is highly embryogenic compared with other genotypes of *M. truncatula* [19]. *M. truncatula* varieties tend to have very low embryogenic capacity unless they belong to a specifically-bred embryogenic seedline.

Equipment and Reagents

- Laminar air flow cabinet or biohazard hood

- Sterile forceps and scalpels (sterilization, as described earlier)
- Chinese food container lids, autoclaved in autoclave bags
- 9 cm Petri dishes containing 25 ml of P4 agar medium (Table 3.1) with 10 μM NAA and 4 μM BAP (P4 10 : 4)
- 9 cm Petri dishes containing 25 ml of P4 agar medium (Table 3.1) with 10 μM NAA, 4 μM BAP and 1 μM ABA (P4 10 : 4 : 1)
- 9 cm Petri dishes each containing 25 ml of P40 agar medium (P40 is P4 medium, Table 3.1, lacking inositol)
- Liquid P40 medium with lower (1% w/v) sucrose concentration
- Sterilized Magenta pots containing two pieces of 9 cm diameter filter paper folded down at sides to create a slightly elevated platform (Figure 3.1g)

Method

1. Collect leaf explant tissue immediately before sterilization. Use healthy glasshouse-grown plants 2–5 months of age as a source of explants. Harvest the youngest expanded trifoliate leaf on a stem as explant source. Place leaves in a small sealed container with some absorbent paper moistened with water to maintain a humid environment around the tissue prior to sterilization.
2. From this point on, work with tissue inside a UV-sterilized laminar flow cabinet or BioHazard hood. Sterilize the leaf tissue as described in Protocol 3.3. While the tissue is in sterilizing solutions, sterilize two forceps and two scalpels and cool as described earlier.
3. Remove an autoclaved Chinese food container lid[a] from autoclave bag, without touching the surface, and place in front of the operator for cutting the tissue.
4. When tissue is sterilized, using forceps, remove a few trifoliate leaves and place onto the sterile surface of the Chinese food lid. Cut explants from the leaves using a scalpel and forceps, using sterile technique, in the following manner (also shown schematically in Figure 3.2):

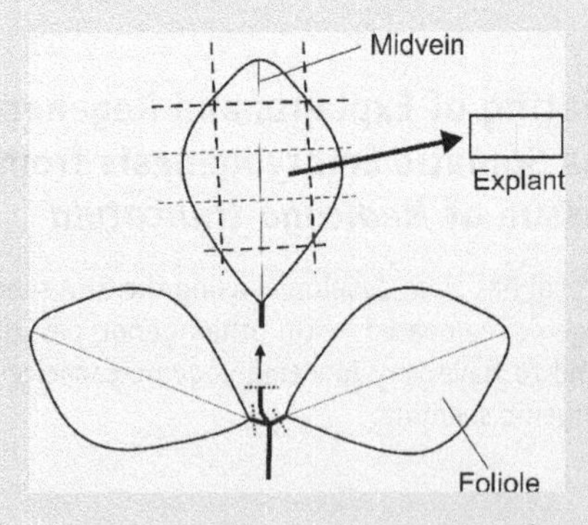

Figure 3.2 Schematic diagram showing how to prepare the explants from *Medicago truncatula* leaves used in Protocol 3.4.

(a) Excise each foliole from the trifoliate leaf by cutting through the petiole (dotted lines in Figure 3.2).

(b) On each foliole, trim the edge of the tissue (heavier dashed lines in Figure 3.2) to leave a rectangular piece of leaf in the middle. Discard edges of tissue.

(c) Depending on the initial size of the leaf, cut the rectangular piece of leaf into 2 or 3 smaller rectangular pieces (lighter dashed lines in Figure 3.2), giving small rectangular explants, with the midvein in the center of the explant and a cut surface at the edges. The size of each explant is 8–10 × 3–5 mm.

5 Transfer explants (abaxial side down) to agar plates. The initial plating medium is P4 10:4. Place six explants on each plate. Position each explant firmly on top of the agar, without pushing the explants below the suface[b].

6 Transfer more leaves from the water container to a sterile Chinese food lid and excise and plate explants using the above procedure. Repeat until the required number of explants have been plated.

7 Wrap a strip of Parafilm around the edge of each agar plate to seal. Once explants are plated, and plates are sealed with Parafilm they can be removed from the laminar flow cabinet. Incubate plates in the dark at 27 °C for 3 weeks.

8 After 3 weeks, transfer explants to P4 10:4:1 medium[c] and continue incubating at 27 °C in the dark. Explants must remain on this medium and sub-cultured to new medium every 3–4 weeks. The first embryos usually appear after about 5 weeks.

9 At each subculture, transfer embryos to hormone-free P40 agar medium in 9 cm Petri dishes and transfer to the light (14 h photoperiod with light intensity of 10 μmol/m^2/s). Transfer embryos to new medium every 3–4 weeks. As somatic embryos form shoots they may be transferred to taller (2 cm high) Petri dishes containing P40 agar medium to accommodate growth (Figure 3.1f).

10 Small plants or embryos with more developed shoots are transferred to Magenta pots containing a filter paper bridge soaked with 6–8 ml of liquid P40 medium with a reduced (1% w/v) sucrose concentration (Figure 3.1g). Continue with subculture every 3–4 weeks until plants are ready to transfer to soil.

Notes

[a]Another form of sterile surface (e.g. a white ceramic tile) can be substituted for autoclaved Chinese food lids. Chinese food lids have the advantage of being inexpensive and readily available from packaging stores.

[b]Switch to sterilized, cooled instruments at regular intervals, for example, after each batch of tissue has been transferred to plates and before cutting up the next batch of tissue. Re-sterilize used instruments and leave to cool. As an extra safeguard against contamination, also use a new cutting surface for new batches of tissue.

[c]1 μM ABA is added to the medium at this stage as it has been shown to increase the number of somatic embryos that form in this system [44].

3.2.8 Transfer to soil – the final stage of regeneration

Plants that have been regenerated from culture are accustomed to growing in a sterile, humid environment and need to be acclimatized to the harsher environment outside of culture. Plants from culture tend to have a poorly developed cuticle and are less tolerant to desiccation. It is important to maintain high humidity at first and to gradually reduce the humidity to allow the plants to 'harden' (Figure 3.1h). The lack of cuticle development also makes the plants more susceptible to infection. Therefore, plants should be transferred to sterilized soil in a clean environment.

PROTOCOL 3.5 Transfer of Regenerated Plants to Soil

Equipment and Reagents

- Small pots of sterilized soil and trays
- Bamboo stakes or wooden skewers
- Plastic cling wrap

Method

1 Thoroughly wet the soil (friable potting mix) in the pot with tap water. Gently remove the plant from the culture vessel and, under gently running water, carefully wash any culture medium from the roots[a].

2 Make a hole in the soil and transfer the plant to soil. Gently fill in around the root system.

3 Apply more water to allow the soil to wash into spaces around the roots.

4 Insert stakes or skewers evenly around the edge of each pot to use like poles of a tent.

5 Cover with plastic cling wrap to form an enclosed space and seal around the edge of each pot.

6 Place the pots in the tray with a few millimeters of water in the bottom of the tray.

7 Grow under light conditions (12–16 h photoperiod) in a culture room or similar controlled environment.

8 After a few days, make an opening several cm wide in the plastic wrap to decrease the humidity inside the wrap.

9 Gradually remove the plastic wrap over several days (Figure 3.1h).

Note

[a]The rich nutrients of culture medium attract the fungi or other microorganisms that can damage or kill the regenerated plant.

3.3 Troubleshooting

- Contamination – Use healthy young tissue, preferably grown in a controlled environment. Always maintain good sterile technique and a clean work environment. Ensure all equipment is appropriately sterilized. Spray cabinet with 70% (v/v) ETOH and wipe regularly with paper tissues.

- Preparation of culture media – Always use high grade chemicals and pure water for making media. Check stock solutions for undissolved or precipitated components before use. The formation of roots on cytokinin-containing medium may be an indication that the cytokinin is prepared incorrectly. Instigate a check system to ensure that all components are added to the medium. Make sure the pH of the medium is adjusted correctly. If agar medium fails to set, it may because of a problem with the pH.

- Browning and necrosis of cultures is likely to be due to an accumulation of phenolics excreted by the plant tissue and can be a major problem in some species such as mango [48]. Media additives, such as activated charcoal, various antioxidants (e.g. ascorbic acid, citric acid or polyvinylpyrolidine, or the ethylene inhibitor, silver nitrate) are often employed [49–51] to counteract this effect. Frequent subculture, incubation in shaking liquid culture, reduced culture temperature or the use of etiolated explants, are also methods that have been used to deal with this problem [48, 52]. However, it should be borne in mind that the appearance of brown or necrotic tissue may not necessarily be a negative factor. An example of this is in soybean, where it has been reported that somatic embryos originated on browning, necrotic tissues [12].

- Absence of sustained root development on regenerated plants. With a sterile scalpel make a clean cut at the base of the shoots so that wounded but fresh tissue is exposed. Try growing the regenerated plants without growth regulators, or with a very low concentration of auxin in the culture medium (see Section 3.2.7).

References

1. Garcês HMP, Champagne CEM, Townsley BT, *et al.* (2007) *Proc. Natl. Acad. Sci. USA* **104**, 15578–83.

2. Koltunow AM, Grossniklaus U (2003) *Annu. Rev. Plant Biol.* **54**, 547–74.

3. Haccius B (1978) *Phytomorphology* **28, 74–81.

A classic paper on the definition of a somatic embryo.

*4. Reinert J (1958) *Naturwissenchaften* **45**, 344–345.

Original paper describing somatic embryogenesis.

*5. Steward FC, Mapes MO, Mears K (1958) *Am. J. Bot*. **45**, 705–708.

Original somatic embryogenesis paper.

6. Williams EG, Maheswaran G (1986) *Ann. Bot*. **57, 443–462 – *Developmental analysis of SE*.

7. Zimmerman JL (1993) *Plant Cell* **5**, 1411–1423.

8. Rose RJ 2004 In: *Encyclopedia of plant and crop science*. pp. 1165–1168. Edited by RM Goodman. Marcel Dekker Inc., New York.

9. Feher A, Pasternak TP, Dudits D (2003) *Plant Cell Tissue Organ Cult*. **74, 201–228.

A comprehensive review of SE mechanisms.

10. Rose RJ, Nolan KE (2006) *In Vitro Cell. Dev. Biol.-Plant* **42, 473–481.

Examination of the molecular genetics of SE.

11. Sakhanokho HF, Ozias-Akins P, May OL, Chee PW (2004) *Crop Sci*. **44**, 2199–2205.

12. Ko TS, Nelson RL, Korban SS (2004) *Crop Sci*. **44**, 1825–31.

13. Kita Y, Nishizawa K, Takahashi M, Kitayama M, Ishimoto M (2007) *Plant Cell Rep*. **26**, 439–447.

14. Mandal AKA, Gupta SD, Chatterji AK (2001) *Biol. Plant*. **44**, 503–507.

15. Chernobrovkina MA, Karavaev CA, Kharchenko PN, Melik-Sarkisov OS (2004) *Biol. Bull*. **31**, 332–336.

16. Filippov M, Miroshnichenko D, Vernikovskaya D, Dolgov S (2006) *Plant Cell Tissue Organ Cult*. **84**, 213–222.

*17. Bingham ET, Hurley LV, Kaatz DM, Saunders JW (1975) *Crop Sci*. **15**, 719–721.

Breeding for SE.

18. Nolan KE, Rose RJ, Gorst JE (1989) *Plant Cell Rep*. **8**, 278–281.

19. Rose RJ, Nolan KE, Bicego L (1999) *J. Plant Physiol*. **155**, 788–791.

20. Fambrini M, Cionini G, Pugliesi C (1997) *Plant Cell Tissue Organ Cult*. **51**, 103–110.

21. Yasuda H, Satoh T, Masuda H (1998) *Biosci. Biotechnol. Biochem*. **62**, 1273–1278.

22. Harvey A, Moisan L, Lindup S, Lonsdale D (1999) *Plant Cell Tissue Organ Cult*. **57**, 153–156.

23. Ma R, Pulli S (2004) *Agr. Food Sci*. **13**, 363–377.

**24. Thorpe TA (1995) *In Vitro Embryogenesis in Plants*. Kluwer Academic Publishers, Dordrecht, The Netherlands.

A comprehensive collection of SE chapters by specialist authors.

*25. Gamborg OL, Miller RA, Ojima K (1968) *Exp. Cell Res*. **50**, 151–158.

Formulation of classic basal culture medium.

*26. Murashige T, Skoog F (1962) *Physiol. Plant*. **15**, 473–497.

Formulation of classic basal culture medium.

*27. Nitsch JP, Nitsch C (1969) *Science* **163**, 85–87.

Classic basal culture medium.

*28. Schenk RU, Hildebrandt AC (1972) *Can. J. Bot.* **50**, 199–204.

Classic basal culture medium.

29. Gamborg OL (1970) *Plant Physiol.* **45**, 372–375.

30. Nomura K, Komamine A (1995) In: *In Vitro Embryogenesis in Plants*. Edited by TA Thorpe. Kluwer Academic Publishers, Dordrecht, The Netherlands, pp. 249–266.

31. Linsmaier EM, Skoog F (1965) *Physiol. Plant.* **18**, 100–127.

32. Ramage CM, Williams RR (2002) *In Vitro Cell. Dev. Biol.-Plant* **38, 116–124.

Review of nutrients in basalculture media.

33. Mordhorst AP, Toonen MAJ, deVries SC (1997) *Crit. Rev. Plant Sci.* **16**, 535–576.

34. von Arnold S, Sabala I, Bozhkov P, Dyachok J, Filonova L (2002) *Plant Cell Tissue Organ Cult.* **69**, 233–249.

35. Grossmann K (2003) *J. Plant Growth Regul.* **22**, 109–122.

36. Halperin W (1966) *Am. J. Bot.* **53, 443–53.

A classic paper on auxin and SE.

37. Murthy BNS, Murch SJ, Saxena PK (1998) *In Vitro Cell. Dev. Biol.-Plant* **34**, 267–275.

38. Kamada H, Ishikawa K, Saga H, Harada H (1993) *Plant Tissue Cult. Lett.* **10**, 38–44.

39. Touraev A, Vicente O, Heberlebors E (1997) *Trends Plant Sci.* **2**, 297–302.

40. Nishiwaki M, Fujino K, Koda Y, Masuda K, Kikuta Y (2000) *Planta* **211**, 756–759.

41. Jiménez VM (2005) *Plant Growth Regul.* **47**, 91–110.

42. Dalton CC, Iqbal K, Turner DA (1983) *Physiol. Plant.* **57**, 472–476.

43. Baweja K, Khurana JP, Gharyalkhurana P (1995) *Curr. Sci.* **68**, 544–546.

44. Nolan KE, Rose RJ (1998) *Aust. J. Bot.* **46**, 151–160.

45. Hutchinson MJ, Senaratna T, Sahi SV, Saxena PK (2000) *J. Plant Biochem. Biotechnol.* **9**, 1–6.

46. Tremblay L, Tremblay FM (1991) *Plant Sci.* **77**, 233–242.

47. Kintzios SE, Taravira N (1997) *Plant Breed.* **116**, 359–362.

48. Krishna H, Singh SK (2007) *Biotechnol. Adv.* **25**, 223–243.

49. Teixeira JB, Sondahl MR, Kirby EG (1994) *Plant Cell Rep.* **13**, 247–250.

50. Zhong D, Michauxferriere N, Coumans M (1995) *Plant Cell Tissue Organ Cult.* **41**, 91–97.

51. Anthony JM, Senaratna T, Dixon KW, Sivasithamparam K (2004) *Plant Cell Tissue Organ Cult.* **78**, 247–252.

52. Phoplonker MA, Caligari PDS (1993) *Ann. Appl. Biol.* **123**, 419–432.

53. Thomas MR, Johnson LB, White FF (1990) *Plant Sci.* **69**, 189–198.

4
Haploid Plants

Sant S. Bhojwani and Prem K. Dantu
Department of Botany, Dayalbagh Educational Institute (Deemed University), Dayalbagh Agra, India

4.1 Introduction

Haploid plants are characterized genetically by the presence of only one set of chromosomes in their cells. In nature, haploids arise as an abnormality when the haploid egg or a synergid forms an embryo without fertilization. Haploids are sexually sterile and, therefore, doubling of the chromosomes is required to produce fertile plants, which are called double haploids (DHs) or homozygous diploids. Haploids and DHs are of considerable importance in genetics and plant breeding programmes. The major advantages of haploids are: (a) the full complement of the genome, including recessive characters, are expressed at the phenotypic level and plants with lethal mutations and gene defects are eliminated, and (b) homozygous diploids can be produced in one generation by doubling of the chromosomes of haploids. The best known application of haploids is in the F_1 hybrid system for the fixation of recombinations to produce homozygous hybrids, allowing easy selection of phenotypes for qualitative and quantitative characters. The doubled haploid method reduces the time needed to develop a new cultivar by 2–4 years, in comparison to conventional methods of plant breeding. This technique is being used routinely in crop improvement programmes and has aided the development of several improved varieties [1].

Natural haploid embryos and plants were first discovered in *Datura stramonium* by Blakeslee *et al.* [2]. To date, naturally occurring haploids have been reported in about 100 species of angiosperms [3]. However, there is no reliable method for experimental production of haploids under field conditions. Therefore, the report of Guha and Maheshwari in 1964 [4] of the direct formation

Plant Cell Culture Edited by Michael R. Davey and Paul Anthony

of pollen embryos in anther cultures of *Datura innoxia* generated considerable interest amongst geneticists and plant breeders, as it offered a potential technique for the production of large numbers of haploids and DHs. In 1967, Bourgin and Nitsch [5] described the formation of haploid plants in anther cultures of *Nicotiana tabacum* and *N. sylvestris*. Since then, this technique has been refined and applied to about 200 species of dicotyledonous and monocotyledonous plants, including several major crop plants [6].

In angiosperms, the haploid state of cells arises when the diploid cells undergo meiosis to form male and female spores. This phase is very short; fertilization of the egg re-establishes the diploid sporophytic phase. It has been possible to raise haploids by inducing the haploid pollen (androgenesis) and egg cells (gynogenesis) to develop into sporophytes without the stimulus of fertilization. Another experimental approach followed routinely to produce haploids of some cereals, is of wide/distant hybridization, followed by embryo culture. In this technique, fertilization occurs normally, but the chromosomes of one of the parents are selectively eliminated during early embryogenesis, resulting in an embryo with only one set of chromosomes. The resulting haploid embryo fails to attain full development *in vivo*, but can be rescued and maintained *in vitro*, to develop into a haploid plant. Indeed, this technique is being used routinely to raise haploids of wheat and barley. However, gynogenesis and distant hybridization techniques have limited application in haploid production. *In vitro* androgenesis remains the major technique for large scale haploid production of a wide range of crop plants. In this article, the technique for the production of androgenic haploids is described in detail. The other two techniques of haploid production are also briefly introduced.

4.2 Methods and approaches

4.2.1 Androgenesis

In androgenesis, immature pollen grains are induced to follow the sporophytic mode of development by various physical and chemical stimuli. There are two methods for *in vitro* production of androgenic haploids, namely anther culture and pollen culture.

Anther culture

This is a relatively simple and efficient technique requiring minimum facilities. Flower buds, with pollen grains at the most labile stage, are surface sterilized and the anthers, excised from the buds under aseptic conditions, are cultured on semi-solid or in liquid medium. In some cases, where the flower buds are small, whole buds or inflorescences enclosing the anthers at the appropriate stage of pollen development are cultured. The cultures are exposed to pretreatments, such as low or high temperature shock, osmotic stress or nutrient starvation, before incubation at 25 °C in the dark. Depending on the plant species and, to some extent, the culture medium, the androgenic pollen either develops directly into embryos (e.g. *Nicotiana* spp., *Brassica* spp.; Figure 4.1a–e) or proliferates to form callus tissue (e.g. rice, wheat; Figure 4.2a, b). After 2–3 weeks, when pollen embryos or calli become visible,

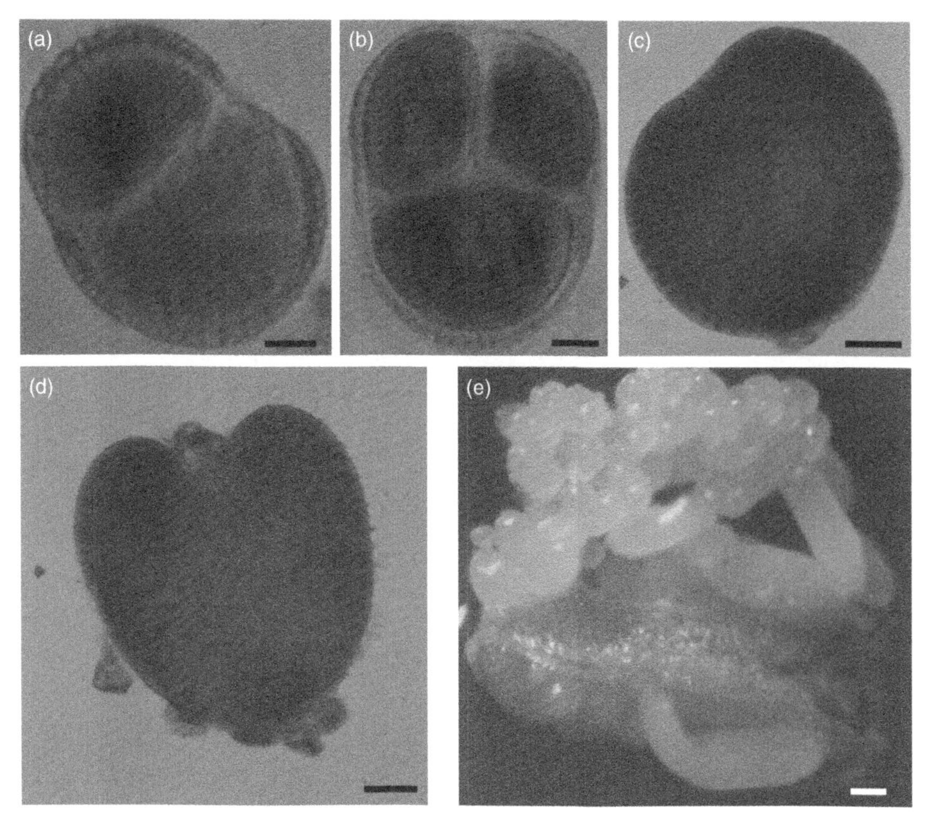

Figure 4.1 Pollen embryogenesis in *Brassica juncea*. (a) Two celled pollen after the first sporophytic division. (b) A three-celled androgenic grain. (c, d) Early and late heart-shaped pollen embryos. (e) A cultured anther that has burst open to release pollen embryos at different stages of development. Bars = 10 μm (a, b); 100 μm (c, d); 1 mm (e).

the cultures are transferred to light for their further development and organogenic differentiation (Figure 4.2c–e), respectively. The shoots regenerated from callus often require transfer to another medium for rooting to form complete plants.

Pollen culture

It is now possible to achieve androgenesis in cultures of mechanically isolated pollen of several plants, including tobacco, *Brassica* species and some cereals. In addition to the culture medium and pretreatment, the plating density (number of pollen grains per unit volume of medium) is a critical factor for the induction of androgenesis in cultured pollen. In most of the cereals, pollen culture involves preculture of the anthers for a few days, or coculture of pollen with a nurse tissue, such as young ovaries of the same or a related plant [7]. Treatment of pollen-derived embryos and pollen-derived callus to recover complete plants is the same as in anther culture.

The nutritional requirements of isolated pollen in culture are more complex than those of cultured anthers. However, unlike the earlier belief, pollen culture is less

Figure 4.2 Anther culture of *Oryza sativa* cv. 1R43 (indica rice). (a) A culture dish showing a large number of anthers with pollen-derived calli. (b) Several pollen-derived calli emerging from a burst anther. (c–e) Plants regenerated from pollen-derived calli. A callus may differentiate to give only green plants (c), only albino plants (d) or both green and albino plants (e). Bars = 10 mm (a, c–e); 1 mm (b).

tedious and time consuming than anther culture. The additional advantages of pollen culture over anther culture for haploid plant production are as follows:

1 An homogeneous preparation of pollen at the developmental stage most suitable for androgenesis can be obtained by gradient centrifugation.

2 Isolated pollen can be modified genetically by mutagenesis or genetic engineering before culture, and a new genotype can be selected at an early stage of development.

3 Pollen culture improves considerably the efficiency of androgenesis. In rapid cycling *Brassica napus*, the culture of isolated pollen was 60 times more efficient than anther culture in terms of embryo production.

4 The exogenous treatments can be applied more effectively and their precise role in androgenesis studied as the unknown effect of the anther wall is eliminated.

5 The culture of isolated pollen provides an excellent system to study cellular and subcellular changes underlying the switch from gametophytic to sporophytic development and the induction of embryogenesis in isolated haploid single cells.

Factors affecting in vitro *androgenesis*

In this chapter, detailed protocols for anther and pollen culture of some selected crops are presented which can act as a guide to the reader on the steps involved in raising androgenic haploids. However, there is considerable variation in the requirements for the optimum androgenic response of different species of a genus, or even different genotypes of a species. In practice, it has been observed that two batches of cultures of the same genotype often exhibit considerable variation in their response, probably because of change in the physiology and the growth conditions of the donor plants. Therefore, it is advisable to manipulate the published protocols when dealing with a new system to optimize the response. Some of the factors that have a profound effect on the fate of pollen in culture are the genotype and the physiological state of the donor plants, the developmental stage of pollen at the time of culture, pretreatments, and the culture medium. Before giving general protocols for anther and pollen culture the effects of these factors on androgenesis are described to facilitate modification of the available protocols to optimize the androgenic response of any specific system.

Genotype The androgenic response is influenced considerably by the plant genotype. The observed interspecific and intraspecific variation is often so great that while some lines of a species are highly responsive, others are extremely poor performers or completely non-responsive.

In general, indica cultivars exhibit poorer response as compared to japonica cultivars of rice [8–10]. Similarly, amongst the crop brassicas, substantial inter- and intraspecific variation has been reported for androgenesis [11, 12]. *Brassica napus* is more responsive than *B. juncea*. Optimum culture conditions may also vary with the genotype. For example, the optimum concentration of ammonium nitrogen for indica rice is almost half of that for japonica rice.

Since plant regeneration from pollen is a heritable trait, it is possible to improve the androgenic response of poor performers by crossing them with highly androgenic genotypes [13–17].

Physiological status of the donor plants The environmental conditions and the age of the plant, which affect the physiology of the plants, also affect their androgenic response. Generally, the first flush of flowers yields more responsive anthers than those borne later. However, in *B. napus* [18] and *B. rapa* [19], pollen from older, sickly looking plants yielded a greater number of embryos than those from young and healthy plants. Similarly, the late sown plants of *B. juncea* yielded more androgenic anthers than the plants sown at the normal time [20]. Application of ethrel, a feminizing hormone [20, 21], or fernidazon-potassium, a gametocidal agent [22], to the donor plants enhanced the androgenic response in some individuals.

Pretreatments Application of a variety of stresses, such as temperature shock, osmotic stress and sugar starvation at the initial stage of anther or pollen culture have proved promotory or essential for the induction of androgenesis. However, the type, duration and the time of application of these pretreatments may vary with the species or even the variety [7].

Of the various treatments, the application of a temperature shock has been most common. In many species, incubation of anther/pollen cultures at low temperature (4–13 °C) for varying periods before incubation at 25 °C enhanced the androgenic response. In practice, the excised panicles of rice are cold treated before removing the anthers for culture [7, 23–25]. The duration of cold treatment is critical to obtain high frequency green plants of pollen origin. For indica rice, cold treatment at 10 °C is essential for the induction of androgenesis, but cold treatment for longer than 11 days, although increasing the androgenic response, adversely affected the frequency of production of green plants [26, 27].

In some plants, such as *Capsicum* [28, 29] and some genotypes of wheat [30], an initial high temperature shock has proved essential or beneficial for androgenesis. A heat shock of 30–35 °C for 2 h to 4 days is a prerequisite for inducing pollen embryogenesis in most *Brassica* species [31]. However, the optimum requirement of high temperature pretreatment varies with different species. The time lapse between isolation of pollen and high temperature treatment can radically affect embryo induction. For example, embryogenesis was completely inhibited when the pollen of *B. napus* was held for 24 h at 25 °C before the application of heat shock [32].

In barley [33] and indica rice [24] pollen cultures, an application of 0.3 M or 0.4 M mannitol to the anthers to induce stress before culture proved better than cold pretreatment. Initial starvation of developing pollen of important nutrients, such as sucrose [34, 35] and glutamine [36, 37] favoured androgenesis in tobacco and barley.

Stage of pollen development The competence of pollen to respond to the various external treatments depends on the stage of their development at the time of culture. Generally, the labile stage of pollen for androgenesis is just before, at, and immediately after, the first pollen mitosis. During this phase, the fate of the pollen is uncommitted because the cytoplasm is cleaned of the sporophyte-specific information during meiosis, and the gametophyte-specific information has not been transcribed by this time. However, it is important to appreciate that the most vulnerable stage of pollen for responding to exogenous treatments may vary with the system.

Culture medium Most of the species require a complete plant tissue culture medium with some growth regulators for *in vitro* androgenesis. The most widely used media for this purpose are those of Murashige and Skoog (MS; [38]) and Nitsch and Nitsch [39] with any modifications. In general, the requirement of isolated pollen in culture is more demanding than that of cultured anthers. Some of the media developed for anther and pollen culture of tobacco, *Brassica* and rice are listed in Table 4.1. Apparently, all the media specifically developed for androgenic haploid production are low salt media as compared to MS-based medium, which is most commonly used in plant tissue culture studies.

Regeneration of androgenic plants may occur directly via embryogenesis from pollen or *via* callus development from pollen, followed by organogenesis. In the latter case, androgenesis is a two-step process, each step requiring different media and culture conditions. Anther cultures of many cereals are very sensitive to inorganic nitrogen, particularly in the form of NH_4^+. Based on this observation, Chu [40] developed N_6 medium which is used extensively for cereals. Indica rice anther and pollen cultures are even more sensitive to the concentration of NH_4^+ in the culture medium.

Sucrose is an essential constituent of media for androgenesis and it is used mostly at 2–4% (w/v). However, some plants require a greater concentration of sucrose to exhibit an optimum response. For potato [41] and some cultivars of wheat [42], sucrose at 6% (w/v) was superior to this carbohydrate at 2% (w/v). Anther and pollen cultures of all crop Brassicas require 12–13% sucrose for androgenesis.

For several cereals, maltose has proved superior to sucrose as the carbon source [10, 25, 43–46]. Substitution of sucrose by maltose in the medium in *ab initio* pollen cultures of wheat allowed genotype-independent plant regeneration [47] and promoted direct pollen embryogenesis [48].

4.2.2 Diploidization

Haploid plants are sexually sterile. In the absence of homologous chromosomes, meiosis is abnormal and, as a result, viable gametes are not formed. In order to obtain fertile homozygous diploids, the chromosome complement of the haploids must be duplicated. In some plants, spontaneous duplication of the chromosome number occurs at a high rate (>50%). This is especially true for the plants where androgenesis occurs via pollen callusing, as in many of the cereals, including wheat, barley and rice. Where the frequency of spontaneous doubling of the chromosomes exceeds 50%, there is no requirement for any special treatment to obtain fertile homozygous diploids. In other cases, the pollen plants should be treated with 0.1–0.4% colchicine solution to diploidize them. Different methods have been followed to diploidize the haploid plants. Generally, the pollen-derived plants, with three to four leaves, are soaked in an 0.5% aqueous solution of colchicine for 24–48 h and, after washing with distilled water, transferred to a potting mixture for hardening and further growth. For *Brassica* species, the roots of the pollen-derived plants, in a bunch of 25–30 plants, are immersed in 0.25% (w/v) colchicine solution for 5 h in the light. After rinsing the treated roots with distilled water, the plants are transferred to a potting mix for hardening and further growth.

Table 4.1 Composition of some media used for androgenic haploid production (concentrations in mg/l).

Constituents	MS[a]	B_5[b]	N&N[c]	AT3[d]	KA[e]	NLN-13[f]	N_6[g]	M-019[h]
KNO_3	1900	2527.5	950	1950	2500	125	2830	3101
NH_4NO_3	1650		725					
$NaH_2PO_4.2H_2O$		150			150			
KH_2PO_4	170		68	400		125	400	540
$(NH_4)_2.SO_4$		134		277	134		463	264
$MgSO_4.7H_2O$	370	246.5	185	185	250	125	185	370
$CaCl_2.2H_2O$	440	150	166	166	750		166	440
$Ca(NO_3)_2.4H_2O$						500		
$FeSO_4.7H_2O$	27.8		27.8			27.8	27.8	27.85
$Na_2.EDTA.2H_2O$	37.3		37.3			37.3	37.3	37.25
Sequestrene 330Fe		28			40			
Fe-EDTA				37.6				
KI	0.83	0.75		0.83	0.75		0.8	0.83
H_3BO_3	6.2	3	10	6.2	3	10	1.6	6.2
$MnSO_4.H_2O$		10				25	4.4	
$MnSO_4.4H_2O$	22.3		25	22.3	10			22.3
$ZnSO_4.7H_2O$	8.6	2	10	8.6	2	10	1.5	8.6
$Na_2MoO_4.2H_2O$	0.25	0.25	0.25	0.25	0.25	0.25		0.25
$CuSO_4.5H_2O$	0.025	0.025	0.025	0.025	0.25	0.025		0.025
$CoCl_2.6H_2O$	0.025	0.025		0.025	0.025	0.025		
Myoinositol	100	100	100	100	100	100		100
Thiamine HCl	0.1	10	0.5	10	10	0.5	1	2.5
Pyridoxine HCl	0.5	1	0.5	1	1	0.5	0.5	2.5
Nicotinic acid	0.5	1	5	1	1	0.5	0.5	2.5
Glycine	2		2			2	2	2
L-Glutamine				1256	800	800		
Glutathione						30		
L-Serine					100	100		
Folic acid			5			0.5		
Biotin			0.5			0.5		
NAA					0.1	0.5		2.5
2,4-D					0.1		2	0.5
BAP						0.05		

Table 4.1 (*continued*).

Constituents	MS[a]	B_5[b]	N&N[c]	AT3[d]	KA[e]	NLN-13[f]	N_6[g]	M-019[h]
Kinetin							0.5	0.5
MES[i]				1950				
Sucrose	30 000	20 000	20 000		10 0000	13 0000	50 000–12 0000	
Maltose				90 000				90 000
Agar	8000	8000	8000		8000		8000	

[a][38]; Murashige & Skoog medium.
[b][55]; B_5 Medium.
[c][39]; for anther culture of tobacco.
[d][56]; for isolated microspore cultures of tobacco; medium is filter sterilized.
[e][57]; for anther culture of *Brassica*.
[f][58]; for isolated microspore culture of *Brassica*.
[g][40]; for rice anther culture.
[h][24]; for microspore culture of indica rice; for japonica rice the NAA is omitted and the concentration of 2,4-D is raised to 2 mg/l.
[i]2-(*N*-morpholino)ethanesulfonic acid.

PROTOCOL 4.1 Anther Culture to Produce Androgenic Haploids of *Nicotiana tabacum* [49]

Equipment and Reagents

- Sterile laminar air flow cabinet, for aseptic manipulations
- Incubators or growth chambers with temperature and light control, to grow experimental plant material
- Refrigerated centrifuge, for cleaning pollen suspensions
- Autoclave, for steam sterilization of media and glassware
- Electronic pH meter, to adjust the pH of media
- Electronic micro- and macrobalances, to weigh chemicals for media and other stock solutions
- Magnetic stirrer, to dissolve chemicals and to isolate pollen
- Light microscope with fluorescence lamp, to observe microscopic preparations, e.g. to determine the developmental stage of pollen and early stages of pollen embryogenesis
- Inverted microscope, to observed cultures in Petri dishes
- Refrigerator, to store chemicals and stock solutions and to give cold pre-treatment
- Millipore filtration unit with filter membranes of pore sizes 0.22 μm and 0.45 μm, to filter sterilize solutions and liquid media
- Haemocytometer to count pollen

- Vernier caliper to measure the size of buds
- Gas or spirit burner, to sterilize instruments used during inoculation, by dipping in alcohol and flaming
- Tea eggs (infusers) to surface sterilize small buds
- Waring blender to macerate anthers to isolate pollen for culture
- Common laboratory glassware, plasticware (e.g. beakers, centrifuge tubes, culture tubes, measuring cylinders, Petri dishes, pipettes, of different sizes).
- Parafilm to seal Petri dishes
- Aluminium foil
- Chemicals to prepare media and stains

Method

1. Grow the plants of *Nicotiana tabacum* in a glasshouse at 20–25 °C under a 16 h photoperiod with a light intensity of 210–270 μmol/m^2/s provided by sodium lamps (400 W).
2. Harvest the flower buds from the first flush of flowers, and transfer to the laboratory in a non-sterile Petri dish.
3. Classify the buds according to their corolla length. Excise anthers from one of the buds of each category and crush them in acetocarmine to determine the stage of pollen development. Identify and select the buds (ca. 10 mm) with pollen just before, at, and immediately after the first pollen mitosis.
4. Incubate the buds at 7–8 °C for 12 days in a sterile Petri dish; seal with Parafilm.
5. Surface sterilize the chilled buds with a suitable sterilant [0.1% (w/v) mercuric chloride for 10 min or 5% (w/v) sodium hypochlorite for 10 min].
6. Rinse the buds three to four times in sterile, double distilled water in a laminar air flow cabinet.
7. Using forceps and a needle, flame-sterilized and cooled, tease out the buds and excise the anthers in a sterile Petri dish. Carefully detach the filament and place the anthers on MS medium (Table 4.1), supplemented with 2% (w/v) sucrose and 1% (w/v) activated charcoal in Petri dishes (five anthers from a bud per 50 mm × 18 mm dish containing 5 ml of medium). Seal the Petri dishes with Parafilm, and incubate the cultures at 25 °C in the dark or dim light (10–15 μmol/m^2/s).
8. After 3–4 weeks, when the anthers have burst to release the pollen-derived embryos, transfer the cultures to a 16 h photoperiod and light intensity of 50 μmol/m^2/s provided by cool white fluorescent tubes. At this stage, if the responding anthers are crushed in acetocarmine (0.5–1.0%) and observed under the microscope, different stages of pollen embryogenesis can be seen which are asynchronous.
9. Complete green plants will develop after 4–5 weeks of culture.
10. Isolate the plantlets emerging from the anthers and transfer them to MS basal medium with 1% (w/v) sucrose and 1% (w/v) activated charcoal to allow root development.

During this period, incubate the cultures under continuous light (3.6 μmol/m^2/s) illumination from Cool White fluorescent tubes.

11 When the plants attain a height of about 5 cm, transfer them to potting mix in small pots or polythene bags and maintain under high humidity. Gradually reduce the humidity and transfer the plants to the field.

PROTOCOL 4.2 Pollen Culture to Produce Androgenic Haploids of *Nicotiana tabacum* [49]

Method

1 Follow steps 1–6 as in Protocol 4.1.

2 Squeeze the anthers from 10 buds in a glass vial (17 ml) with about 3 ml of medium B (37), containing (in mg/l) KCl (49), $CaCl_2.2H_2O$ (147), $MgSO_4.7H_2O$ (250), KH_2PO_4 (136) and mannitol (54 700).

3 Place a magnetic bar in the vial and stir for 2–3 min at maximum speed until the medium becomes milky.

4 Collect the suspension of pollen and debris using a Pasteur pipette and filter it through a 40–60 μm pore size metal or nylon sieve.

5 Centrifuge the filtrate for 2–3 min at 250 g. Discard the supernatant and the upper green pellet using a 200 μl or 1000 μl pipette.

6 Suspend the lower whitish pellet in 2–10 ml of medium B and centrifuge again. Repeat the fifth step two to three times until there is no green layer above the white pellet.

7 Suspend the white pellet, comprised of purified pollen, in the B-medium and dispense the suspension in a presterilized Petri dish. Seal the Petri dish with Parafilm and incubate in the dark at 33 °C for 5–6 days. The induction of androgenesis occurs during this starvation stress treatment.

8 After the pretreatment, transfer the suspension to a screw-capped centrifuge tube and pellet by centrifugation at 250 g for 5 min.

9 Discard the supernatant and suspend the pellet in AT-3 medium (Table 4.1) and dispense into the original dishes (1 ml per dish). Seal the dishes with Parafilm. Incubate the dishes in the dark at 25 °C.

10 After 4–5 weeks, when fully differentiated pollen embryos have developed, transfer the culture dishes to a 16 h photoperiod with cool white fluorescent light (50 μmol/m^2/s).

11 After 1 week, transfer individual embryos to culture tubes or jars containing MS-based medium with 1% (w/v) sucrose and 1% (w/v) activated charcoal for germination and full plant development. Incubate the cultures in the light as above during this period.

12 After the plants attain a height of about 5 cm, transfer them to potting mixture in small pots or polythene bags and maintain them under high humidity. Gradually reduce the humidity, and finally, transfer the plants to the field.

PROTOCOL 4.3 Pollen Culture to Produce Androgenic Haploids of *Brassica juncea* [12]

Method

1. Sow the seeds in 20 cm pots containing an artificial potting mixture, such as Agropeat PV, and maintain them at 25 °C under natural light.
2. At the bolting stage, move the plants to a growth chamber at 10 °C/5 °C day/night temperatures and with 16 h photoperiod and 150–200 μmol/m^2/s of light intensity from Cool White fluorescent tubes.
3. After 2 weeks, when 2–3 flowers have opened, collect the young green inflorescences and transfer to the laboratory in non-sterile Petri dishes.
4. Classify the buds into two to four categories on the basis of their length (2.7–2.9 mm, 3.0–3.1 mm, 3.2–3.3 mm and 3.4–3.5 mm) using a Vernier caliper.
5. Determine the stage of pollen development in the buds of the different categories by staining with DAPI (4,6-diamino-2-phenylindole; 2–4 μg/ml of McIlavaine buffer of pH 7; McIlavaine buffer: mix 18 ml of 0.1 M citric acid with 82 ml of 0.2 M $Na_2HPO_4.2H_2O$) and observe under UV light using a fluorescence microscope. Select the buds at the late uninucleate stage for culture, when the nucleus has migrated to one side. Hereafter, all operations must be performed under axenic conditions in a laminar flow cabinet.
6. Transfer the selected buds to a tea egg; immerse in 0.1% (w/v) mercuric chloride or 2% (w/v) sodium hypochlorite solution, with a drop of Tween 20 or Teepol for 10–12 min with continual shaking.
7. After three rinses each of 5 min in cold sterile distilled water, transfer the buds (maximum 20) to an autoclaved 25 ml beaker containing 7 ml of cold liquid B_5 medium with the salts reduced to half strength and 13% (w/v) sucrose (½B_5-13; Table 4.1).
8. Homogenize the buds by crushing them with the aid of an injection piston, applying turning pressure movement to release the pollen. Wash the piston with ½B_5-13 medium.
9. Filter the pollen suspension through a double layer of nylon (Nytex 63 μm pore size top and 44 μm bottom) in a 15 ml sterilized, screw cap centrifuge tube. Rinse the nylon sieve with 2 ml of ½B_5-13 medium and adjust the volume to 10 ml with medium.
10. Wash the pollen twice with ½B_5-13 medium by pelleting at 100 g for 3 min in a refrigerated centrifuge precooled to 4 °C.
11. Wash the pollen in NLN-13 medium containing 0.83 mg/l KI (NLN-13-KI; Table 4.1).
12. Suspend the pellet in 1 ml of NLN-13-KI medium and determine the density of pollen using a haemocytometer. Adjust the density to 1×10^4 pollen grains/ml using NLN-13-KI medium.
13. Dispense the suspension into sterile Petri dishes (3 ml per 60 mm dish) as thin layers. Seal the dishes with Parafilm and incubate at 32 °C or 35 °C in the dark.

14 After 3 days, transfer the culture dishes to 25 °C in the dark.

15 After another 3 weeks of culture, transfer individual embryos to B_5 medium with 2% (w/v) sucrose for germination. Place the dishes in a culture room with a 16 h photoperiod (50–100 μmol/m^2/s provided by cool white fluorescent tubes) at 25 °C. If necessary, after 4–5 days, reorientate the embryos in a vertical plane to facilitate their germination.

16 After 2 weeks, transfer the plants to culture tubes with their roots immersed in 1–2 ml colchicine solution [0.1–0.2% (w/v)] and leave overnight. Wash the roots with sterile distilled water and transfer them to a 1 : 1 (v : v) mixture of Agropeat and soil in Hycotrays (Sigma); maintain in a glasshouse under high humidity. Gradually move the plants to areas of decreasing humidity. The plants should be ready after another 3 weeks for transfer to the field.

PROTOCOL 4.4 Anther Culture to Produce Androgenic Haploids of *Oryza sativa* [50]

Method

1 Collect, at 8–9 a.m., the tillers from glasshouse-grown plants with the central florets at the middle to late uninucleate stage of the pollen.

2 Wipe dry and wrap the spikes in aluminium foil and store at 8–10 °C for 8 days.

3 Rinse the spikes in 70% (v/v) ethanol for 30 s before surface sterilizing them with 2% (v/v) Chlorax (a commercial bleach with 5.2% $NaOCl_2$) containing a drop of Teepol for 20 min. Carry out all further steps in a laminar air flow cabinet.

4 Rinse the spikes three times in sterile distilled water.

5 To excise and culture the anthers, cut the base of the florets just below the anthers with sharp scissors. Pick the floret at the tip with forceps and tap on the rim of the Petri dish so that the anthers fall in the dish containing N_6 medium (Table 4.1) supplemented with 5% (w/v) sucrose, 0.5–2.0 mg/l 2,4-D (callus induction medium). About 60 anthers may be cultured in a 55 mm diameter Petri dish containing 6 ml of callus induction medium.

6 Incubate the cultures at 25 °C in the dark.

7 After 4–5 weeks, transfer the pollen-derived calli (each 2–3 mm in diam.) to MS-based regeneration medium with 0.5–4.0 mg/l kinetin and incubate the cultures in the light (12 h photoperiod with 50–100 μmol/m^2/s provided by Cool White fluorescent tubes) at 25 °C.

8 Transfer the regenerated shoots to MS-based rooting medium lacking growth regulators.

PROTOCOL 4.5 Pollen Culture to Produce Androgenic Haploids of Rice [24]

Method

1 Follow steps 1, 3 and 4 of Protocol 4.4.

2 Collect 150 axenic anthers in a 55 mm diameter Petri dish containing 0.4 M mannitol solution, following the procedure described in step 5 of Protocol 4.4. Incubate in the dark at 33 °C.

3 Simultaneously isolate the unfertilized ovaries from the same batch of florets and culture in Petri dishes, each containing 3 ml of M-019 medium (Table 4.1), to condition the medium for pollen culture. Culture 30 ovaries per dish and incubate the dishes in the dark at 25 °C.

4 After 4 days of osmotic stress in the mannitol solution, some of the pollen grains will be liberated from the anthers into the pretreatment medium. Transfer the pollen suspension with the pretreated anthers to a small beaker and stir at slow speed for 2–3 min using a Teflon-coated magnet to release the remaining pollen.

5 Filter the above suspension through a nylon/metallic sieve (40–60 μm pore size), pipette out the filtrate, transfer it to a screw cap centrifuge tube and centrifuge at 500 rpm for 2–3 min. Discard the supernatant and suspend the pellet in new mannitol solution and wash again by centrifugation. Give the final wash in M-019 medium. Finally, suspend the pollen in M-019 medium conditioned by the cultivation of unfertilized ovaries (1 ml suspension per 3.5 cm dish) for 4 days. Transfer 10 ovaries into each dish. Seal the Petri dishes with Parafilm and incubate the cultures in the dark at 25 °C.

6 After 4 weeks from initiation of the pollen cultures, transfer the embryo-like structures (ELS) or calli, each measuring 2–3 mm in size, to semi-solid MS-based regeneration medium (Table 4.1) supplemented with benzylaminopurine (BAP) (2.0 m/l), kinetin (1.0 mg/l), naphthaleneacetic acid (NAA) (0.5 mg/l) and gelled with 0.6% (w/v) agarose (Sigma). Incubate the cultures under a 12 h photoperiod (50–100 $\mu mol/m^2/s$ illumination provided by cool white fluorescent tubes). After 7–10 days, more ELS/calli from the induction medium may be transferred to regeneration medium.

7 Transfer the regenerated plants to hormone-free 1/4 strength MS-based medium containing 2% (w/v) sucrose and gelled with 0.25% (w/v) Phytagel (Sigma) in culture tubes.

8 When the plants attain a height of about 15 cm, transfer them to liquid 1/10 strength MS-based medium without sucrose, vitamins or hormones for hardening, before transfer to pots.

4.3 Troubleshooting

- The quality of plants and the stage of pollen at the time of culture are of utmost importance. The correlation between the stage of pollen development and the external morphological markers, such as bud/corolla length, changes with the age and growth conditions of the anther donor plants. Therefore, the experimental plants should be grown under controlled conditions and the anthers/pollen for culture should be taken from the first one to two flushes of flower buds for reproducible results.

- The cultures should be raised in the morning (8–11 a.m.) and the time lapse between picking the buds and subjecting them to a pretreatment or the initiation of cultures should be a minimal. This is essential for optimum results.

- Anther and pollen cultures should always be maintained in the dark. Light is detrimental for the induction of androgenesis.

- Anther culture and, more recently, isolated pollen culture, have become a practical approach to haploid production of crop plants. Androgenic haploids are being used routinely in crop improvement programmes. The major advantage of androgenesis is the availability of a large number of haploid cells (pollen) which can be induced to form haploid plants. However, there are some serious problems associated with this technique, since (a) it is highly genotype specific, (b) all efforts to produce androgenic haploids of some crop plants have been unsuccessful, and (c) in most of the cereals, a large number of pollen plants are albinos (Figure 4.2d, e), which are of no value in breeding programmes. Sometimes, the frequency of albinos may exceed 80% [51, 52]. To overcome these problems, alternative methods of haploid production have been developed.

The most effective method to produce green haploid plants of wheat is to cross this cereal with maize, followed by *in vitro* culture of the embryos [53]. Similarly, the best method to produce green haploid plants of barley is to cross it with *Hordeum bulbosum*, a wild relative of barley, and culture the embryos on artificial medium. In these distant crosses, fertilization occurs normally, but the chromosomes of maize and *bulbosum*, respectively, are selectively eliminated during early embryogenesis. The resulting haploid embryos, which abort prematurely *in situ*, form complete plants in culture.

For some plants, such as onion, sunflower and mulberry, where it has not been possible to induce androgenesis, haploids may be produced by *in vitro* cultivation of unfertilized ovules, ovaries or flower buds [54]. Interestingly, the gynogenic haploids of cereals are, to a large extent, green, unlike androgenic haploids.

References

*1. Maluszynski M, Kasha KJ, Forster BP, Szarejko L (eds) (2003) *Doubled Haploid Production in Crop Plants*. Kluwer Academic Publishers, Dordrecht, The Netherlands.

A volume devoted to haploid production in a range of crop plants.

2. Blakeslee AF, Belling J, Farnham ME, Bergner AD (1922) *Science* **55**, 646–647.

3. Vasil IK (1997) In: *In Vitro Haploid Production in Higher Plants*. Edited by SM Jain, SK Sopory and RE Veilleux. Kluwer Academic Publishers, Dordrecht, The Netherlands. pp. vii–viii.

4. Guha S, Maheshwari SC (1964) *Nature* **204, 497.

The first report of *in vitro* androgenesis in Datura.

5. Bourgin JP, Nitsch JP (1967) *Ann. Physiol. Veg*. **9**, 377–382.

6. Kott LS (1998) *Agbiotech* **10**, 69–74.

7. Datta SK (2001) In: *Current Trends in the Embryology of Angiosperms*. Edited by SS Bhojwani and WY Soh. Kluwer Academic Publishers, Dordrecht, The Netherlands, pp. 471–488.

8. Miah MAA, Earle ED, Khush GS (1985) *Theor. Appl. Genet*. **70**, 113–116.

9. Cho MS, Zapata FJ (1990) *Plant Cell Physiol*. **31**, 881–885.

*10. Raina SK (1997) *Plant Breed. Rev*. **15**, 141–186.

A detailed review of the literature on androgenesis in rice.

11. Duijs JG, Voorrips RE, Visser DI, Custers JBM (1992) *Euphytica* **60**, 45–55.

12. Chanana NP, Dhawan V, Bhojwani SS (2005) *Plant Cell Tissue Organ Cult*. **83**, 169–177.

13. Rudolf K, Bohanec B, Hansen N (1999) *Plant Breed*. **118**, 237–241.

14. Cloutier S, Cappadocia M, Landry BS (1995) *Theor. Appl. Genet*. **91**, 841–847.

15. Foroughi-Wehr B, Friedt W, Wenzel G (1982) *Theor. Appl. Genet*. **62**, 233–239.

16. Petolino JF, Jones AM, Thompson SA (1988) *Theor. Appl. Genet*. **76**, 157–159.

17. Barloy D, Dennis L, Beckert M (1989) *Maydica* **34**, 303–308.

18. Takahata Y, Brown DCW, Keller WA (1991) *Euphytica* **58**, 51–55.

19. Burnett L, Yarrow S, Huang B (1992) *Plant Cell Rep*. **11**, 215–218.

20. Agarwal PK, Bhojwani SS (1993) *Euphytica* **70**, 191–196.

21. Wang CC, Sun CS, Chu ZC (1974) *Acta Bot. Sin*. **16**, 43–54.

22. Picard, E, Hours C, Gregoire S, Phan TH, Meunier JP (1987) *Theor. Appl. Genet*. **74**, 289–297.

23. Ogawa T, Hagio T, Ohkawa Y (1992) *Japan J. Breed*. **42**, 675–679.

24. Raina SK, Irfan ST (1998) *Plant Cell Rep*. **17**, 957–962.

25. Pande H, Bhojwani SS (1999) *Biol. Plant*. **42**, 125–128.

26. Pande H (1997) PhD Thesis, University of Delhi, India.

27. Gupta HS, Borthakur DN (1987) *Theor. Appl. Genet*. **74**, 95–99.

28. Dumas de VR, Chambonnet D, Sibi M (1982) In: *Variability in Plants Regenerated from Tissue Culture*. Edited by ED Earle and Y Demarly. Praeger Publishers, NJ, USA, pp. 92–98.

29. Rincs HW (1983) *Crop Sci*. **23**, 268–272.

30. Li H, Qureshi JA, Kartha KK (1988) *Plant Sci*. **57**, 55–61.

*31. Custers JBM, Cordewener JHG, Fiers MA, Maassen BTH, van Lookeren Campagne MM, Liu CM (2001) In: *Current Trends in the Embryology of Angiosperms*. Edited by SS Bhojwani and WY Soh. Kluwer Academic Publishers, Dordrecht, The Netherlands, pp. 451–470.

Discusses the cellular and subcellular basis of androgenesis in anther and pollen cultures of *Brassica napus*.

32. Pechan PM, Bartels D, Brown DCW, Schell J (1991) *Planta* **184**, 161–165.

33. Roberts-Oehlschlager SL, Dunwell JD (1990) *Plant Cell Tissue Organ Cult*. **20**, 235–240.

34. Aruga K, Nakajima T, Yamamoto K (1985) *Jpn J. Breed*. **35**, 50–58.

35. Wei ZM, Kyo M, Harada H (1986) *Theor. Appl. Genet*. **72**, 252–255.

36. Kyo M, Harada H (1985) *Plant Physiol*. **79**, 90–94.

37. Kyo M, Harada H (1986) *Planta* **168**, 427–432.

38. Murashige T, Skoog F (1962) *Physiol. Plant*. **15**, 473–497.

39. Nitsch JP, Nitsch C (1969) *Science* **163**, 85–87.

*40. Chu CC, Wang CC, Sun CS, *et al.* (1975) *Sci. Sin*. **18**, 659–668.

Authors formulated the N_6 medium widely used for cereal anther culture.

41. Sopory SK, Jacobsen E, Wenzel G (1978) *Plant Sci. Lett*. **12**, 47–54.

42. Ouyang T, Hu H, Chuang C, Tseng C (1973) *Sci. Sin*. **16**, 79–95.

43. Datta SK, Schmid J (1996) In: *In Vitro Haploid Production in Higher Plants*. Edited by SM Jain, SK Sopory and RE Veilleux. Kluwer Academic Publishers, Dordrecht, The Netherlands, Vol. **2**. pp. 351–363.

44. Xie J, Gao M, Cai Q, Cheng X, Shen Y, Liang Z (1995) *Plant Cell Tissue Organ Cult*. **42**, 245–250.

45. Letini Z, Reyes P, Martinez CP, Roca WM (1995) *Plant Sci*. **110**, 127–138.

46. Last DI, Brettell RIS (1990) *Plant Cell Rep*. **9**, 14–16.

47. Mejza SJ, Morgant V, DiBona DE, Wong JR (1993) *Plant Cell Rep*. **12**, 149–153.

48. Navarro-Alvarez W, Baenziger PS, Eskridge KM, Shelton DR, Gustafson VD, Hugo M (1994) *Plant Breed*. **112**, 53–62.

49. Touraev A, Heberle-Bors E (2003) In: *Doubled Haploid Production in Crop Plants*. Edited by M Maluszynski, KJ Kasha, BP Forster and L Szarejko. Kluwer Academic Publishers, Dordrecht, The Netherlands, pp. 223–228.

50. Zapata-Arias FJ (2003) In: *Doubled Haploid Production in Crop Plants*. Edited by M Maluszynski, KJ Kasha, BP Forster and L Szarejko. Kluwer Academic Publishers, Dordrecht, The Netherlands, pp. 109–116.

51. Serazetdinova LD, Lörz H (2004) In: *Encyclopedia of Plant and Crop Science*. Edited by RM Goodman. Marcel Dekker, NY, USA, pp. 43–50.

*52. Babbar SB, Narayan JP, Bhojwani SS (2000) *Plant Tissue Cult*. **10**, 59–87.

Reviews the problem of albinism in pollen-derived plants of cereals.

53. Inagaki MN (2003) In: *Doubled Haploid Production in Crop Plants*. Edited by M Maluszynski, KJ Kasha, BP Forster and L Szarejko. Kluwer Academic Publishers, Dordrecht, The Netherlands, pp. 53–58.

*54. Bhojwani SS, Thomas TD (2001) In: *Current Trends in the Embryology of Angiosperms*. Edited by SS Bhojwani and WY Soh. Kluwer Academic Publishers, Dordrecht, The Netherlands, pp. 489–507.

Reviews the technique of gynogenesis for haploid production.

55. Gamborg OL, Miller RA, Ojima K (1968) *Exp. Cell Res*. **50**, 151–158.

56. Touraev A, Heberle-Bors E (1999) In: *Methods in Molecular Biology*. Edited by RD Hall. Humana Press Inc., NJ, USA, Vol. III. pp. 281–291.

57. Keller WA, Armstrong KC (1977) *Can. J. Bot*. **55**, 1383–1388.

58. Polsoni L, Kott LS, Beversdorf WD (1988) *Can. J. Bot*. **66**, 1681–1685.

5 Embryo Rescue

Traud Winkelmann[1], Antje Doil[2], Sandra Reinhardt[3] and Aloma Ewald[3]

[1]*Institute of Floriculture and Woody Plant Science, Leibniz University Hannover, Hannover, Germany*

[2]*University of Applied Sciences and Research Institute for Horticulture, Weihenstephan, Freising, Germany*

[3]*Institute of Vegetable and Ornamental Crops, Kuehnhausen, Germany*

5.1 Introduction

Hybridization is the driving force in plant breeding in order to create genetic variability. As long as sexual crosses are performed within a species, seeds develop containing viable embryos. However, if hybridization is carried out between species or even genera [1], hybrids often cannot be obtained *in situ*, because different barriers prevent these crosses. Often these barriers act after the fusion of pollen and egg cell – postfertilization. In many cases, young embryos abort, because they are no longer nourished by the endosperm, which starts to degenerate at some time during seed development. If this is the case, embryo rescue is a suitable strategy to permit these wide crosses. This technique involves culturing the embryo *in vitro* on a nutrient medium.

Thus, embryo rescue techniques aim to generate wide crosses. The applications of these hybridizations are various. The main objectives are the introgression of genetic material into a species, to create new hybrids as novel ornamentals, reduction of the breeding cycles in species with long seed dormancy, fundamental research in embryo and endosperm development, and cytological as well as molecular phylogenetic studies.

Plant Cell Culture Edited by Michael R. Davey and Paul Anthony

In the literature the term embryo culture is often used synonymously with embryo rescue [2]. However, to be accurate, embryo rescue means that under normal conditions the cultivated embryo would not develop naturally. Embryo culture should be utilized in a broader sense for all kinds of cultured embryos, for example, to shorten the breeding cycle [3] in hybridizations which would also lead to seed set *in vivo*, but to a lesser rate of success or in a longer period of time. Today, as in the past, embryo rescue plays an important role in plant breeding programmes, and is undertaken by many private breeders and research institutes. In the future, it is expected to retain or even broaden its significance, since plants obtained by embryo rescue do not have to be considered as genetically modified (transgenic) organisms.

The scope of this chapter is to document the steps in embryo rescue, starting from determination of the type and time of hybridization barriers, selecting and isolating explants, culture media and conditions, to plant regeneration and verification of the hybrid state.

5.2 Methods and approaches

Before entering into detail regarding the different sections of this topic, some important general remarks must be mentioned on pollination and culture of the seed and pollen parents. It is important to expend care in the culture of the partners to be crossed regarding environmental conditions, namely light, humidity and temperature, plant health and nutrition according to best agricultural or horticultural practices. In addition, pollen viability should be checked by staining [4] or *in vitro* germination [5] assays. Staining with fluorescein diacetate (FDA) [6] or with 2,5-diphenyl tetrazolium bromide (MTT) [7] has been found to give reliable results for assessments of pollen viability in different species such as *Streptocarpus, Cyclamen, Hydrangea* and *Primula*. In some instances, the flowering time of both parents cannot be synchronized, and storage at $-18\,^{\circ}C$ of fresh pollen, after drying at room temperature for 24 h, may be useful. It is strongly recommended that as many pollinations (see Protocol 5.1) as possible are performed and all observations documented, since only a small proportion of embryos will develop and can be rescued.

5.2.1 Identification of the time and type of barrier in hybridization

The first step should be the careful monitoring of pollen tube growth in order to determine the type and point in time of the hybridization barrier. Embryo rescue techniques can only be pursued if fertilization can be observed. In the case of prefertilization barriers, other techniques have to be taken into consideration, such as *in vitro* pollination of isolated ovules or even egg cells [8–10], cut or grafted style pollination [11], physical or hormonal treatments (as suggested in reference [12]), or the application of mentor pollen [13].

If the pollen tubes can be monitored to penetrate the ovules and reach the egg cells (see Protocol 5.2, Figure 5.1), the barrier is most likely postzygotic and embryos can be rescued. Therefore, it is essential to study pollen tube development thoroughly, starting a few hours after pollination and taking samples every 12

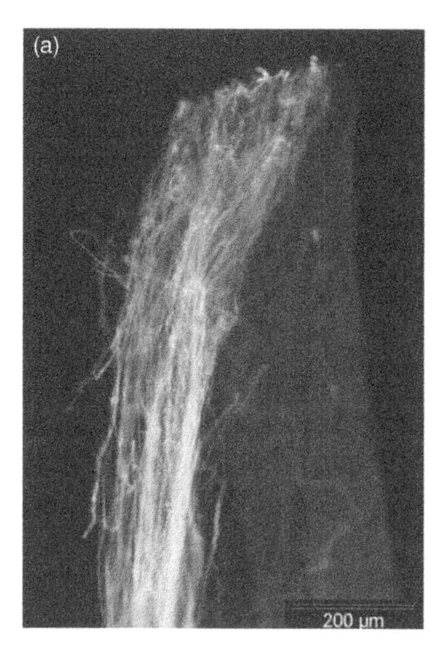

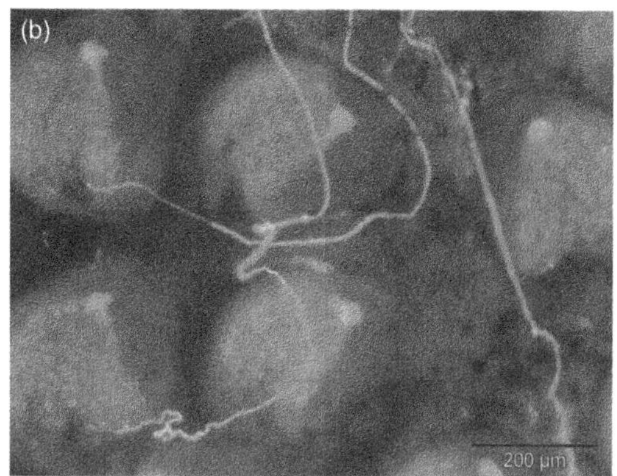

Figure 5.1 Pollen tube growth in the combination *Cyclamen persicum* × *Cyclamen purpurascens*. (a) In the upper part of the pistil 3–4 days after pollination. (b) Pollen tubes entering the micropyle of ovules 6–7 days after pollination.

to 24 h. Since, in some incompatible combinations pollen tube growth is retarded severely, sampling should be continued for 5–6 days or even longer.

The development of seed capsules should be monitored accurately, and histological analyses of endosperm and embryo development should accompany the first experiments as, for example, shown for interspecific crosses between *Cyclamen persicum* and *C. purpurascens* [14], or within the genus *Trifolium* [15]. If the first differences occur in the development of hybrid embryos compared to embryos from compatible crosses, this point in time is often the best to rescue the embryos.

In many, if not all species, strong genotypic or specific cross combination differences have been observed in the success of embryo rescue, for example, in *Cicer* [16], *Rhododendron* [17] and *Dianthus* [18]. Therefore, it is crucial: (i) to test a number of different genotypes of both parental species, and (ii) to perform the reciprocal crosses, since unilateral incompatibilities have often been reported as, for example, in *Cyclamen* [19], *Hibiscus* [20] and *Dianthus* [18]. The recent molecular and genetic achievements in understanding embryo and endosperm development and genome interactions [21, 22] will allow more systematic selection of parental plants in the future. These two factors may be far more important than culture conditions or culture media. Finally, the ploidy of both seed and pollen parent, can have an effect on the outcome of interspecific hybridizations [23, 21].

5.2.2 Isolation of plant material after fertilization

Three types of culture can be distinguished with respect to explant material: these being (i) ovary culture, (ii) ovule culture and (iii) embryo culture. Combinations have also been exploited, for example, commencing with ovary or ovule culture and isolating the embryos after a few weeks [24]. The benefits of ovary and ovule culture are that the risk of wounding the embryos is minimal, since the embryos are protected and surrounded by maternal tissue. The risk of precocious germination, which would result in malformed embryos and plantlets, is also reduced [25]. Drawbacks might be that degenerating maternal tissue could inhibit embryo development, that poorer diffusion of nutrients through this tissue could result in slower growth, and that maternal tissue could give rise to callus which could inhibit embryo growth, or from which non-hybrid plants can regenerate. Finally, the isolation of the embryo will be advantageous, if inhibiting substances are present in either endosperm or testa. The size and accessibility of the embryo will also be an important factor in making the choice of culture method. To date, for the present authors, ovary culture has given better results than isolated ovules in interspecific hybridizations with *Cyclamen* (see Protocols 5.3 and 5.4). In contrast, in *Tulipa* hybrids, ovule culture was reported to be superior to isolated embryo culture [26], while in *Cuphea*, half ovules containing the embryo were appropriate explants [24]. Isolated embryos were found to be ideal starting material in species with relatively large embryos, as in the case of *Trifolium* [27], wheat × *Agropyron* [28] or *Prunus* [29].

In general, embryos should be prepared as late as possible, to allow them to develop for the longest possible time on the plant. Many studies have shown that the frequency of success in embryo rescue increased with age and size of the prepared embryos [2, 30, 31]. Conversely, the degenerating endosperm or disturbances in embryo development can have severe, deleterious effects on the success of culture, so that the isolation of the embryos must not be too late. Embryos at the heart stage, or later in development, have often been reported to be successful in embryo rescue. If fruit abortion takes place very early, treatment of the pollinated flowers with plant growth regulators such as gibberellic acid, auxins and cytokinins, can prolong the time the developing fruits remain on the plants [2, 30].

5.2.3 Culture conditions and media

Amongst the physical culture conditions, light and temperature have been listed as important factors influencing the growth and development of rescued embryos. However, it is extremely dependent on the species and it is not something to generalize which conditions are recommended. It should be taken into consideration whether the first days or weeks of culture are performed in the dark, as the latter was found to be beneficial for *Cyclamen* [14, 19], *Rosa* [32, 33] and several other species [2, 30]. In some cases, photoperiods of 12–16 h resulted in success in *Cuphea* [24] and *Cucumis* [34], for example.

Regarding temperature, the conditions known for germination or micropropagation of the respective species should be tested initially. In some genera like *Tulipa* [26], *Rosa* [35], or *Prunus* [36], cold treatments of 4–5 °C for several weeks improved significantly the success in the culture of rescued embryos. Again, the natural requirements for germination of the species provide considerable information for starting experiments.

Culture media formulation may also have a strong impact, especially when embryos are cultured in their very early stages of development. Immature embryos have been considered to express a type of heterotrophic growth [37], meaning that these young embryos depend upon a richer medium. Mainly organic compounds, which are provided naturally by the endosperm, are needed in these cases as, for example, amino acids, vitamins, sugar alcohols, casein hydrolysate, malt extract, coconut water (which itself is liquid endosperm) and nitrogenous compounds [2]. Sharma (2004; [30]) reviewed several hundred reports on embryo rescue and came to the conclusion that the vast majority of protocols used the composition of macro- and microelements in the culture medium according to Murashige and Skoog (1962; [38]). Again, from the authors' experience, any information available on tissue culture of the target species should serve as a basis for the first embryo rescue trials in relation to medium formulation.

One component of media for embryo rescue that has received particular attention, is the carbohydrate source and its concentration. Sucrose is most commonly used in concentrations of 1.5–6% (w/v). Besides nutritional effects, sucrose as well as other sugars, has an impact on the osmotic potential of the medium. During development, zygotic embryos are subjected to decreasing osmotic potential of the surrounding endosperm. Therefore, it could be important to commence culture with elevated sugar concentrations and to reduce subsequently these concentrations.

Although liquid culture systems, as well as two-layer systems [24], have been suggested for the culture of rescued embryos, most of the protocols use semisolid culture medium (see Figure 5.2). The application of plant growth regulators seems to be strongly dependent on the species and, also, on the age of the isolated embryos. While many reports were successful with hormone-free culture media [37], others report the use of plant growth regulators to be essential. Mainly auxins and cytokinins in low concentrations, and gibberellic acid have been recommended in some species [37]. Callus formation may result in negative effects on embryo germination as shown for grape [23]. In rare cases, embryogenic or organogenic cultures have been induced from rescued embryos. Gibberellic acid not only leads to elongation of cells and, in consequence, of the whole embryo, but could also help to overcome dormancy, as shown for *Rhododendron* [17].

5.2.4 Confirmation of hybridity and ploidy

Since accidental pollination and apomictic or parthenogenetic origin of developed plantlets cannot be excluded completely, the hybrid nature of the plants has to be confirmed. Ideally, this confirmation should be done as early as possible to save labour and space for cultivation of non-hybrid plants. Different possibilities exist for proof of hybridity, namely morphological characters, cytological analyses, flow cytometric analyses, isozymes and molecular markers, including random amplified polymorphic DNAs (RAPDs), amplified fragment length polymorphisms (AFLPs) and simple sequence repeats (SSRs). The choice of the method should be based on cost, accuracy, time required and availability. If one or more paternal characters can be detected, the respective plant may be considered as a hybrid plant. In the

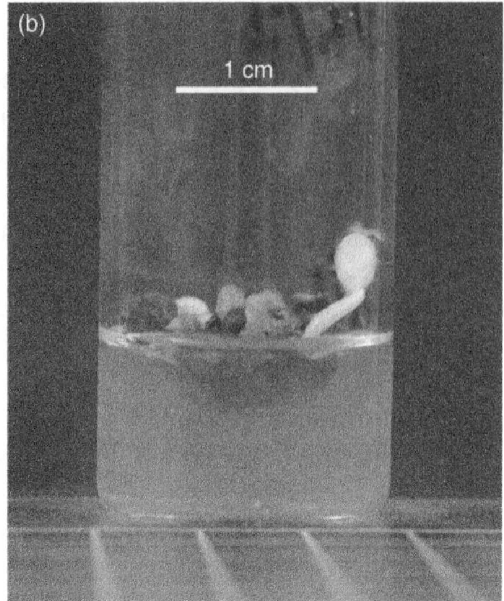

Figure 5.2 Ovary culture after interspecific hybridization in *Cyclamen*. (a) Cultured ovary (*C. persicum* × *C. graecum*), isolated 28 days after pollination and cultured for 6 weeks; (b) Ovary with a seedling (*C. persicum* × *C. graecum*) 13 weeks after pollination; (c) Development of hybrid seedlings (*C. persicum* × *C. purpurascens*) 38 weeks after pollination.

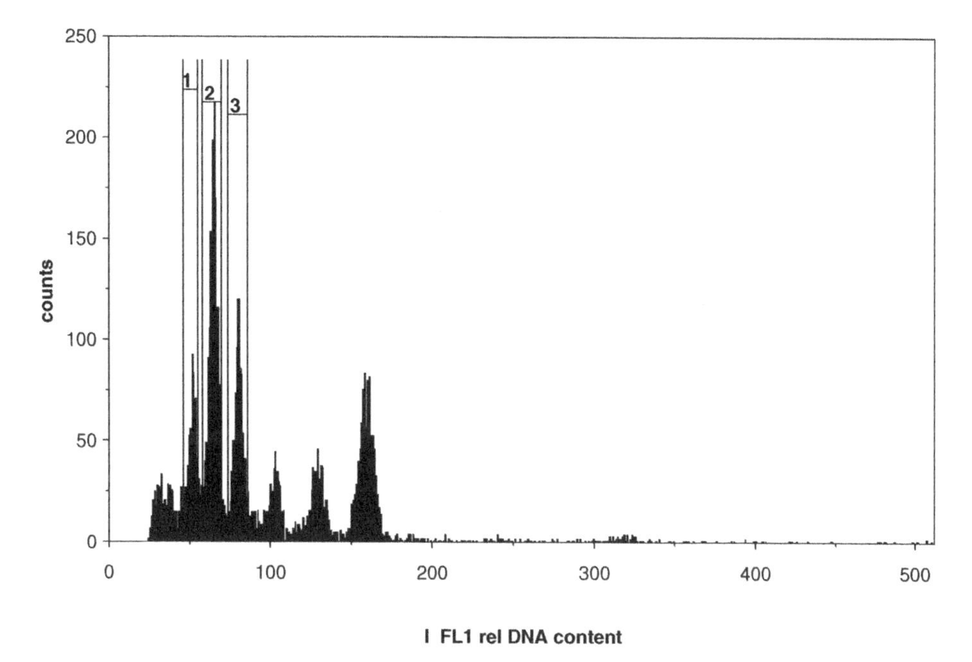

Figure 5.3 Flow cytometric measurement of relative DNA contents of parents and an interspecific hybrid. Peak 1 – *Cyclamen persicum* (seed parent); peak 2 – interspecific hybrid; peak 3 – *C. hederifolium* (pollen parent).

authors' hands [19], and also for other species [18], flow cytometric confirmation (see Protocol 5.5, Figure 5.3) of interspecific hybrids was found to be reliable, fast and inexpensive. However, it is only possible to use this technique if the genome sizes of the parental species are sufficiently different.

Cytological information on the plants obtained may be important with regard to their further use in breeding programmes. Chromosome counting is one way to verify that the hybrids contain the complete genomes of both parents, since chromosomal losses are often observed, especially when the genetic distance is large. Modern staining techniques, such as fluorescence *in situ* hybridization (FISH) [39], or genomic *in situ* hybridization (GISH) [40] or their combination, allow observation of the fate of parental chromosomes and may reveal interesting information regarding the balance of genomic components.

Amongst the molecular markers, RAPDs [41] have often been used to verify the hybrid state of plants (See Protocol 5.6, Figure 5.4). The advantages of this type of marker are that it is simple, rapid and inexpensive and does not need any species-specific sequence information. One important drawback is their limited reproducibility. Conversely, AFLPs [42] are very reliable markers giving the additional advantage of large amounts of information on individual gels. This allows an estimation of the genomic composition of hybrids, especially their maternal and paternal contributions. However, AFLPs require more sophisticated equipment, greater amounts of DNA of high quality and, moreover, the patent considerations may result in additional costs. Microsatellites or SSRs [43] are more elaborate in

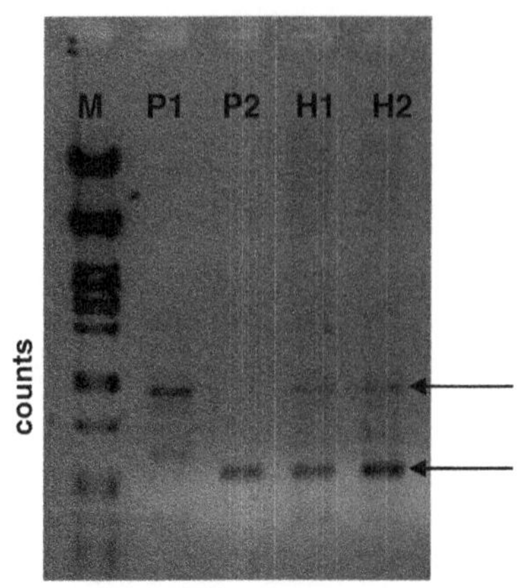

Figure 5.4 Identification of interspecific *Streptocarpus* hybrids by RAPD markers. M = size marker: λDNA digested with Pst, P1 = *Streptocarpus glandulosissimus* (seed parent), P2 = *S. caulescens* (pollen parent), H1 and H2 = interspecific hybrids. Arrows indicate specific bands of both parents which are present in the hybrids (Photo by R. Afkhami-Sarvestani).

their development but, if available, they are very well suited for hybrid identification because they are highly reproducible and very informative.

5.2.5 Conditions for regeneration of embryos to plants

Once the rescued embryos have developed cotyledons and roots, conditions have to be established that support their further growth and propagation. Sometimes it is advisable to multiply the hybrids obtained by axillary shoot formation in order to minimize the risk of losing genotypes during acclimatization. If cytokinins have been applied during embryo rescue, subsequent transfer of the shoots to auxin-supplemented rooting media may be necessary. Again, any general recommendation is difficult, but all information available on the respective species regarding their requirements and growth conditions in nature, as well as tissue culture protocols, should be taken into account. Acclimatization (see Protocol 5.7) can be handled the same way as for any other micropropagated plant.

The protocols described in this chapter use *Cyclamen* and *Streptocarpus* as the examples.

PROTOCOL 5.1 Emasculation and Pollination in Cyclamen

Equipment and Reagents

- Fine forceps
- Petri dishes (6 cm diameter)
- Aluminium foil
- Tag labels

Method

1 Carry out crosses under controlled conditions, in growth chambers (12 h photoperiod with 250 µmol/m^2/s high pressure sodium lamps, 20/16 °C day/night temperatures, 60% relative humidity), or in the glasshouse (18 °C heating, 22 °C ventilation temperature).

2 Emasculate flowers about 3 days before anthesis by removing the corolla to which the anthers are attached. If all anthers are not detached at once, use forceps to eliminate all residual anthers. Collect the anthers in Petri dishes or in other suitable vessels, if the pollen is needed for further pollinations.

3 Dry anthers of the pollen parent at room temperature overnight in unsealed Petri dishes; after 24 h, release pollen from the anthers by taking hold of the anthers with fine forceps and beating them onto the bottom of the Petri dish[a]. Store in a sealed vessel at 4 °C in the short term, or at −18 °C for longer.

4 Cover stigmas from emasculated flowers with pollen by dipping them carefully into the pollen in the Petri dishes; isolate the flowers with covers of aluminium foil. Label the flowers individually with parents and date.

5 Repeat the pollinations after 3 and 7 days, because ovules mature gradually.

Note

[a]Study the viability of pollen by staining with FDA (reference [6], modified by [44]) or MTT [7] to be sure that pollinations are performed with pollen of high quality. Analyse at least 300 pollen grains from each sample.

PROTOCOL 5.2 Aniline Blue Staining of Pollen Tubes

Equipment and Reagents

- 99% (v/v) ethanol
- Lactic acid (Carl Roth GmbH)
- 1 M NaOH
- Aniline blue ($C_{32}H_{25}N_3Na_2O_9S_3$) (Serva)
- $K_3PO_4xH_2O$
- Aniline blue staining solution: Dissolve 100 mg aniline blue and 767.6 mg K_3PO_4 in 100 ml distilled water. Keep this solution for 24 h under natural light until its blue colour turns yellow; store at 4 °C in the dark
- 1.5 ml microfuge tubes
- Glycerine: 10% (v/v) solution
- Glass slides; cover slips

- Deionized water
- Microscope with fluorescence facilities

Method

1. Cut off the pollinated carpels and fix them in 1.5 ml microfuge tubes with ethanol : lactic acid (2 : 1, v : v) immediately to terminate pollen tube growth.
2. After storage for a minimum of 24 h at room temperature[a], rinse the carpels three times in deionized water.
3. Macerate tissue by incubation in 1 M NaOH at 60 °C for 45 min[b]; rinse three times in deionized water.
4. Incubate the carpels in aniline blue staining solution for at least 30 min at room temperature.
5. Place each carpel on a microscope slide in a drop of aniline blue staining solution or glycerine (10% v/v) and squash carefully under a cover slip.
6. Observe pollen tube growth under a fluorescence microscope with the following filter combination: excitation filter BP 340–380, dichromatic mirror 400, suppression filter LP425. Callose is visible as a bright yellow to green fluorescence, indicating pollen tube walls.

Notes

[a]The carpels can be stored in this fixing solution at 4 °C in the refrigerator for several months without loss of quality.

[b]This step can be omitted if the tissue is already softened during a longer storage period in the fixative.

PROTOCOL 5.3 Ovary Culture in *Cyclamen*

Equipment and Reagents

- 70% (v/v) ethanol (EtOH); it is not necessary to use pure ethanol for sterilization; denatured ethanol is adequate
- Sodium hypochlorite (NaOCl, 3% active chlorine): Dilute commercially available NaOCl solution (12–14% active chlorine) with tap water 1 : 4 (v : v), plus one drop of Tween 20 (detergent) per 400 ml. Care should be taken with this bleach and caustic solution, which should be prepared as required
- Deionized sterile water
- Ovary culture medium: macro- and micronutrients of Murashige and Skoog (MS; [38]) at full strength, 100 mg/l myoinositol, 2.0 mg/l glycine, 0.5 mg/l nicotinic acid, 0.1 mg/l thiamine HCl, 0.5 mg/l pyridoxine HCl, 30 or 60 g/l sucrose, and 2.5 g/l Gelrite (Duchefa), pH 5.8

- Glass vessels: 10 cm in height, 45 ml volume (welted glasses, Carl Roth GmbH), sealed with two layers of aluminium foil
- Germination medium: Nitsch medium [45] supplemented with 30 g/l sucrose and 2.5 g/l Gelrite
- Proliferation medium: Nitsch medium [45] supplemented with 30 g/l sucrose, 1.5 mg/l BA (benzyladenine), 1.0 mg/l IAA (indoleacetic acid), 120 mg/l adenine and 2.5 g/l Gelrite
- Rooting medium: Nitsch medium [45] supplemented with 20 g/l sucrose, 0.5 mg/l NAA (1-naphthyleneacetic acid), and 2.5 g/l Gelrite

Method

1. Using the method of Ishizaka and Uematsu (1992; [46]) excise flowers/seed capsules 14, 21, 28, 35 and 42 days after pollination.
2. Surface sterilize the flowers in 70% (v/v) EtOH for 30 s followed by 3% hypochlorite solution for 20 min; rinse three times for 5 min each with sterile deionized water. Leave the flowers in the last water wash until required.
3. Remove the ovary wall and isolate the central placenta containing the ovules.
4. Place the cut surfaces of the explants on the ovary culture medium in glass culture vessels ensuring good contact of the cells with the medium.
5. Incubate the cultures at 20–24 °C in the dark.
6. Transfer germinating embryos to germination medium lacking plant growth regulators[a] for rapid growth, or to proliferation medium on which plants can be multiplied to minimize the risk of loosing important genotypes. In the latter case, proliferating shoot cultures need a special rooting medium.
7. Incubate plantlets, with a height of about 2 cm, under a 16 h photoperiod (cool fluorescent illumination; 20–40 μmol/m^2/s).

Note

[a]Other media formulations such as MS [38] at half strength, but with FeEDTA at full strength and 250 mg/l peptone, or U-medium [47], both with 30 g/l sucrose and 3.7 g/l Gelrite, have also been used for plantlet development.

PROTOCOL 5.4 Ovule Culture in *Cyclamen*

Equipment and Reagents

- 70% (v/v) ethanol (EtOH). It is not necessary to use pure ethanol for sterilization; denatured ethanol is adequate
- Sodium hypochlorite (NaOCl, 3% active chlorine) solution: See Protocol 5.3

- Ovule culture medium composed of macro- and micronutrients of MS [38] medium at full strength, 100 mg/l myoinositol, 2.0 mg/l glycine, 0.5 mg/l nicotinic acid, 0.1 mg/l thiamine HCl, 0.5 mg/l pyridoxine HCl, 60 g/l sucrose and 2.5 g/l Gelrite, pH 5.8
- Glass vessels, 10 cm in height, volume 45 ml (welted glasses; Carl Roth GmbH), sealed with two layers of aluminium foil
- Preparation needles
- Stereo microscope

Method

1. Excise flowers/seed capsules every 7 days from 21–56 days after pollination.
2. Surface sterilize the flowers, as in Protocol 5.3.
3. Carefully dissect the ovules from the ovary under a stereo microscope.
4. Preparation of ovules is best done using two needles; one to fix the peduncle, and the other to gently touch the individual ovules. They will adhere to the needle and can be placed on ovule culture medium.
5. Place 25 ovules in a vessel and incubate at 20–24 °C in the dark.
6. Transfer germinating embryos[a] to new culture medium of the same composition.
7. Incubate plantlets, with a height of about 2 cm, under a 16 h photoperiod (cool fluorescent illumination; 20–40 µmol/m^2/s).

Note

[a]If germination becomes visible, it is recommended to isolate the embryos from the ovule and to culture them on MS [38] medium at half strength, but with FeEDTA at full strength, 250 mg/l peptone, 30 g/l sucrose and 3.7 g/l Gelrite.

PROTOCOL 5.5 Flow Cytometric Analyses of Putative Hybrids

Equipment and Reagents

- Razor blades
- Petri dishes (6–9 cm diameter)
- Sieves with 40 µm mesh (Partec)
- 5 ml reaction tubes (Partec)
- CyStain UV Precise P kit (Partec)
- Flow cytometer (Cell analyser CAII or PA; Partec)

Method

1 Excise small leaf segments (each about 0.5 cm^2) and chop with a sharp razor blade in a plastic Petri dish with 0.5 ml nuclei extraction buffer of the CyStain UV Precise P kit.

2 Filter the suspension through 40 µm sieves and collect the filtrate in 5 ml tubes.

3 After 2 min, add 2 ml of the staining buffer of the CyStain UV Precise P kit.

4 After 2 min, analyse the DNA content of the released nuclei with the flow cytometer[a].

5 Evaluate the position of the peaks in relation to those of the parental plants[b].

Notes

[a]Adjust the sensitivity (gain) so that the parent with the smallest DNA content reveals a peak position at about one fifth to one tenth of the scale.

[b]It is recommended that analysis is carried out on mixed samples containing the nuclei of both parents and the putative hybrid (see Figure 5.3).

PROTOCOL 5.6 RAPD Analysis of Putative Hybrids of *Streptocarpus*

Equipment and Reagents

- DNeasy Plant Mini Kit (Qiagen)
- *Taq* polymerase (5 U/µl stock; Invitek GmbH)
- dNTPs (stock solution 1 mM each; Carl Roth GmbH)
- 10 × Williams buffer for PCR: 100 mM Tris pH 8.3, 500 mM KCl, 20 mM $MgCl_2$, 0.01% gelatin
- Decamer primer (Roth) (5 pmol/µl dilution of 100 pmol/µl stock solution)
- 200 µl thin wall reaction tubes (Sarstedt)
- Thermocycler (Biometra T3)
- 10 × loading buffer: 2% (w/v) bromphenol blue in 34.5% (v/v) glycerol
- Agarose (SeaKem, LE Agarose; Cambrex Inc.)
- Gel electrophoresis and documentation equipment
- TAE buffer: 0.04 M Tris acetate, 1 mM EDTA, pH adjusted to 8.44 with acetic acid
- Liquid nitrogen
- Sterile distilled water

Method

1 Isolate DNA from 100 mg of leaf tissue (*in vitro* or *ex vitro* plants); grind the leaves in liquid nitrogen according to the manual of the DNeasy kit.

2 Mix the following at a reaction volume of 25 µl: 5–20 ng of genomic DNA, 2 µl of decamer primers, 2 µl of dNTPs, 2.5 µl of 10 × Williams buffer for PCR, 0.2 µl of *Taq* polymerase; add sterile distilled water to the final volume.

Conduct PCR in 200 µl thin wall tubes in a thermocycler with the following programme:

No. of cycles	Programme
1	94 °C for 5 min
40	92 °C for 1 min, 35 °C for 1 min, 72 °C for 1 min
1	72 °C for 10 min
1	Hold at 20 °C

3 Mix the whole volume of PCR product with 2.5 µl of 10 × loading buffer and transfer into an agarose gel containing 1.0% (w/v) agarose with 0.29 µg/ml ethidium bromide in 1 × TAE buffer.

4 Electrophorese samples at 3 V/cm for 5 min and at 4.5 V/cm until the front of the loading buffer reaches the middle of a gel.

5 Document gels under UV (320 nm) illumination.

PROTOCOL 5.7 Acclimatization of Cyclamen *in vitro*-derived Plants to *ex vitro* Conditions

Equipment and Reagents

- Compost mixture: Einheitserde P, Einheitserde (Sinntal-Jossa) : perlite (1 : 1, v : v)
- Trays, multicell plates or 6 cm diam. pots
- Foil tunnel: length 2–3 m, width 1.2–1.5 m, height 0.5–0.8 m[a]

Method

1 Remove developed plants with a tuber, two to three leaves and roots, from the culture medium; wash carefully in lukewarm water and, if necessary, reduce the roots to 2 cm in length.

2 Place the plants into trays or multicell plates or 6 cm pots in the compost : perlite mixture, taking care that one third of the tuber is above the surface of the potting medium.

3 After thorough watering[b], keep the plants under a foil tunnel[a] at 90–95% relative humidity and 22–24 °C. Prevent direct sunlight in the first 2–3 weeks.

4 Open the foil gradually after 7–10 days.

Notes

[a]Hardening can be facilitated with heating mats covered with sand on glasshouse benches. Place the trays with the plants on the sand-covered mats; cover the plants with transparent foil.

[b]Spraying or watering with antimicrobial compounds such as 8-hydroxyquinoline (0.1%) or a fungicide prevents losses during the first few days after transfer to compost.

5.3 Troubleshooting

- One of the most striking problems in establishing embryo rescue protocols is between fact that only a limited number of pollinations can be performed. Therefore, multifactorial experimental designs can rarely be realized. Moreover, the adoption of existing protocols is not possible and adapting them is time and labour consuming.
- The first steps, involving emasculation and pollination, require special attention to avoid self pollination. One should be aware that the anthers must be detached very carefully and that flowers have to be isolated immediately thereafter. Sometimes pollen does not adhere to the stigmatic surface. In this case, increased relative humidity or the use of more pollen may help. The isolation and culture of very young embryos is still difficult, not only regarding the preparation itself, but also with respect to the development of an appropriate culture medium. One phenomenon which has often been reported [17, 20] is the occurrence of albinism, which might be the result of imbalanced nuclear and chloroplast genomes. It has been recommended to test combinations of the reciprocal crossing and other parental genotypes to overcome this problem.
- The following general remarks may assist in planning and conducting embryo rescue experiments. Perform the most pollinations possible and prepare as many ovaries, ovules or embryos, respectively, as are manageable. Since, in many species, especially woody plants, flowering takes place only once a year, very thorough planning and design of the combinations to be crossed, the timing of pollination and the initiation of culture are essential. The viability and germination of pollen should be tested to ensure that viable pollen is used. Finally, much important information can be obtained from careful observations, macroscopic as well as microscopic, of all the details during the development of capsules and embryos.

References

1. Sharma HC (1995) *Euphytica* **82**, 43–64.

*2. Sharma HC, Kaur R, Kumar K (1996) *Euphytica* **89**, 325–337.

Comprehensive review on embryo rescue techniques in plants, covering culture details and applications.

3. Goetze BR (1979) Dissertation, Akad. Landwirtsch.-Wiss., Berlin.

*4. Rodriguez-Riano T, Dafni A (2000) *Sex. Plant Reprod.* **12**, 241–244.

Comparison and assessment of different pollen viability assays.

5. Brewbaker JL, Kwack BH (1963) *Am. J. Bot.* **50**, 859–865.
6. Heslop-Harrison J, Heslop-Harrison Y, Shivanna KR (1984) *Theor. Appl. Genet.* **67**, 367–375.
7. Khatum S, Flowers TJ (1995) *J. Exp. Bot.* **46**, 151–154.
8. Deverna JE, Myers JR, Collins GB (1987) *Theor. Appl. Genet.* **73**, 665–671.
9. Kranz E, Bautor J, Loerz H (1991) *Sex. Plant Reprod.* **4**, 12–16.

*10. Zenkteler M (1990) *CRC Crit. Rev. Plant Sci.* **9**, 267–279.

Review on methods to obtain embryos after *in vitro* pollination.

11. Van Tuyl J, van Dien M, van Creij M, van Kleinwee T, Franken J, Bino R (1991) *Plant Sci.* **74**, 115–116.
12. Bhat S, Sarla N (2004) *Genet. Res. Crop Evol.* **51**, 455–469.
13. Singsit C, Hanneman RE (1991) *Plant Cell Rep.* **9**, 475–478.
14. Ishizaka H, Uematsu J (1995) *Euphytica* **82**, 31–37.
15. Repkova J, Jungmannova B, Jakesova H (2006) *Euphytica* **151**, 39–48.
16. Clarke HJ, Wilson JG, Kuo I, *et al.* (2006) *Plant Cell Tissue Organ Cult.* **85**, 197–204.
17. Eeckhaut T, de Keyser E, van Huylenbroeck J, de Riek J, van Bockstaele E (2007) *Plant Cell Tissue Organ Cult.* **89**, 29–35.
18. Nimura M, Kato J, Mii M, Morioka K (2003) *Theor. Appl. Genet.* **106**, 1164–1170.
19. Ewald A (1996) *Plant Breed.* **115**, 162–166.
20. van Laere K, van Huylenbroeck J, van Bockstaele E (2007) *Euphytica* **155**, 271–283.

*21. Burke JM, Arnold ML (2001) *Ann. Rev. Genet.* **35**, 31–52.

This review deals with the genetic basis of hybrid sterility and inviability.

*22. Orr HA, Presgraves DC (2000) *BioEssays* **22**, 1085–1094.

Review article on the genetics of sterility and inviability of interspecies hybrids.

23. Yang D, Li W, Li S, Yang X, Wu J, Cao Z (2007) *Plant Growth Regul.* **51**, 63–71.
24. Mathias R, Espinosa S, Roebbelen G (1990) *Plant Breed.* **104**, 258–261.

*25. Bridgen MP (1994) *HortScience* **29**, 1243–1246.

This compact review lists practical aspects of plant embryo culture.

26. Custers JBM, Eikelboom W, Bergervoet JHW, van Eijk JP (1995) *Euphytica* **82**, 253–261.

27. Phillips GC, Collins GB, Taylor NL (1982) *Theor. Appl. Genet.* **62**, 17–24.

28. Sharma H, Ohm H (1990) *Euphytica* **49**, 209–214.

29. Liu W, Chen X, Liu G, Lian Q, He T, Feng J (2007) *Plant Cell Tissue Organ Cult.* **88**, 289–299.

**30. Sharma HC (2004) *Recent Research Developments in Genetics and Breeding*. Research Signpost, Trivandrum, Vol. 1 Part II, pp. 287–308.

Literature review summarizing 651 publications on embryo rescue and underlining important factors for success.

31. Roy AK, Malaviya DR, Kaushal B, Kumar D, Tiwari A (2004) *Plant Cell Rep.* **22**, 705–710.

32. Marchant R, Power JB, Davey MR, Chartier-Hollis J (1994) *Euphytica* **74**, 187–193.

33. Mohapatra A, Rout GR (2005) *Plant Cell Tissue Organ Cult.* **81**, 113–117.

34. Sisko M, Ivancic A, Bohanec B (2003) *Plant Sci.* **165**, 663–669.

35. Gudin S (1994) *Euphytica* **72**, 205–212.

36. Zagaja SW, Hough LF, Bailey CH (1960) *Proc. Am. Soc. Hort. Sci.* **75**, 171–180.

**37. Rhagavan V (1980) In: *Perspectives in Plant Cell and Tissue Culture. Int. Rev. Cytol.* 11B, edited by IK Vasil. Academic Press, New York, pp. 209–240.

Review with comprehensive information on media components and their effects on cultured embryos.

38. Murashige T, Skoog F (1962) *Physiol. Plant.* **15**, 473–497.

39. Kamstra SA, Ramanna MS, de Jeu MJ, Kuijpers AGJ, Jacobsen E (1999) *Heredity* **82**, 69–78.

*40. Schwarzacher T, Leitch AR, Bennett MD, Heslop-Harrison JS (1989) *Ann. Bot.* **64**, 315–324.

First report dealing with fluorescent techniques for chromosome identification in wide hybrids.

41. Williams JGK, Kubelik AR, Livak KJ, Rafalski JA, Tingey SV (1990) *Nucl. Acids Res.* **18**, 6531–6535.

42. Vos P, Hogers R, Bleeker M, *et al.* (1995) *Nucl. Acids Res.* **23**, 4407–4414.

43. Tautz D (1989) *Nucl Acids Res.* **17**, 6463–6471.

44. Kison HU (1979) Dissertation, Akad. Landwirtsch.-Wiss., Berlin.

45. Nitsch JP (1969) *Phytomorphology* **19**, 389–404.

46. Ishizaka H, Uematsu J (1992) *Jpn. J. Breed.* **42**, 353–366.

47. Haensch KT (1999) *Gartenbauwissenschaft* **64**, 193–200.

6

In vitro Flowering and Seed Set: Acceleration of Generation Cycles

Sergio J. Ochatt[1] and Rajbir S. Sangwan[2]

[1] *Laboratoire de Physiologie Cellulaire, Morphogenèse et Validation (PCMV), Centre de Recherches INRA de Dijon, Dijon, France*

[2] *Laboratoire AEB, Universite de Picardie Jules Verne, Amiens, France*

6.1 Introduction

Plant breeding is the basis of efficient agricultural production and involves the recovery of novel, agronomically interesting genotypes, as rapidly as possible, so that they may be registered for commercial cultivation. This process, however, takes 10–12 generations before interesting traits, that may have been incorporated into breeding lines through crosses or introduced by biotechnological approaches, can be fixed in the genome and become stable. In this respect, at best, only two generations per year are feasible in the field with crops such as protein legumes. Frequently, this is possible only when planting in opposite hemispheres [1]. Two to three generations/year may also be obtained under glasshouse conditions, but at an extra cost that prevents this approach for some crops.

It is therefore of value to accelerate generations by shortening each cycle, and to induce flowering and seed set *in vitro*, particularly for rare and valuable genotypes where the initial number of seeds is limited [2–4]. Additionally, this would favour a more rapid fixation of new traits when regenerated shoots are difficult to root [5], or when establishing regenerated plants is difficult, as in legumes [6]. Since seeds

Plant Cell Culture Edited by Michael R. Davey and Paul Anthony

are harvested *in vitro*, this avoids the frequent and significant glasshouse losses and the production of sterile plants or plants with reduced fertility [5].

Efforts in this area [7] now permit seven to nine generation cycles/year in field pea, and 3–5 cycles/year with some neglected and underutilized protein legumes, including grass pea [8, 9] and bambara groundnut [10–12] (see Protocols 6.1–6.3). The technique is being extended to other major grain legumes, including lentil, lupin and chickpea. More recently, this same strategy has been adopted for *Arabidopsis thaliana* where, depending on genotype, 15–19 generations are feasible each year (see Protocol 6.4) [13]. *In vitro* flowering and seed set holds considerable potential in crop breeding as a reliable tool for the rapid follow up of the introgression of traits into progeny, as revealed by a kinetic genomic *in situ* hybridization (GISH) analysis of successive generations, by flow cytometry, or through immunolabelling. It has been used for this objective in hybrids of field pea and some of its wild relatives [14]. It may be useful for single seed descent (SSD) studies for a faster generation of novel genotypes of interest in crop science. An additional, potential application of this strategy is when transgene fixation in the genome of genetically modified organisms (GMOs) might prove politically difficult or too costly. *In vitro* flowering is an attractive procedure to carry out those tests in an environmentally riskless and politically correct manner.

This chapter describes the general strategy conducive to the induction of flowering and seed set *in vitro*. It provides guidelines for the specific modifications to this general method in order to adapt it for different species.

6.2 Methods and approaches

6.2.1 Protein legumes [7]

Three methods have been devised aimed at reducing the generation cycles applied to a range of genotypes of pea, grass pea and bambara groundnut.

1. In the glasshouse (see Protocol 6.1): Six genotypes of pea (*Pisum sativum* L.), were tested, including the spring protein types Baccara and Terese, the winter protein types Cheyenne and Victor, and the winter forage types Champagne and Winterberger. Four landraces of bambara groundnut (*Vigna subterranea* L.) were also studied; two from Ghana (GB1 and GB2) and two from Mali (MB1 and MB2).
2. An intermediate methodology involving *in vivo* plus *in vitro* stages (see Protocol 6.2): Victor, Frisson and Terese peas, and all four landraces of bambara groundnut were tested over a 2-year period with 12 successive generations, using a strategy modified from that reported by Stafford and Davies [15], as described below.
3. *In vitro* only, with all stages up to and including seed set occurring *in vitro* (see Protocol 6.3). Five pea (Frisson and its hypernodulating mutants [16] P64, P79 and P90 and Terese) and three grass pea (*Lathyrus sativus* L.) genotypes (L3 and L12, with coloured flowers and wrinkled, coloured seeds, and LB with white

flowers and smooth, flat, white seeds) were used. Shoots were compared derived either from excised embryo axes germinated on modified B5 [17] medium [6], or regenerated *in vitro* from hypocotyl explants of pea [6] and grass pea [9, 18, 19], or from leaf protoplasts of pea [20].

PROTOCOL 6.1 Glasshouse Strategy

Equipment and Reagents

- Mature, dry seeds of the genotypes to be studied, preferably harvested not more than 2 years before use
- Nutrient solution containing 14.44 mM NO_3, 3.94 mM NH_4, 15.88 mM Ca, 17.9 mM K_2O, 4 mM MgO, 2.46 mM P_2O_5, 2.00 mM SO_3 as macroelements and the microelements as in Murashige and Skoog [21] medium, i.e. (in mg/l) 0.025 $CoCl_2.6H_2O$, 0.025 $CuSO_4.5H_2O$, 0.25 $Na_2MoO_4.2H_2O$, 0.83 KI, 6.2 H_3BO_3, 8.6 $ZnSO_4.7H_2O$, 16.9 $MnSO_4.H_2O$. Prepare these as two stock solutions, concentrated 10 × for the macroelements and 1000 × for the microelements, so as to add, respectively 100 ml and 1 ml per litre of medium. Keep these stock solutions at 4 °C in the dark until use, or renew every month for the macro-elements and once a year for the microelements
- Perlite (SA Sonofep)
- Flurprimidol 2-methyl-1-pyrimidine-5-yl-1-(4-trifluoromethoxyphenyl)propane-1-ol (Topflor; Dow-Agrosciences)

Method

1 Sow seeds at a density of 230 seeds/m^2, with perlite as substrate.

2 Water plants by capillarity with nutrient solution.

3 For pea, control the temperature at 20 °C/16 °C day/night, with a maximum of 26 °C. Adapt the photoperiod according to genotype, i.e. use a 16 h photoperiod from 400 W sodium lamps or continuous illumination for 16 h/day, but supplement with incandescent bulbs (8 h/day) to complete the far-red supply and thus permit floral initiation of genotypes which are sensitive to photoperiod.

4 For bambara groundnut, use 27 ± 1 °C/25 ± 1 °C (day/night) and a 10 h photoperiod (cf. 3. above).

5 In pea the commercial antigibberellin, Flurprimidol, may be used (0.5% w/v) to reduce internode elongation; spray three times every 10 days from the three-leaf stage[a].

6 Cease watering and providing nutrients when pods are whitish in colour (50–60% seed dry matter content) to hasten plant maturation. Perlite favours plant dehydration.

7 Harvest at full maturity to preserve maximum germination; resow seeds immediately following the same procedure. The number of seeds/pod is reduced compared to pods produced following standard procedures[b].

8 The duration of each generation is the time between sowing date and pod harvest date (Figure 6.1).

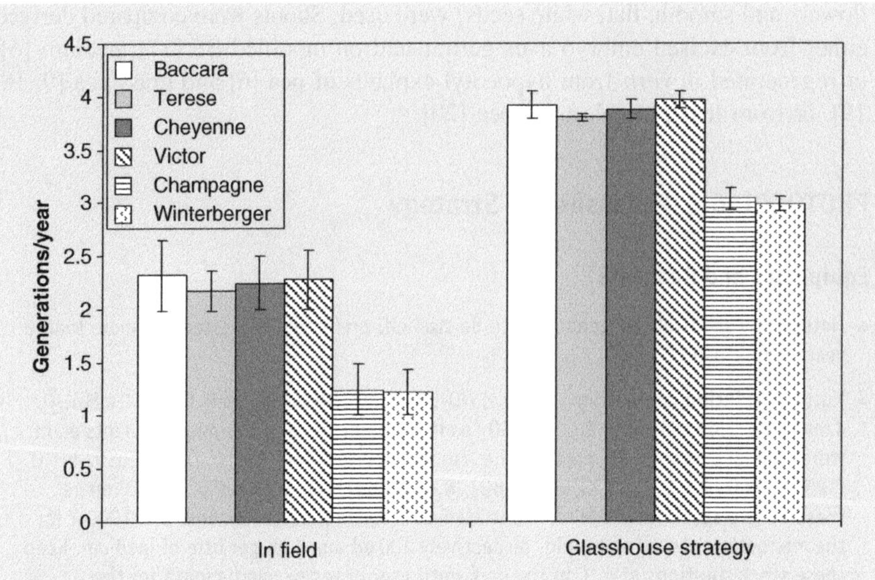

Figure 6.1 The number of generations in one year for pea using the glasshouse strategy. Mean ± SE data from two consecutive years.

Notes

[a]The final goal of this work is to integrate this technique into a SSD selection scheme. Therefore, sow plants at a high density. Since *P. sativum* is naturally of an indeterminate type, it is essential to obtain plants with reduced vegetative development to be able to shorten generations. Using Flurprimidol, plants of Baccara, Terese, Cheyenne and Victor are 20–25 cm in height (versus 70–120 cm) and Champagne and Winterberger 25–35 cm (instead of 150–200 cm) at maturity, with no significant effect of photoperiod on plant height.

[b]The mean number of seeds/plant is reduced by Flurprimidol, but not by photoperiod, to 2.5 and 6.2 seeds/plant for protein and forage pea genotypes, respectively. This is of little consequence for SSD, where one or two seeds per plant suffice.

PROTOCOL 6.2 *In vitro* Plus *In vivo* Strategy

Equipment and Reagents

- 3 l plastic pots with Vermiculite (SA Sonofep)
- Glasshouse nutrient solution: see Protocol 6.1
- Deionized water
- Ethanol 70% (v/v)

- $Ca(ClO)_2$ at 35 g/l and 50 g/l
- Murashige and Skoog (MS) culture medium [21] consisting of macroelements, full-strength MS microelements, Fe-ethylene diaminetetraacetic acid (EDTA) and MS vitamins, plus 15 g/l sucrose (pH 6), semisolidified with 6 g/l agar for pea
- Bambara medium (BM) containing MS macroelements, microelements and vitamins of Nitsch and Nitsch [22], 2% (w/v) sucrose, plus growth regulators (NAA, IBA) at various concentrations (0.0, 0.5, 1.0 mg/l), semisolidified with 6 g/l agar
- Transparent plastic vessels (50 mm diam. × 100 mm height, 110 ml capacity, screw-capped, autoclavable; Falcon, Dutscher)
- Petri dishes with vents (100 × 20 mm; CellStar Greiner bio-one)
- 25 × 150 mm glass culture tubes, autoclavable (Dutscher)
- Laminar air flow cabinet; dissection instruments

Method

1. Sow the seeds in vermiculite in the pots and water with nutrient solution (see Protocol 6.1) every 7 days throughout the experiment and with deionized water once or twice every 7 days, according to plant development.
2. During growth and until seed production, maintain the plants under a 16 h photoperiod at 24 °C/20 °C (see Protocol 6.1), with 70% relative humidity.
3. After 2 months, detach yellowing pods with mature undried seeds and surface-sterilize the unopened pods in 70% (v/v) ethanol (1 min), $Ca(ClO)_2$ at 35 g/l (20 min) for pea, and at 50 g/l (30 min) for bambara groundnut.
4. Open the pods aseptically and excise the embryos from the 3 central (pea) or randomly chosen (bambara) seeds.
5. Culture the embryos on hormone-free semisolid (6 g/l agar) MS-based medium, as above, for pea[a], or on BM medium for bambara[b].
6. For seed germination of bambara groundnut, use only half strength BM medium[b].
7. Pour the media into: (i) transparent plastic vessels (30 ml medium/vessel) and close the lids loosely to favour gas exchange, (ii) Petri dishes (20 ml medium/dish), or (iii) culture tubes (15 ml medium/tube).
8. Maintain the cultures at 24 °C/22 °C, under a 16 h photoperiod for pea, and in an environment room under short days (10 h photoperiod) as in Protocol 6.1 for bambara.
9. Within 14–21 days, transfer 4–5 cm tall pea plants *ex vitro* under the conditions in Protocol 6.1, into large containers with vermiculite; retain the plants until new pods are mature enough for extraction of the seed for the next generation[a].
10. Similarly, transfer bambara plants to the glasshouse for seed set when 3–4 cm in height[b].

Notes

[a]Several points emerge from preliminary experiments with pea:

1. The presence of the integuments delays root growth and germination by several days.
2. Optimum germination (90–100%) occurs *in vitro*, which avoids fungi and desiccation, and justifies the use of excised embryos for early plant growth.
3. A nutrient solution should be simple, inexpensive and effective. This was half-strength MS [21] macroelements, full-strength microelements, Fe-EDTA and vitamins, with 15 g/l sucrose, 6 g/l agar, at pH 6.0, in sterile containers and stored at 4°C until use (many months without deterioration).
4. With pea, optimum results are obtained through successive generations from seed-to-seed by alternating the first step *in vitro* for germination, with a second step *ex vitro* for full development. Under such conditions, the mean time for one generation ranges from 67 ± 5 days in Frisson (with a mean field cycle of 143 ± 3 days, which allows for two generations/year at best), to 76 ± 6 days in Terese. When looking at the duration of each generation cycle over a 2-year period under artificial conditions, some seasonal fluctuations can be observed, with spring generally being more favourable than autumn and winter. This phenomenon was more evident in Terese and Victor than in Frisson.
5. Management of plant development by removing heads to keep only the first two flowering nodes, optimizes the number of cycles/year (Figure 6.2). However, results may be improved.

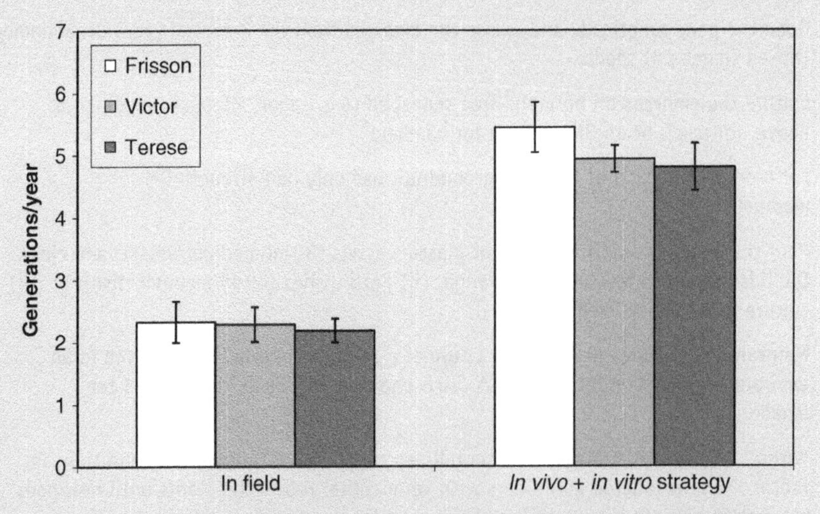

Figure 6.2 The number of generations in one year for pea using the glasshouse plus *in vitro* strategy. Mean ± SE data from two consecutive years.

[b]For bambara groundnut, responses were as follows:

1 Germination starts by day 7 for peeled seeds, while unpeeled controls take 14 days. However, by 21 days, the percentage of germination and plant morphology for peeled and unpeeled seeds are comparable.

2 Root growth and plant development are optimum and faster with an auxin (0.5–1 mg/l 1-naphthaleneacetic acid; NAA) than on hormone-free BM medium.

3 Embryo axes germinate more uniformly and faster than peeled or unpeeled seeds, but plants from embryos are significantly smaller by 28 days of culture, probably due to the reserves in the cotyledons in peeled/unpeeled seeds. Embryo axes have no cotyledons. However, this has none or little effect on the duration of flowering or seed set. All bambara landraces give low pod yields in the glasshouse, with small differences between landraces in terms of mean leaf number per plant, leaf canopy and pod dry weight.

4 In the glasshouse, seed-to-seed cycles for the genotype MB2 last 160 ± 8 days, similar to plants grown in the field in Mali, allowing for two generations per year at best. However, by removing the seed coat/integuments, germination can be accelerated and the duration of the cycle reduced. As with pea, over a 2-year period, some seasonal fluctuations are also observed, and best results in bambara are obtained by alternating a first step *in vitro* for germination and a second step *ex vitro* for full development (applicable to breeding programmes), whereby the mean time span for one generation is approximately halved (Figure 6.3).

5 Plants obtained are morphologically normal and fertile, as are their progenies. Thus, for breeding bambara groundnut the *in vitro* plus *in vivo* approach is the best in terms of efficiency, ease of execution and cost.

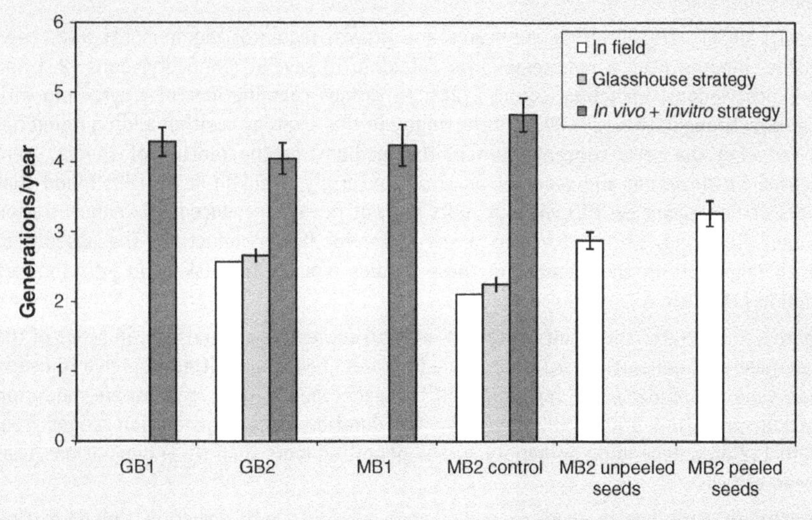

Figure 6.3 The number of generations in one year for bambara groundnut using various strategies and seed treatments. Mean ± SE data from two consecutive years.

PROTOCOL 6.3 *In vitro* only Strategy

Equipment and Reagents

- ≥1 cm tall *in vitro* shoots of any origin (explants, callus, cell suspensions, protoplasts)
- Hormone-free, half- and full-strength, MS-based medium [21]
- NAA. Prepare a stock solution at 1 mg/ml in ethanol, store at 0–5 °C until use and renew regularly (minimum every 12 months)
- Laminar air flow cabinet; dissection instruments

Method

1. Transfer shoots (≥1 cm tall and of any origin), comprising two internodes, onto hormone-free MS medium for elongation, flowering and seed set[a].
2. Alternatively, transfer shoots to half-strength MS medium without hormones or with 1 mg/l NAA [6, 18–20] for rooting, prior to flower and fruit production[b].
3. Harvest and resow immature seeds on the same medium as above (hormone-free MS) and repeat the procedure[c].
4. The number of generations feasible/year is defined as the number of d between transfer of the initial 1 cm tall shoots onto the medium and the harvest of seeds for the first generation (R1). For the R2 and subsequent generations, the duration of each generation is the number of d from *in vitro* seed germination to seed set *in vitro*[d].

Notes

[a]Flowering and seed set is obtained *in vitro* for all genotypes and without any previous need to root shoots.

[b]Reports on *in vitro* flowering are scarce and growth regulator requirements have been variable, ranging from a requirement for cytokinin in several monocotyledons [23] and some dicotyledons, including legumes [24], to various combinations of a cytokinin with other growth regulators [25–27]. Interestingly, in this strategy, neither adding hormones nor reducing the salts concentration in the medium, or the rooting of shoots, were essential for flowering and seed set *in vitro*. Conversely, Franklin *et al.* [28] found that shoots of *P. sativum* cv. PID without roots did not flower, a reduced NH_4 concentration favoured flowering, while auxin was a key factor for flower induction. The absence of growth regulators in the medium in these studies reduces the risk of *in vitro*-induced variation [18–20].

[c]Figure 6.4 illustrates the results obtained, over 10 successive generations, in terms of the mean number of generation cycles per year. In protein pea, it permitted from five to nearly seven generations per year, depending on the genotype. In grass pea, where field crop duration varies from 150 to 180 days [29], the duration of each generation ranged from 104 to 112 days depending on genotype, and permitted more than three generations/year instead of two.

[d]In previous work, the recovery of explant-derived plants with flowering, pod formation and viable seed production, was 12–14 weeks in pea [6] and 17–21 weeks of culture in grass pea [9, 18, 19]. In pea, the process takes 12–15 months from leaf protoplasts [20].

A more efficient exploitation of such approaches for breeding (e.g. for stress resistance) can be envisaged by using the methods reported here, as time-spans may be reduced further, the rooting step no longer being required with regenerated shoots (generation R_1), or with any subsequent generation. Indeed, this strategy has been exploited to accelerate generations involving hybrids of pea with *P. fulvum* which, taken to generations F_{12}–F_{14} [14], are now cultivated in the field to assess their reaction *vis-à-vis Aphanomyces euteiches*, responsible for root rot, to which the wild pea parent is reportedly resistant. For pea and grass pea, this strategy is the most appropriate for breeders.

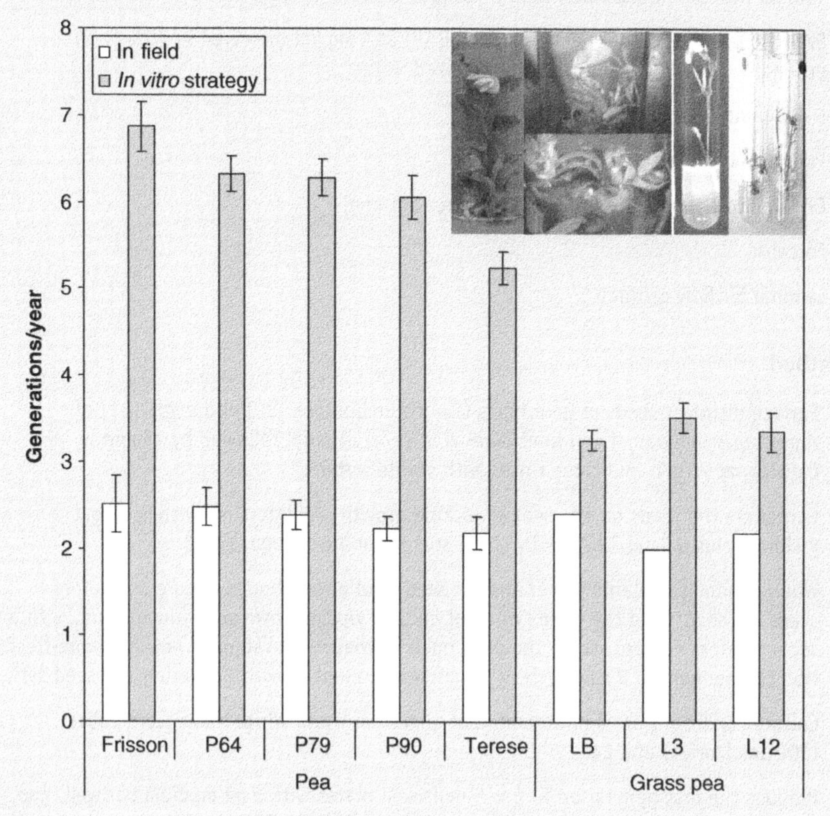

Figure 6.4 The number of generations in one year for pea and grass pea using the *in vitro* only strategy. Mean ± SE data from two consecutive years. Insert – flowering *in vitro* shoots of pea and grass pea, and of pods and seeds formed in pea.

6.2.2 Arabidopsis thaliana [13]

PROTOCOL 6.4 *In Vitro* Strategy

Equipment and Reagents

- Mature seeds of the target genotypes

- Ethanol 70% (v/v)
- $Ca(OCl)_2$ at 70 g/l
- Sterile water
- MS based medium
- Stock solution of Picloram at 1 mg/ml dissolved in 70% (v/v) ethanol and made to volume with deionized water
- Stock solution of 6-benzylaminopurine (BAP) at 1 mg/ml dissolved in 70% (v/v) ethanol and made to volume with deionized water
- 5 × 5 multi-well dishes (Sterilin)
- Plastic Petri dishes (10 cm diam.)
- Glass culture tubes (25 × 150 mm; Greiner-bio one)
- Forceps
- Laminar air flow cabinet

Method

1. Surface disinfect seeds of genotypes C24, Columbia, *hoc* [30] and *amp1* [31] of *Arabidopsis thaliana*; 1 min in ethanol 70% (v/v) ethanol followed by 15 min in $Ca(OCl)_2$ at 70 g/l, and three rinses with sterile water.
2. Germinate the seeds on MS medium lacking growth regulators, or with 0.1 mg/l Picloram plus 0.5 mg/l BAP with 30 g/l sucrose and 6 g/l agar (pH 5.6)[a].
3. When germinated plants flower and set seed, and once siliques mature but before seeds are shed, hold the plants or plant clusters upside down with forceps and, with a second pair of forceps, crush the pods open so that seeds fall onto new hormone-free MS [21] medium for a new cycle of germination, plant growth, flowering and seed set[b].
4. Culture conditions are a photoperiod of 16 h from Warm White fluorescent tubes (100 μmol/m^2/s) and 24 ± 2 °C.
5. Produce the first generation in 5 × 5 multiwell plates with 2 ml medium per well, and, subsequently, in culture tubes with 15 ml medium/tube.
6. Alternatively, lay immature siliques on hormone-free MS medium in 10 cm Petri dishes and leave the seed to germinate inside the siliques. This simplifies the procedure. Recover the seedlings and treat as above.
7. Determine the number of seeds per silique and calculate the number of feasible generations per year, as the mean (± SD) number of days elapsed between successive seed sowings, i.e. the number of days from *in vitro* seed germination to seed set by the resulting seedlings[c].

Notes

[a]Twenty five to 45 days elapse between germination of original seeds and flowering and seed set, depending on the medium used. Growth regulators in the medium do not affect

germination, but flowering and seed set occur significantly faster than on hormone-free medium. Use of a medium with an auxin (Picloram) and a cytokinin (benzylaminopurine) does not affect seedlings being true-to-type.

[b]With the four genotypes of *Arabidopsis* tested using this simple strategy, it is possible to obtain fertile, flowering and fruiting seedlings of successive generations from the F_2 generation within 17–22 day/cycle, depending on the genotype (Figure 6.5). This is about the same duration required for the development of *A. thaliana* seeds alone [32], and allows more than 10 generations (up to 19) per annum.

[c]The fertility of plants is reduced, with 80–100 seeds/silique, which is significantly less than the number of seeds/pod produced *in vivo*. However, since such seeds are nearly all capable of germination, this has no effect on the production of sizeable progeny from each silique. This simple and efficient strategy for fast cycling should fulfil its promise when coupled with genetic studies.

Figure 6.5 Flowering and seed set of *Arabidopsis thaliana in vitro*. (a) Individual seeds sown on culture medium. (b) Developing siliques with one at the optimum stage showing direct germination of seeds on the medium. (c) A rosette seedling with flowers. (d) A cluster of unrooted shoots with flowers. Bars = 1 cm.

6.3 Troubleshooting

- The main goal of the glasshouse experiments with legumes, i.e. to produce four generations per calendar year, can be achieved for the protein pea genotypes listed, regardless of the specific genotype or photoperiod. However, there is a significant seasonal effect, with shorter cycles in spring and summer. In the case of the forage genotypes, Champagne and Winterberger, only three successive generations can be completed in 1 year (Figure 6.1).

- For embryo excision, open the pods carefully and discard the wet outer and inner integuments from each seed, avoiding damage to the cotyledons and embryo axis. Take care to avoid breaking the root tip with its cap (essential for rapid germination).

- The efficiency and need to use excised embryos and to work under *in vitro* conditions has been verified by comparing entire seeds and excised embryos (without integuments) extracted from surface sterilized pods, which were sown *in vitro* directly onto vermiculite, but under non-sterile conditions. In pea, this was done over four successive generations. In bambara groundnut, entire seeds, peeled seeds and embryo axes excised from unpeeled seeds were also compared.

- For pea, MS [21] and B5 [17] media were compared at full, half and quarter strength of macro- and microelements, with or without vitamins, with a range of concentrations of glucose and sucrose (0–40 g/l), agar (5–8 g/l) and a pH of 5.5–6.5.

- Rooting lengthens each cycle, by 15–30 days, and affects flowering, particularly for grass pea.

- Optimum flowering occurs on growth regulator-free medium, while growth regulators systematically reduce it; halving the salts concentration may reduce seed set, and is coupled with a lower germination competence of the seeds produced.

- Pod dehiscence and seed germination were sometimes observed on shoots *in vitro*, but was restricted to Frisson and its mutants, and occurred only on rooted shoots. Similar results have been observed in amaranths [4] but, somewhat surprisingly, not in pea by other workers [29].

In vitro flowering has been reported in a number of species [24, 25, 27, 28] but complex mixtures of growth regulators were employed. Conversely, in the protocols described in this chapter, a very simple strategy is adequate, with growth regulators not being essential for flowering and seed set. Growth regulators have no influence on the duration of each generation as a carry-over effect on germination of initial seeds. Also, with few exceptions [4, 25, 29], previous reports of *in vitro* flowering did not proceed to seed set.

The protocols described have considerable potential for several breeding schemes, including marker-assisted selection, SSD, and the analysis of introgression of transgenes within the progeny of primary transformants. In this context, highly cost-effective methods to accelerate generations, such as those reported here, should be useful for plant breeding companies and research institutes.

References

1. Roumet P, Morin F (1997) *Crop Sci*. **37**, 521–525.

2. Al-Wareh H, Trolinder NL, Goodin JR (1989) *Hort Sci*. **24**, 827–829.

3. Dickens CWS, Van-Staden J (1988) *South Afr. J. Bot*. **54**, 325–344.

4. Tisserat B, Galleta PD (1988) *Hort Sci*. **23**, 210–212.

5. Bean SJ, Gooding PS, Mullineaux PM, Davies DR (1997) *Plant Cell Rep*. **16**, 513–519.

6. Ochatt SJ, Pontécaille C, Rancillac M (2000) *In Vitro Cell. Dev. Biol. Plants* **36**, 188–193.

7. Ochatt SJ, Sangwan RS, Marget P, Assoumou Ndong Y, Rancillac M, Perney P (2002) *Plant Breed*. **121, 436–440.

The original publication on the acceleration of generation cycles in protein legumes using protocols as described in this chapter.

8. Campbell CG (1997) *Grass pea*, *Lathyrus sativus* L. IPGRI, Rome/Gatersleben.

9. Ochatt SJ, Abirached-Darmency M, Marget P, Aubert G (2007) In: *Breeding of Neglected and Under-utilized Crops, Herbs and Spices*. Edited by SJ Ochatt and SM Jain. Science Press, Plymouth, USA, pp. 41–60.

10. Heller J, Begemann F, Mushonga J (1997) Bambara groundnut, *Vigna subterranea* (L.) Verdc. IPGRI, Rome/Gatersleben.

11. Koné M, Patat-Ochatt EM, Conreux C, Sangwan RS, Ochatt SJ (2007) *Plant Cell Tissue Organ Cult*. **88**, 61–75.

*12. Sanwan RS, Adu-Dapaah HK, Bretaudeau A, Ochatt SJ (2007) In: *Breeding of Neglected and Under-utilized Crops, Herbs and Spices*. Edited by SJ Ochatt and SM Jain. Science Press, Plymouth, USA, pp. 81–94.

Description of the method for acceleration of generation cycles in Bambara groundnut.

13. Ochatt SJ, Sangwan RS (2008) *Plant Cell Tissue Organ Cult*. **93, 133–137.

The original publication on the acceleration of generation cycles in *Arabidopsis* using the protocol described in this chapter.

*14. Ochatt S, Marget P, Benabdelmouna A, *et al.* (2004) *Euphytica* **137**, 353–359.

The strategy described in Protocol 6.3 was used in this publication to accelerate generations of *P. sativum* × *P. fulvum* hybrids for breeding for disease resistance.

15. Stafford A, Davies DR (1979) *Ann. Bot*. **44**, 315–321.

16. Duc G, Messager A (1989) *Plant Sci*. **60**, 207–213.

17. Gamborg OL, Miller RA, Ojima K (1968) *Exp. Cell. Res*. **50**, 151–158.

18. Ochatt S, Durieu P, Jacas L, Pontécaille C (2001) *Lathyrus Newsl*. **2**, 35–38.

19. Ochatt SJ, Muneaux E, Machado C, Jacas L, Pontécaille C (2002) *J. Plant Physiol*. **159**, 1021–1028.

20. Ochatt S, Mousset-Déclas C, Rancillac M (2000) *Plant Sci*. **156**, 177–183.

21. Murashige T, Skoog F (1962) *Physiol. Plant*. **15**, 473–497.

22. Nitsch, JP & Nitsch C (1969) *Science* **163**, 85–87.

23. Zhong H, Srinivasan C, Sticklen MB (1992) *Planta* **187**, 490–497.

24. Narasimhulu SB, Reddy GM (1984) *Theor. Appl. Genet*. **69**, 87–91.

25. Tepfer SS, Karpoef AJ, Greyson RI (1966) *Am. J. Bot*. **53**, 148–157.

26. Rastogi R, Sawhney VK (1987) *J. Plant Physiol*. **128**, 285–295.

27. Peeters AJM, Proveniers M, Koek AV, *et al.* (1994) *Planta* **195**, 271–281.

28. Franklin G, Pius PK, Ignacimuthu S (2000) *Euphytica* **115**, 65–73.

29. Swarup I, Lal MS (2000) *Lathyrus sativus and Lathyrism in India*. Surya Offset Printers, Gwalior.

30. Catterou M, Dubois F, Smets R, *et al.* (2002) *Plant J*. **30**, 273–287.

31. Chaudhury AM, Letham S, Craig S, Dennis ES (1993) *Plant J*. **4**, 907–916.

32. Baud S, Boutin JP, Miquel M, Lepiniec L, Rochat C (2002) *Plant Physiol. Biochem*. **40**, 151–160.

7

Induced Mutagenesis in Plants Using Physical and Chemical Agents

Chikelu Mba, Rownak Afza, Souleymane Bado and Shri Mohan Jain*

Plant Breeding Unit, International Atomic Energy Agency, Laboratories Siebersdorf, Vienna International Centre, Vienna, Austria

**Current address – Department of Applied Biology, University of Helsinki, Helsinki, Finland*

7.1 Introduction

Mutation, the heritable change to the genetic make up of an individual, occurs naturally and has been the single most important factor in evolution as the changes that are passed on to offspring lead to the development of new individuals, species and genera. The first reported cases of artificial induction of mutations, that is, the creation of genomic lesions above the threshold observable in wild types, were in the 1920s with work on *Drosophila*, maize and barley. Since these pioneering activities, induced mutagenesis has become widespread in the biological sciences, primarily for broadening the genetic base of germplasm for plant breeding and, more recently, as a tool for functional genomics.

Mutations are induced in plants by exposure of their propagules, such as seeds and meristematic cells, tissues and organs, to both physical and chemical agents with mutagenic properties [1]. In some instances, whole plants are also exposed. Physical mutagens are mostly electromagnetic radiation such as gamma rays, X-rays, UV light and particle radiation, including fast and thermal neutrons, beta and alpha particles. Chemical mutagens include alkylating agents (such as the commonly used ethyl methane sulfonate – EMS), intercalating agents (such as ethidium bromide)

Plant Cell Culture Edited by Michael R. Davey and Paul Anthony

and base analogues (such as bromouracil that incorporate into DNA during replication in place of the normal bases). Other chemical agents cause a myriad of genome lesions, including the formation of triesters and depurination as a result of alkylation, and even gross chromosomal damage. Mba *et al.* [2] and the United Nations Organization [3] listed the commonly used chemical and physical mutagens and their modes of action. In general, these agents bring about changes in DNA sequences and, consequently, change the appearance, traits and characteristics of the treated organism.

In the past, irradiation was carried out in either of two ways, these being chronic or acute irradiation. While the former refers to exposure at relatively low doses over extended periods of time of weeks or even months, the latter refers to single exposures at higher doses over very short periods of time (seconds or minutes). The prevailing opinion then was that acute irradiation resulted in greater mutation frequencies. Currently, this reasoning is that in practice, such differences have had no discernible impact on the outcomes of induced mutagenesis, with most induction being of the acute type.

At the Seibersdorf, Austria Laboratories of the International Atomic Energy Agency, a cobalt-60 source (Gammacell Model No. 220, Atomic Energy of Canada, Ottawa, Ontario, Canada) is used routinely for gamma irradiation of seeds and other plant propagules. The facility also provides cost-free irradiation services and additional information on this service can be obtained via e-mail from <official.mail@iaea.org>.

In general, the development and dissemination of validated protocols for induced mutagenesis, especially for less studied plant species, have not progressed apace with the enthusiastic use of mutation induction to create novel alterations in the genome. This chapter seeks to redress the dearth of information on appropriate methodologies for inducing mutations by providing guidance on protocols for determining the optimal doses, and methods relevant to the use of both physical and chemical mutagens. It illustrates these procedures in both seed and vegetatively propagated plants. The use of *in vitro* propagules in induced mutagenesis, a strategy for mitigating the confounding effects of chimeras and for achieving homozygosity rapidly, is also included in the protocols.

7.2 Methods and approaches

7.2.1 Determination of the optimal doses of mutagens for inducing mutations

The dose of a mutagen that achieves the optimum mutation frequency with the least possible unintended damage, is regarded as the optimal dose for induced mutagenesis. For physical mutagens, this is estimated by carrying out tests of radiosensitivity (from radiation sensitivity), a term described as a relative measure that gives an indication of the quantity of recognizable effects of radiation exposure on the irradiated subject [4]. Its predictive value therefore guides the researcher in the choice of optimal exposure dosage depending on the plant materials and desired outcome.

Procedures for determining radiosensitivity and carrying out bulk induced mutation treatments using seeds, and *in vitro* nodal segments are described in the following sections. When using chemical mutagens, optimal doses are also inferred using the same underlying principles of quantifying observed damage.

PROTOCOL 7.1 Radiosensitivity and Induction of Mutations in a Seed Propagated Crop (Rice) Through the Gamma Irradiation of Seeds

Equipment and Reagents

- Seeds of target plant: e.g. those of rice which should be dry, clean, disease-free and of uniform size
- Gamma radiation source: A source provider is available at official.mail@iaea.org
- Paper seed envelopes (air- and water-permeable standard paper envelopes without wax or lining)
- Vacuum dessicator
- Sterilized soil
- Pots and glasshouse facilities
- Petri dishes (9 cm diam.)
- Whatman No. 1 filter papers (9 cm diam.)
- Sterile water
- Glycerol (60%, v/v)
- Chlorox bleach solution: 20% (v/v); 5.25% (w/v) solution of sodium hypochlorite; Chlorox Co.) with one to two drops of Tween 20 (Sigma)
- Blotting paper (cut to 5 × 11 cm from Gel Blotting Paper; GB002; Schleicher and Schuell BioScience GmbH)
- Racks (see suggestions for the construction of racks under the 'Sandwich blotter method' below)
- Plastic trays (any plastic tray for holding water to a depth of 5 cm)

Method

This involves preirradiation handling, irradiation, postirradiation handling of the seeds, data collection and analyses.

1 Preirradiation handling of seeds:

   (a) Make 36 batches each of 40–50 of the sorted seeds. These correspond to three replications of the untreated seeds and each of the 11 doses (see below). Place each batch inside an air and water permeable seed envelope and label the envelopes accordingly.

(b) Place the packed seeds in a vacuum dessicator over glycerol (60% by volume) and leave at room temperature for 5–7 days. This equilibrates the seed moisture content to 12–14%, the ideal moisture condition for achieving efficient induction of mutation.

2 Irradiation of seeds:

Expose the seeds to gamma irradiation in the source, taking care to observe all safety precautions. Figure 7.1(a) shows a cobalt-60 source while Figure 7.1(b) is a close-up showing the elevated loading stage with rice grains in a Petri dish. Successful irradiation is dependent on having the precise dosimetry data, as this is used to calculate the exposure time given by the formula:

$$\text{Exposure time in seconds} = \text{Desired dose/dose rate}^{a}$$

Where there is a dedicated gamma source operator, the above step is not necessary and the seeds are submitted through established procedures.

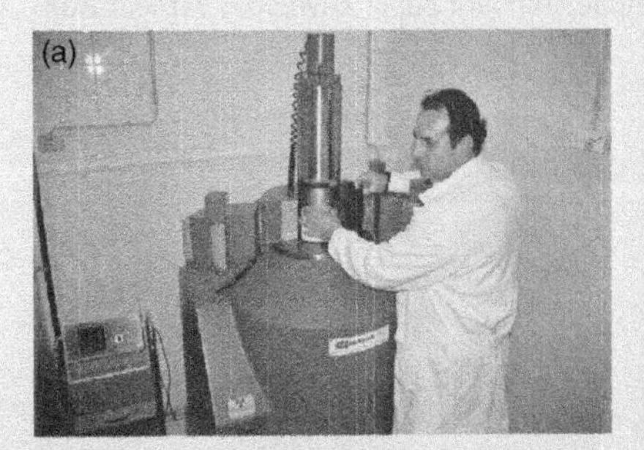

Figure 7.1 (a) A cobalt-60 gamma source with a raised loading stage (credit: IAEA). (b) Close-up of the raised loading stage of a cobalt-60 gamma source with rice grains in a Petri dish (credit: IAEA).

Note

[a]Depending on the genotype, gamma ray dosages of 100–400 Gy have been reported to be optimal for inducing mutations in rice seeds. In practice, it is advisable to carry out a pilot study by exposing batches of similar seeds to irradiation doses around this range, staggered by 50 Gy (i.e. 11 doses of 100, 150, 200, 250, 300, 350, 400, 450, 500, 550 and 600 Gy).

3 Postirradiation handling of seeds

(a) In order to minimize additional damage, sow seeds as soon as possible after irradiation. If a delay is necessary, store seeds at room temperature for a maximum of 4 weeks. Beyond this period, storage should be in dry conditions, with a minimum of oxygen (in airtight vials or bags, in the dark and at low temperature (2–5 °C). These conditions minimize metabolic activity and prevent additional lesions to the genome.

(b) Sow seeds exposed to the same dosage simultaneously with the batches arranged in a manner to permit easy visual comparison of the different treatments. The different laboratory or glasshouse methods for sowing the seeds that permit the determination of germination or emergence, seedling height and leaf spotting, all indicative of the extent of damage caused by the mutagenic treatment are presented below. Other subsequent observations, such as fertility and survival, are carried out in pots in the glasshouse. Field observations are discouraged on account of the influence of environmental factors.

4 Flat method
In the glasshouse, sow the seeds in rows in trays containing adequately moistened and well-drained heat- or steam-sterilized soil. Plant the seeds in order of increasing dose with replications sown in different trays. Alternatively, sow the seeds in pots or individual cells of compartmentalized trays[a].

5 Petri dish method
Place the seeds on wet, preferably sterile, filter paper in Petri dishes; keep the filter paper continually moist[b].

Notes

[a]As much as is practical, ensure that all the environmental factors and sowing depth are uniform for all the treatments.

[b]Fungal attack may compromise the data to be collected. To control this, in addition to sterile filter papers, it is strongly recommended to disinfect the seeds (e.g. surface sterilization in 20% (v/v) Chlorox bleach, 5.25% (w/v) NaOCl active ingredient, for 20 min) and to use sterile Petri dishes and sterile water.

6 Sandwich blotter method[a]
Presoak the seeds and place them between two wet blotting papers which are pressed together and supported vertically in racks. Place the racks in plastic trays with water. This simple and robust method proposed by Myhill and Konzak [5] and described more recently by Martınez *et al.* [6] has been used for more than 40 years at the laboratories of the IAEA. The method consists of sowing seeds between the edges of one end of two wet blotting papers (cut to 5 × 11 cm from Gel Blotting Paper, GB002; Schleicher and Schuell BioScience GmbH) that are held together to form a 'sandwich'. With the edge containing the seeds uppermost, the 'sandwich' is held upright in a plastic rack. One way of making a rack is to link two plastic combs together with a firm horizontal support such that their teeth face upwards. A blotter sandwich is supported by wedging it within aligned grooves of these two combs. This arrangement (sandwich and comb) is placed inside a plastic tray containing distilled water with the lower end of the sandwich dipping into the water (up to half the height of the blotting paper). At the IAEA laboratories, racks are constructed from 4 mm thick plastic bars measuring 140 mm in length by 50 mm in height. Each groove measures 35 mm in depth, is 2 mm wide, and is separated from the next by a ridge 5 mm wide. At the centre is a non-grooved solid portion. Two combs are held together by a similar plastic bar measuring 120 mm and with two grooves near both ends. This is done by sliding the two combs at their solid middle portions into each of the two grooves of the 'bridge' such that the two grooves form interlocks with the solid middle portions of the two combs. A convenient plastic tray for holding four of these racks (two combs and one bridge each) measures approx. 305 mm in length with a width and depth of 255 mm and 55 mm, respectively. Figure 7.2 shows a typical rack and the parts.

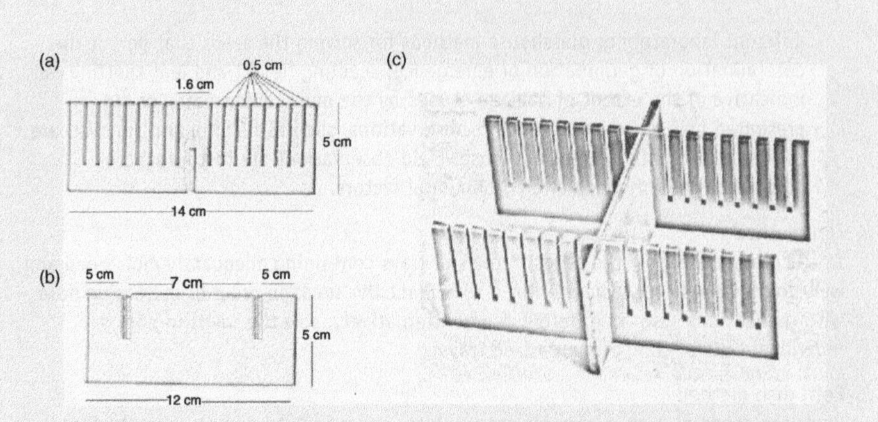

Figure 7.2 Sketch of the rack used for holding the sandwich of wet filer papers and seeds upright. (a) The 'comb'; (b) the 'bridge'; (c) the assembled plastic rack.

Note

[a]This method, in addition to the advantage of saving on labour, provides accurate data. However, it requires additional equipment, such as a plastic film-covered growth cabinet and a humidifier. At the IAEA laboratories, the growth chamber is constructed locally and consists of a cubic frame with all sides left open except for the bottom. The metallic frames and base are cut and welded together by machinists. To further control the environment, the structure is covered by a plastic sheet with the edges tucked beneath the base of the chamber. High humidity is maintained by pumping air through a vessel of water into the chamber.

7 Data collection and analyses

Collect data on the following parameters:

- Germination rate
- Seedling height[a]
- Survival rate
- Chlorophyll mutation
- Number of tillers
- Seed set
- Fertility test in the M_2 (second) generation

Note

[a]Data collection can be adapted to any type of plant germination (monocotyledons, and both epigeal and hypogeal types in dicotyledons). For measurement of seedling height (usually the most common parameter assayed), the methodologies are outlined in Table 7.1.

Table 7.1 Methods for measuring seedling height in dicotyledons and monocotyledons[a].

Germination type	Plant parts measured
Monocotyledons (e.g. cereals)	In pots: from soil level to the tip of the first or secondary leaf
	In Petri dishes and sandwich blotter: from the origin of the root to the tip of the first or secondary leaf[a]
Dicotyledons: epigeal germination (e.g. *Phaseolus*)	The length of the epicotyl is measured i.e. the region between the point of attachment of the cotyledons to the tip of the primary leaves or to the stem apex.[b]
	Alternatively, seedling height can be taken from soil level to the tip of the primary leaves or to the stem apex, without compromising the data.
Dicotyledons: hypogeal germination (e.g. *Pisum*)	In pots, the length from soil level to the tip of the primary leaves (longest leaf) or to the stem apex.
	In the Petri dish and sandwich blotter methods, measure the distance between the origin of the roots and the stem apex.

[a]NB For cereals, the leaf that emerges through the coleoptile is the first true leaf; seedlings with only the coleoptile emerging and no true leaf are not included in measurements.
[b]NB The hypocotyl region is relatively insensitive to radiation and is therefore not measured.

8 Data handling:

Create a spreadsheet[a] and enter the mean data for each treatment and control (wild-type, untreated). Calculate the differences between each treatment and control and express these as percentages (see sample below). Plot a graph of the absorbed doses against these percentage differences for each parameter (see sample, Figure 7.3).

Note

[a]The percentage in reduction of plant height is a good parameter for estimating the damage due to mutagenic treatment. By inserting the 'line of best fit' and reading off the dose corresponding to 50% reduction, the so-called lethal dose 50, written as LD_{50}, is obtained. This, and values corresponding to other percentages, can be read from the line of best fit or, more precisely, calculated using the straight line equation, $y = mx + c$. The LD_{50} is an appropriate dose for irradiation but, in practice, a range of doses around it is used. Tables 7.2 and 7.3 are suggested formats for data collection sheets for determining optimal irradiation and EMS doses, respectively, for inducing mutations. They could be used as a guide for the treatment conditions and useful for collecting data.

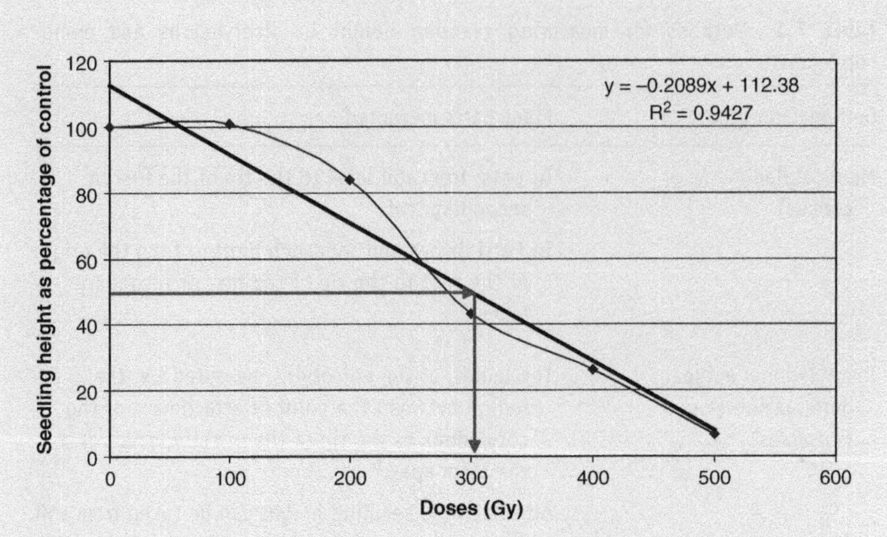

Figure 7.3 Percentage reduction in plant height of seedlings from seeds exposed to gamma irradiation (compared with seedlings from untreated seeds), plotted against gamma irradiation dosage.

Table 7.2 Suggested format for data collection sheet on treatment conditions for determining the optimal conditions for irradiation-mediated mutagenesis.

Absorbed doses (Gy)	Average measurements	Percentage of control (%)
0		
50		
100		
150		
250		
300		
350		
400		
450		
500		

Table 7.3 Suggested format for data collection sheet on treatment conditions for determining the optimal conditions for EMS-mediated mutagenesis[a].

Concentration of EMS (M)	Treatment temperature (°C)	Treatment duration (h)	Average measurements	Percentage of control (%)
0.050	30	0.5		
		1		
		1.5		
		2		
	32.5	0.5		
		1		
		1.5		
		2		
	35	0.5		
		1		
		1.5		
		2		
0.075	30	0.5		
		1		
		1.5		
		2		
	32.5	0.5		
		1		
		1.5		
		2		
	35	0.5		
		1		
		1.5		
		2		
0.100	30	0.5		
		1		
		1.5		
		2		
	32.5	0.5		
		1		
		1.5		
		2		

Table 7.3 (*continued*).

Concentration of EMS (M)	Treatment temperature (°C)	Treatment duration (h)	Average measurements	Percentage of control (%)
	35	0.5		
		1		
		1.5		
		2		
0	0	0		

[a]Adapted from barley experiments at the Plant Breeding Unit, Joint FAO/IAEA Agriculture and Technology Laboratory, Seibersdorf Laboratories of the International Atomic Energy Agency.

PROTOCOL 7.2 Radiosensitivity and Induction of Mutations in a Vegetatively Propagated Crop (Cassava) Through the Gamma Irradiation of *In Vitro* Nodal Segments

The induction of mutations in seed propagated crops compared with vegetatively propagated plants is easier, mostly on account of the relative ease of achieving homozygosity and dissociating the chimeras in the progeny of zygotic embryos through a limited number of cycles of selfing. *In Vitro* strategies are used to mitigate this bottleneck in vegetatively propagated crops such as cassava. Ideally, the most appropriate strategy should involve the exploitation of totipotency through somatic embryogenesis (e.g. friable embryogenic callus) so that plants originate from one or a few irradiated cells. There is a dearth of information on reproducible protocols for somatic embryogenesis for many crops and where they exist [7–9]. Genotypic specificity often prevents the horizontal application of the protocols across species. *In vitro* nodal segments are convenient as starting material for the induction of mutations in cassava [10].

Rapid Micropropagation of Cassava

Equipment and Reagents

- Potted plants with new shoots
- Laminar air flow cabinet
- Bottles or flasks containing sterile water
- Chlorox bleach solution: 20% (v/v) of a 5.25% (w/v) solution of sodium hypochlorite; Chlorox Co.) with one to two drops of Tween 20 (Sigma)
- 75% (v/v) ethanol
- 250 ml flasks
- Liquid Murashige and Skoog (MS; 1962; [11]) basal medium lacking growth regulators

- Sterile distilled water
- Gyrotatory shaker
- Petri dishes (9 cm diam.)
- Parafilm (VWR International GmbH)
- Gelrite (Sigma)

Method

1 Excise actively growing, new shoots and remove the leaves; cut the stems into single- or two-nodal segments (explants).

2 Wash the explants in running water (to remove dirt) for 1 h and, in a laminar flow cabinet, place the segments in a covered bottle or 250 ml flask containing water.

3 Prepare 20% (v/v) Chlorox solution with one to two drops of Tween 20 in 500 ml of water.

4 Rinse the explants in 75% (v/v) ethanol.

5 Add 100 ml of the prepared Chlorox solution to 250 ml flasks containing the explants and place the flasks with their contents on a gyratory shaker. Agitate for 10–20 min at 30 rpm, or agitate by hand every 5 min.

6 Wash the explants three to four times with sterile distilled water and transfer the explants to Petri dishes containing sterile water.

7 Transfer the explants, five to six per flask, to 10 ml of liquid Murashige and Skoog basal medium (see recipe below) with 20 g/l sucrose[a].

8 Maintain the flasks on a horizontal gyratory shaker at 300 rpm at 26 °C under continuous light (65 μmol/m^2/s; Cool White fluorescent tubes, Philips TLP 36/86).

9 After 2–3 weeks, remove the shoots formed from axillary buds and divide each into two-node segments and subculture to new medium, again placing five to six segments in each flask.

10 After about 4 weeks, de-leaf the growing explants and cut into pieces, each containing two nodes. Place 10 explants in each Petri dish containing sterile distilled water and seal the Petri dishes with Parafilm. These are ready for irradiation.

Radiosensitivity test:

11 Irradiate each Petri dish with different doses (5, 10, 15, 20, 25 and 30 Gy). In a laminar flow cabinet, transfer the irradiated explants to sterile labelled conical flasks containing 10 ml of liquid medium.

12 With the control non-irradiated samples, place the flasks on a horizontal gyratory shaker at 300 rpm and allow the explants to grow at room temperature (about 26 °C) under continuous light (65 μmol/m^2/s; Cool White fluorescent tubes, Philips TLP 36/86).

13 After 4–5 weeks, take measurements of the parameters, weight of the explants and the number of nodes.

Data handling:

14 Create a spreadsheet and enter the average data for each treatment and control (wild-type, untreated). Calculate the differences between each treatment and control and express these as percentages (see earlier example from seed propagated crops). Plot a graph of the irradiation doses against these percentage differences for each parameter[b].

Notes

[a]The following growth media have been validated for cassava micropropagation with African and South American cassava clones in the tissue culture facilities of the Plant Breeding Unit of the Joint FAO/IAEA Agriculture and Biotechnology Laboratories, Agency Laboratories, Seibersdorf, Austria.

For one litre of liquid medium, use the following:

- MS basal medium (Sigma) = 4.4 g
- Sucrose (Grade1, Sigma) = 20 g
- Make up to 1 l with sterile, double distilled water
- Adjust the pH to 5.8

For one litre of semisolid medium, use the following:

- As above, but add 1.8 g Gelrite
- Adjust the pH to 5.8

[b]These estimates of the percentage in plant growth reduction are good parameters for estimating the damage due to mutagenic treatment. By inserting the 'line of best fit' and reading off the dose corresponding to 50% reduction, the so-called lethal dose 50 (LD_{50}), is obtained. This (and values corresponding to other percentages) can be read from the line of best fit, or calculated more precisely using the straight line equation i.e. $y = mx + c$. The LD_{50} is an appropriate dose for irradiation but, in practice, a range of doses around this value should be used.

PROTOCOL 7.3 Induction of Mutations in a Seed Propagated Crop Using the Chemical Mutagen EMS Based on Protocols Validated for Barley

Equipment and Reagents

- Supply of seeds of target plants (e.g. barley)

- Polyethylene mesh bags (ca. 11 × 7 cm. in dimension; made from locally available plastic net screens, such as mosquito nets, that are cut to size and formed with a heat sealer)
- Beaker (500 ml)
- Distilled water
- EMS
- Dimethyl sulfoxide (DMSO)
- Collection vessels for EMS waste solution

Method

Preparation of EMS solution:

1 Use only freshly prepared EMS solution. This consists of EMS, the active ingredient, DMSO as the carrier agent and distilled water. Prepare the EMS solution in two phases:

First, mix the required volumes of water and 2% (v/v) DMSO and autoclave at 120 °C for 15 min at 103.5 kPa (15 psi). Leave the mixture to cool to room temperature. This step may carried out in advance, and the sterile mixture used for the preparation of the EMS solution up to 24 h later.
The second phase, which must be carried out in a laminar flow cabinet, involves the addition of EMS to the water–DMSO mixture. When ready to incubate the target materials in the mutagen, use a sterile syringe and a 0.2 μm filter to add the required volume of EMS solution to the sterile water–DMSO mixture. Shake the resulting solution vigorously to give an homogeneous emulsion.

Example: To prepare 200 ml of 0.5% (v/v) EMS with 2% DMSO, mix 4 ml of DMSO and 1 ml of EMS solution (Sigma, d = 1.17 g/ml) in 195.5 ml distilled water.
As a guide for the volume required, prepare 1 ml of solution for every seed to be treated.

Pretreatment handling of seeds and determination of optimal treatment conditions:

2 Select genetically similar and normal shaped seeds that are disease-free, dry and quiescent. The seeds should have good germination. Divide the seeds into 37 batches, each of about 25 seeds. Leave 1 batch untreated as a control, while 26 batches correspond to the possible combinations of concentrations of EMS (range of 0.05 to 0.2 M solution), 2–3 treatment temperatures (range of 30–35 °C) and 2 treatment durations (range of 2–6 h). Table 7.3 can be used as a guide for the treatment conditions and is useful for collecting data to aid the investigator in determining the optimal treatment condition.

3 Place seed batches in appropriately labelled polyethylene mesh bags (ca. 11 × 7 cm in dimension) with tops folded over and secured with plastic paper clips. A common labelling method is to attach marked plastic tags to the mesh bags using cotton strings. Each bag is easily made from locally available plastic net screens such as mosquito nets. Cut the sheets to size and use a heat sealer to form the bags.

4 Soak the seeds by placing the bags in a beaker with distilled (or deionized) water and leave standing for 16–20 h at 20–22 °C. Facilitate aeration by intermittent agitation, or by pumping in air or oxygen to create bubbles.

5 Towards the end of this stage (presoaking), prepare new solutions of EMS according to the desired concentrations (see Protocol 7.4 and Notes).

6 At the end of this presoaking period, remove the bags and shake off excess water.

EMS treatment of seeds:

7 Using a water bath to maintain the desired temperature, soak the seeds in the EMS solutions according to the desired combinations of concentration, temperature and duration.

8 After each treatment, wash the seeds (to remove excess EMS) under running cold tap water for 2–3 h. Dispose of EMS according to local safety rules.

9 Shake off excess moisture and place the seeds on blotting paper for a short period to surface dry the seeds.

Post-treatment handling of seeds:

10 For optimal results, especially in order to prevent the occurrence of artefacts such as unintended lesions after treatment, sow the seeds immediately after treatment on uniform well-prepared seedbeds or soil in pots. If the soil is dry, irrigate immediately after sowing in order to avoid injury due to dry-back when in the soil.

11 If needed, seeds may be stored or transported. For these options, dry the seeds by hanging the bags of seeds in an air current ('dry-back treatment'). After 1–2 days of drying, store the seeds in a refrigerator at 4 °C.

PROTOCOL 7.4 Mutation Induction in a Vegetatively Propagated Crop Using the Chemical Mutagen, EMS, Based on Protocols Validated for Cassava

Equipment and Reagents

- Laminar flow cabinet
- Sterile glass or plastic Petri dishes (9 cm diam.)
- Parafilm (VWR International GmbH)
- Forceps
- Gyrotatory shaker
- Sterile distilled water

- Sterile sieves for washing off excess EMS: metal; 70 mm diam., 70–100 μm pore size (VWR International GmbH)
- Membrane filter unit (sterile) for filtering the EMS solution: 25 mm diam., 0.2 μm pore size (VWR International GmbH)
- Collection vessels for EMS waste solution
- Sterile Whatman filter papers
- EMS
- DMSO
- Distilled water

Method

Preparation of EMS solution:

1 See Protocol 7.3, above.

Pre-treatment handling of explants:

Carry out the following procedures under aseptic conditions, preferably in a laminar flow cabinet:

2 Remove the leaves from the plants (from liquid or semisolid growth medium) and cut the stems into explants each with two nodes.

3 Keep these nodal segments in a sterile plastic or glass Petri dish containing sterile distilled water. Seal the Petri dishes with Parafilm to avoid contamination. If necessary, the explants can be left this way in the air-flow cabinet for about 24 h before EMS treatment.

4 Using sterile forceps, transfer the explants from the water into the homogeneous EMS solution under aseptic conditions in the air-flow cabinet. As a guide, 200 ml of EMS solution can used to treat 50–100 explants (the volume depends on the size of the explants, but it is crucial that the explants are immersed completely in the solution).

5 Leave the explants immersed in EMS solution for the desired, predetermined time. In order to enhance the viability of the explants, the set up should ideally stand on a gyratory shaker (80–120 rpm).

6 After treatment, wash the explants in sterile distilled water under aseptic conditions. The washing is done by passing the explants onto a sterile sieve and transferring into a conical flask or beaker containing sterile water before being shaken. The process, of transferring to a new sterile sieve and washing by thorough shaking in sterile water, is repeated at least three times to remove all traces of EMS.

7 Collect the EMS and the wash solutions for appropriate disposal as hazardous wastes (see below for safe disposal of EMS)[a].

8 Transfer the washed explants into conical flasks containing liquid growth medium (6–10 explants/10 ml medium). Incubate on a horizontal gyratory shaker at 60 rpm under continuous light at 26 °C. After 24 h, transfer the explants into new liquid medium under aseptic conditions. This exchange of medium (effecting additional

washing) may be repeated at least twice to ensure the removal of any residual mutagen, thus avoiding continuous exposure to EMS during growth and development of the plants.

9 If the explants are not to be established in liquid medium (e.g. for shipment), after the last wash in growth medium, transfer the explants to sterile Whatman filter paper to soak up excess liquid growth medium, and transfer to semisolid MS basal growth medium[b].

Notes

[a]Disposal of EMS: EMS is a toxic chemical and must be disposed off according to current safety regulations in the laboratory (check with personnel responsible for toxic materials or local health authority). It may be necessary to use a specially designated sink for toxic chemicals for the washing step.

Detoxify the waste and all unused EMS solution by adding 4% (w/v) NaOH or 10% (w/v) sodium thiosulfate ($Na_2S_2O_3.5H_2O$) in a 3:1 ratio by volume. Pour into a designated container (marked with 'Disposal of suspected carcinogen' in some laboratories) and leave to stand for at least six half lives. As a guide, the half-life of EMS in 4% NaOH is 6 h at 20 °C and 3 h at 25 °C. For EMS in a 10% sodium thiosulfate solution, the half-life is 1.4 h at 20 °C and 1 h at 25 °C. All body parts or laboratory coats contaminated with EMS should be washed thoroughly with water and detergent and further neutralized with 10% (w/v) sodium thiosulfate.

[b]Need for preliminary tests to determine the range of optimal 'dosage'. There is a significant genetic component (even between cultivars of the same species) to the overall mutagenic efficiency of a chemical mutagen that, in turn, combines with the mutagens and prevailing environment to produce effects ascribable to the induced mutagenesis assay. It is usually advisable to carry out a preliminary experiment with different treatment combinations (such as those outlined above) in order to determine the parameters, mutation effectiveness (mutations per unit dose) and mutation efficiency (ratio of mutation to injury or other effect).

Determining the primary injury in M_1 seedlings under glasshouse conditions achieves this purpose efficiently. Primary injury could be ascertained from measuring growth parameters, including seedling height, root length, survival rate and chlorophyll mutation. To determine the optimal treatment condition for specific crops, cultivars, or genotypes, it is advisable to identify the range of the EMS concentration by a combination of treatment-duration, at which treatment growth reduction of about 20–30% occurs. The graphical method for determining LD in induced mutagenesis using physical agents can also be used for this purpose.

7.3 Troubleshooting

7.3.1 Factors influencing the outcome of mutagenesis using chemical mutagens

Factors that are critical to induced mutagenesis assays include the condition of the mutagenic solution, the inherent characteristics of the target tissue and the environment.

- Concentration of mutagen. This is the most critical factor with the results of assays depending to a great extent on the use of optimal concentrations of the mutagen. As a rule, an increase in the concentration of EMS, for instance, normally results in more mutation events, but these are accompanied by a corresponding greater amount of injury to seedlings and lethality.

- Treatment volume. The samples should be immersed completely in the mutagen solution the volume of which must be large enough to prevent the existence of concentration gradients during treatment. This ensures that all seeds (or other samples) are not exposed differently to the active ingredients of the mutagen. As a guide, a minimum of 0.5–1.0 ml of mutagen solution per seed is suitable for most cereals.

- Treatment duration. The treatment should be long enough to permit hydration and infusion of the mutagen to target tissue. The relevant seed characteristics that impact on this include seed size, permeability of the seed coat and cell constituents. Additionally, in order to minimize the unintended effects of EMS hydrolysis (acidic products) and in order to maintain the mutagen concentration, the treatment solution should be buffered or renewed with newly prepared EMS solution when the treatment duration is longer than the half life of the mutagen. For EMS, this is 93 h at 20 °C or 26 h at 30 °C, the time at which half of the initial active ingredient is hydrolysed or otherwise degraded. With practice, it is also possible (and advisable) to reduce the treatment duration when the target seeds have been presoaked.

- Temperature. Related to hydrolysis is the temperature of the environment in which the plant material is treated. Temperature influences the rate of hydrolysis of the mutagenic solution; at low temperatures, hydrolysis rate is decreased, implying that mutagen remains stable for longer. For EMS, the optimal temperature to achieve a half life of 26 h is 30 °C.

- Presoaking of seeds. This enhances the total uptake, the rate of uptake and the distribution of mutagen within the target tissue. With seeds, for example, presoaking leads to the infusion of a maximum amount of mutagen into the embryo tissue within the shortest possible time. This is on account of the fact that embryonic tissues of cereals, for instance, commence DNA synthesis rendering the seeds most 'vulnerable' to mutagenesis and hence resulting in high mutation frequencies, but with relatively less chromosomal aberrations. The duration of pre-soaking depends primarily on the anatomy of the seed; hard and thick seed coats require longer pre-soaking times than soft and thin ones. For barley, a pre-soaking period of 16–20 h is sufficient; the cells of the embryos attain the S-phase of mitotic cell division during this time.

- pH. The hydrogen ion concentration of the solution influences the hydrolysis of EMS. While low pH seems not to be critical for the rate of hydrolysis of EMS, biological systems seem to be more sensitive at low pH values. Buffers are used to control this, and by maintaining the pH of the EMS solution at the optimal vale of 7.0, injury to seeds and explants is minimized.

- Catalytic agents. Certain metallic ions such as Cu^{2+} and Zn^{2+} have been implicated in the enhancement of chromosomal aberrations induced by EMS. It is for this reason that it is recommended to use deionized water to prepare the EMS emulsion.

- Post-treatment handling: The by-products of the incubation process (resulting from hydrolysis) and residual active ingredients should be promptly washed off the incubated target tissues after treatment. This prevents continued absorption of the mutagen beyond the intended duration, so-called dry-back, which leads to lethality.

7.3.2 Factors influencing the outcome of mutagenesis using physical mutagens

- Oxygen. This is the major component of the environment with significant impact on mutagenesis. An electroaffinic agent, its presence in the target tissue is related directly to the number of mutation events. Iodine is another example, while others include chemical agents already identified as mutagens (interfering with DNA metabolism in different ways) as well as antibiotics that have been shown to interfere with DNA repair. The interplay between oxygen (and these other agents) and ionising radiation continue from irradiation to post-irradiation storage.

- Moisture content. Seed moisture content is important. In barley, for instance, it has been shown that at seed moisture content below 14%, there is marked increase in mutation frequencies as the moisture content decreases. It is therefore necessary to equilibrate the seed moisture content prior to ionization.

- Temperature. While low treatment temperatures have not been conclusively established as depressing mutation frequencies, preheating of cell lines has been shown to increase the incidence of mutation events.

- Other physical ionizing agents. The presence of other unintended agents (electromagnetic and ionizing radiation) has been shown to increase mutation frequencies, necessitating a deliberate attempt to exclude all other agents in order to guarantee reproducibly of result.

- Dust and fibres. Particles in the environment including dust and fibres (e.g. from asbestos) have been demonstrated to increase significantly the incidence of mutagenicity of irradiation and should therefore be eliminated in order to ensure reproducibility of results.

- Biological and infectious agents. Complex interaction mechanisms, based on animal studies, exist between hormonal concentrations and the effects of irradiation. While clear-cut inference is difficult to reach, it is advised that extraneous sources of hormones be excluded from irradiation set-ups in order to prevent confounding of results. Infectious agents (both viral and bacterial) have been shown to elevate radiosensitivity.

7.3.3 Facts about induced mutations

- Reproducibility of results. Induced mutations are random events, implying that even adherence to published irradiation conditions might not result in the same mutation events. A way of mitigating this uncertainty is to rely on statistical probability and to work with large population sizes. A guide is to target the production of an M_2 population of a minimum of 5000–10 000 individuals. A corollary to this is that estimates of radiosensitivity are so specific to the genotypes (and conditions in the reporting laboratory) that it is strongly advised that, whenever feasible, some preliminary tests are carried out with the experimental materials destined for induced mutagenesis.
- Dormancy. It is important to overcome dormancy before induced mutagenesis treatments. Preliminary seed viability tests, to detect whether or not the seeds are dormant, are usually recommended before treatment of seeds so that other underlying factors do not confound the estimates of radiosensitivity. For dormant seeds, efforts must be made to break the dormancy. Prechilling, heating and several forms of scarification (chemical and mechanical) have been established as ways of breaking dormancy [12].
- Safety. Radioactivity is potentially injurious to health (mutagenic and carcinogenic). Radioactive sources should therefore be operated only by trained and authorized personnel. Local regulations are usually explicit. Also, EMS is highly toxic (mutagenic and carcinogenic). In addition to the strict observance of good laboratory practices (e.g. no ingestion of foods and drinks, correct labelling of reagents, the use of laboratory coats and gloves), extra precautions should be observed when handling this chemical. The avoidance of contact with skin or any body parts should be strictly enforced. All procedures involving this biohazard should be carried out in a functional fume chamber, or, in exceptional cases, only if the experimenter is wearing a face mask. The bench surface should be covered with disposable absorbent paper with all spills correctly removed with absorbent paper or sawdust.

References

*1. IAEA (1977) *Technical report series* No. 119, 289 pp. International Atomic Energy Agency, Vienna, Austria.

Provides protocols on many aspects of induced mutations in plants.

2. Mba C, Afza R, Jain SM, *et al.* (2007) In: *Advances in Molecular Breeding Towards Drought and Salt Tolerant Crops*. Edited by MA Jenks, PM Hasegawa and SM Jain. Springer-Verlag, Berlin, Heidelberg, pp. 413–454.

*3. United Nations Organization (1982) United Nations Scientific Committee on the Effects of Atomic Radiation (UNSCEAR). 1982 Report to the General Assembly.

Very informative as it provides background information on the state of knowledge on different types of ionizing radiation.

4. Van Harten AM (1998) *Mutation Breeding: Theory and Practical Applications*. Cambridge University Press, Cambridge, UK.

The most recent compendium on induced mutations as a crop improvement strategy.

5. Myhill RR, Konzak CF (1967) *Crop Sci.* **7**, 275–276.

Describes the blotter method for measuring damage (growth reduction) due to irradiation.

*6. Martínez AE, Franzone PM, Aguinaga A, *et al.* (2004) *Envir. Expt. Bot.* **51**, 133–144.

Describes the blotter method for measuring damage (growth reduction) due to irradiation.

7. Raemakers CJJM, Amati M, Staritsky G, Jacobsen E, Visser RGF (1993) *Ann. Bot.* **71**, 289–294.

8. Côte FX, Domergue R, Monmarson S, *et al.* (1996) *Physiol. Plant.* **97**, 285–290.

9. Taylor NJ, Edwards M, Kiernan RJ, *et al.* (1996) *Nat. Biotechnol.* **14**, 726–730.

10. Owoseni O, Okwaro H, Afza R, *et al.* (2007) *Plant Mutation Rep.* **1**, 32–36.

11. Murashige T, Skoog F (1962) *Physiol. Plant.* **15**, 473–497.

*12. Kodym A, Afza R (2003) In: *Methods in Molecular Biology*. Edited by E Grotewold. Humana Press, Totowa, NJ, USA, Vol. 236. pp. 189–203.

In addition to providing the theoretical bases for many aspects of induced mutagenesis practice, it is also a source of valuable references.

8

Cryopreservation of Plant Germplasm

E.R. Joachim Keller and Angelika Senula
Genebank Department, Leibniz Institute of Plant Genetics and Crop Plant Research (IPK), Gatersleben, Germany

8.1 Introduction

Although most methods of plant cell culture are aimed at fundamental research or supporting other methods in biotechnology to create new genetic combinations, cell culture also has promising potential with respect to conservation strategies. This becomes increasingly important in view of the destruction of natural habitats and genetic erosion. Plant germplasm is maintained *in situ* in its natural surroundings and *ex situ* in living plant collections (genebanks). The propagules of higher plants are their seeds which are the main storage material in genebanks (orthodox seed). Many species, however, do not develop seeds that survive dry or cold periods, and therefore cannot be stored as seeds (recalcitrant seed). In genebanks, such plants must be maintained vegetatively. Similarly, plants which do not set seeds at all, or whose genotype is not truly represented by seeds, must be maintained vegetatively. The latter is the case in many varieties and hybrids. Shoot tips of these plants, excised embryos or embryo axes, callus, cell suspensions and pollen, are materials for which cell culture methods have been developed for conservation [1–3]. Temperature reduction is crucial in the storage of many items. This can be achieved by reducing the temperature, but maintaining it above 0 °C, to slow down developmental processes (so-called slow-growth culture) or exploiting very low temperatures, as in cryopreservation.

Plant Cell Culture Edited by Michael R. Davey and Paul Anthony

8.2 Methods and approaches

8.2.1 Main principles

Cryopreservation involves storage of biological material in liquid nitrogen (LN) at −196 °C, or above LN at −150 to −196 °C. 'Cryo' comes from the Greek word *κρυοσ* which means 'cold, frost, freezing'. All molecular processes are temperature dependent. Therefore, at such ultra-low temperatures, biochemical reactions do not occur and stored material does not undergo decay or genetic changes. A number of critical points are common to cryopreservation in reaching ultralow temperature and returning to warm conditions [4].

Several methods have been developed, which have been improved to avoid critical steps and to minimize their risks. Three main risk factors are common to cryopreservation: (1) size of the object to be cryopreserved, (2) its water content and (3) the speed of temperature transitions. These factors are tightly connected and interact with each other. The size of the object is crucial, because any local temperature transition will cause mechanical tensions within the material. If the object is too large, these tensions result in cracking. When a cryoprotecting chemical does not enter the object sufficiently, its concentration gradient may lead to over-accumulation, especially in the outer cell layers. Many cryoprotectants are poisonous compounds. The object to be stored must be sufficiently small to permit successful cryopreservation. The second factor is the water content. Although water is the basis of all life activities, its changes during cooling and warming are the most critical for the biological material because water is the main component of tissues, being 50–98% of its total mass. Water influences tissues in two ways during cooling. Ice formation in the cell wall (extracellular ice) leads to osmotic imbalances, removing water from the inner cell spaces in the course of the osmotic equilibration process. This causes plasmolysis. Intracellular ice crystals destroy the cellular organelles mechanically; both processes may interact. The speed of the temperature changes may be critical since ice formation requires time and, if the changes are very rapid, ice formation can be avoided. Two procedures are adopted to overcome these destructive forces. The first is slow freezing involving extracellular ice formation and increase of intracellular solute concentration, protecting the inner cell space from freezing. The other is rapid cooling, in which the contents of cryoprotective substances are so high that the viscosity of the solutions leads to their amorphous solidification during rapid cooling. This process is called glass transition or 'vitrification', as derived from the Latin word *vitrum* (glass; [5]). The risk is very high when cooling and warming speeds are slow, since in the heterogeneous cell compartments, some solution clusters remain that are of lower concentration and act as ice crystallization centres. Therefore, the temperature transitions have to be rapid to avoid general crystallization. In the ideal situation, no solution remains that can crystallize. It is, however, also possible to maintain cellular integrity when very small ice crystals are formed (microcrystalline ice).

Overall, the aim is the same with all methods, namely, safe storage of living plant material for very long periods. Several main procedures have been reported, including slow, two-step freezing [6], vitrification [7], encapsulation–dehydration [8] and dimethyl sulfoxide (DMSO) droplet freezing [9]. Recently, several combinations of methods have been described, such as droplet–vitrification [10] and encapsulation–vitrification [11]. Simplified methods can be used in the case of cold-hardened buds [12] and orthodox seeds [13, 14], while storage of pollen [15] and spores [16] requires specific procedures.

Cryopreservation techniques are under intense development as recent surveys document [17], since protocols must be modified for any given species and type of tissue. All methods comprise dehydration that may be detrimental if tissues are not pre-adapted. Therefore, several dehydrating steps occur prior to cryopreservation proper, including preculture periods with low or alternating temperatures to cold-adapt the target plants [18], dehydration using solutions of high osmotic pressure, or air desiccation.

As cryopreservation imposes harsh stress on biological materials, not all specimens survive. Conditions need to be optimized to maximize regeneration. After rewarming, some adaptive culture steps may be needed, such as stepwise reduction of the osmotic pressure of the medium and culture in the dark or under reduced light intensity to avoid photo-oxidation injury to the tissue. Methods are available to confirm survival of cells and organs. The most exploited are staining with triphenyl tetrazolium chloride (TTC) and fluorescein diacetate (FDA) for cells and callus [19], and peroxidase and 3-(4,5-dimethyldiazol-2-yl)-2,5-diphenyltetrazolium bromide (MTT) for pollen grains [20]. Shoot tips are often assessed for their survival 1–2 weeks after recovery from cryopreservation. Survival is defined as the existence of green structures with swelling and, sometimes, with callus production. However, regrowth or shoot regeneration observed after 4–8 weeks, depending on the species, can be regarded as the only reliable measure of the success of cryopreservation. The same is true for pollen, which can be cultured in hanging drops of medium, but the final assessment must be pollen tube germination and its ability to fertilize egg cells. A survey of the various possible steps of the methods and the target object is given in Figure 8.1.

The main cryopreservation methods were developed and, initially, optimized empirically. For further refinement and the development of new procedures, fundamental research is needed into various factors of the cryopreservation process. Water in the tissue can be analysed by differential scanning calorimetry (DSC) using its thermal behaviour [21]. Sugars acting as cryoprotectants, polyamines, membrane lipids and components with antioxidative effect can be analysed biochemically. Proteins, involved in gene expression of signal-transducing chains of cold-adaption and injury repair (e.g. stress proteins), may be analyzed by two-dimensional-gel electrophoresis and mass spectrometry (proteomics; [22]). Finally, histological and ultrastructural investigations may give insight into damage and repair mechanisms and pathways of regeneration [23]. Changes in the cytoskeleton may be of value in this respect [24]. Several other methods have been used to investigate freezing and thawing processes, including nuclear magnetic resonance [25] and cryomicroscopy [26].

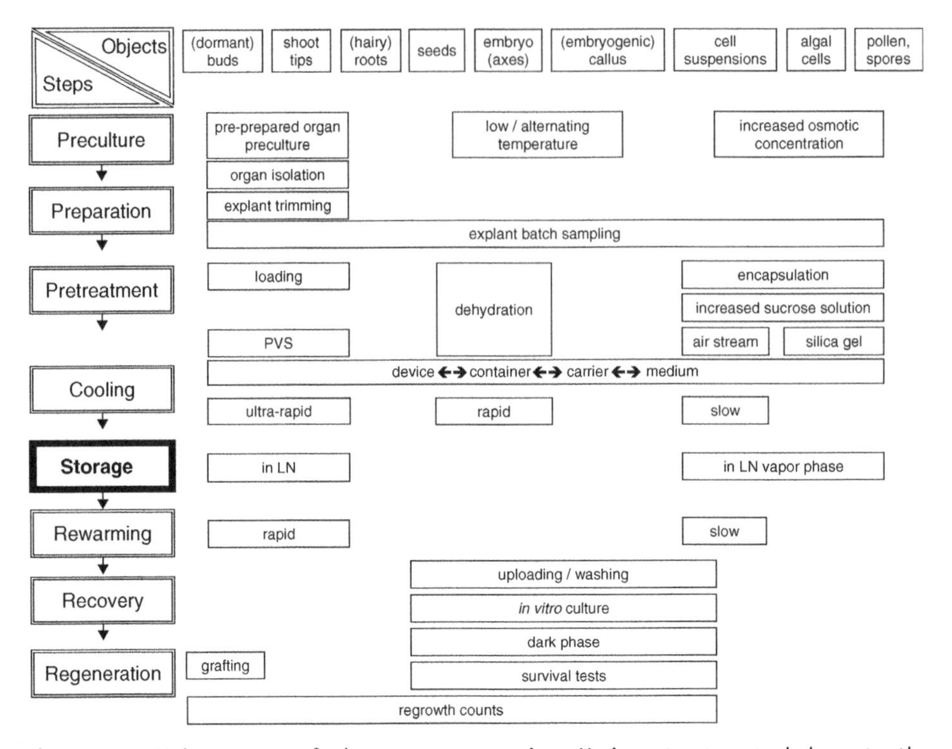

Figure 8.1 Main aspects of plant cryopreservation. Various treatments belong to the respective main steps of the method. No strict sequence arrangements are given because, in most cases, various options are possible for the different steps and biological materials.

8.2.2 Slow (two-step) freezing

Defined, slow cooling procedures (about 0.1–0.5 °C/min) may cause stepwise concentration of intracellular solutes with increased viscosity, through osmotically driven water efflux caused by extracellular ice formation. The use of cryoprotective compounds and artificial induction of early ice crystallization outside cells ('seeding') may be beneficial, so that intracellular spaces do not supercool to very low temperatures with subsequent spontaneous ice crystallization. This takes place about −40 °C, which is the critical temperature for spontaneous ice crystallization in pure water. The speed of the decrease in temperature must be fixed very precisely. Thus, programmable freezers are used, which gently add aliquots of LN to reaction vessels. Thermal behaviour below −40 °C is not as critical as above this temperature. Material can be plunged directly into LN. The name 'two-step cooling' is derived from these different velocities above and below the critical temperature. Simplified equipment may be used in the case of less sensitive material [2].

8.2.3 Vitrification

The glassy state of cellular liquids is obtained by sufficiently high concentrations of solutes (cryoprotectants) and rapid cooling. The change of osmotic conditions is much stronger than in slow freezing. Therefore, several steps of dehydration may be necessary, which are sometimes differentiated as loading (weaker solution) and dehydration (stronger solution). Depending on their molecular size, solutes penetrate into the intracellular space (e.g. DMSO, glycerol), or remain in the cell wall (e.g. sucrose and other carbohydrates). Accordingly, they affect vitrification processes in different compartments of the tissue. The optimum equilibrium is reached when a mixture of cryoprotective substances is used. Some standard mixtures are published and are employed extensively, many of which are called PVS (plant vitrification solution). The most commonly used mixture is PVS2, consisting of 0.4 M sucrose, 3.2 M glycerol, 2.4 M ethylene glycol, 1.9 M DMSO in liquid culture medium as appropriate for the respective plant [7]. As certain components of PVS may be toxic, some protocols utilize low temperature pretreatments at 0 °C using an ice bath. After dehydration in PVS, samples are again transferred into PVS solution, and tubes containing the solution with the samples are plunged immediately into LN, thus enabling high cooling rates (~300 °C/min). It is essential that rewarming of the samples is rapid as well to avoid re-crystallization of cellular solutes. Very high osmotic values of PVS and the poisonous character of some components, necessitate washing treatments with stepwise decreasing concentrations [7].

8.2.4 Encapsulation–dehydration

Some of the principles of this method resemble natural seed development, where a dehydration step occurs during ripening. Dehydration is also included, once the encapsulation process is complete. At this stage, the explants are also used directly for agricultural purposes in a similar way to seeds. Consequently, encapsulated shoot tips are also termed 'artificial seeds'. Encapsulation uses the chelating potential of alginates in the presence of bivalent ions (mainly Ca^{2+}) to produce gel capsules, so-called beads, in which the explants are embedded. Alginates are extracted from bacteria or marine algae and consist of long carbohydrate chains that are gelled by ion bridges. For bead production, explants are sampled in liquid medium devoid of calcium ions, but containing alginate solution. The liquid with floating explants is transferred drop-wise by a pipette into calcium chloride solution. As soon as the drops come into contact with this solution, they are transformed to gelatinous globules (beads). Beads containing explants are then further dehydrated in liquid medium with greater sucrose contents and dried over silica gel, or in the air stream of a laminar flow cabinet. Finally, they are transferred rapidly into LN. Rewarming can be slow or rapid. Regenerating plantlets grow out of the beads, or may be excised from the latter.

8.2.5 DMSO droplet freezing

Ways have been found to increase the speed of temperature changes even more than in some previously described methods. Since aluminium is a good heat conductor, aluminium foils are used as carriers for explants. Simple solutions, like 10% DMSO, are sufficient as a cryoprotectant in this procedure. Small droplets of cryoprotectant solutions are placed on aluminium foil strips, and explants placed into these droplets. Droplets containing explants adhere to the foils; the latter are transferred immediately into vials containing LN. Both aluminium and the direct transfer increase the temperature change velocity up to 8000–12 000 °C/min. This much higher speed is the basis for the term 'ultra-rapid freezing'. Vitrification is not complete in this method, and small amounts of freezable water can be found in the tissue. However, there may not be enough time for formation of larger ice crystals, and the small-sized crystals do not damage cell structure. It is also necessary to ensure ultra-rapid rewarming, which may be achieved by plunging the foils with the adhering explants into sterile culture medium at room temperature.

8.2.6 Combined methods

As various methods are developed, more combinations of protocols are published. Thus, in the encapsulation–vitrification method, alginate beads are produced as in encapsulation–dehydration. However, after culture with increased concentrations of sucrose, the beads are transferred into PVS solution, as in the vitrification procedure, and treated accordingly. In droplet–vitrification, the advantageous influence of the aluminium foils is combined with the use of PVS solutions instead of simple DMSO solutions, thus combining the heat conducting effect of the foil with complete vitrification.

8.2.7 Freezing of cold-hardened buds

Woody plants, adapted to the conditions of temperate zones where they have to survive subzero temperatures during winter, have developed mechanisms of cryoprotection to withstand freezing injury. Thus, using twigs of these plants in winter after adaptation mechanisms have established cold-hardiness, may facilitate cryopreservation. Twigs are cut into short pieces and stored in appropriate containers, and buds excised after rewarming. In some cases, they can be used directly as scions and grafted onto rootstocks, as in conventional grafting procedures. Alternatively, the buds can be grown *in vitro* after rewarming.

8.2.8 Freezing of orthodox seeds

Orthodox seeds are ones that can be dried and stored for long periods at reduced temperatures and under low humidity [13]. Such seeds reduce their water content during ripening. Therefore, they are normally stored easily at subzero temperatures. Storage in seed genebanks is usually performed at temperatures of −15 to −20 °C. There are, however, some seed collections where personnel have accumulated experience in the cryopreservation of orthodox seeds. The seeds are placed into appropriate containers and stored in the vapour phase above LN. However, not all orthodox seeds are storable; seeds rich in oil components may be difficult to cryopreserve [3].

8.2.9 Freezing of pollen and spores

Pollen grains are also carriers of genetic information and can be stored to preserve germplasm. However, a difficulty is their small size. A possibility is to store pollen within ripe, bud-enclosed anthers and to exploit one of the cryopreservation methods mentioned earlier. Anther tissue has to be removed carefully after rewarming. Pollen can also be placed in special containers such as cryotubes, gelatin capsules, butter paper, or tightly sealed aluminium pouches. These containers have to be transferred directly, or after a precooling phase, into LN. Protocols exist for rewarming, using fast or slow temperature changes.

PROTOCOLS – General Equipment and Reagents for all Methods

- Laminar air flow cabinet
- Dewar vessels, 1 l volume (KWG Isotherm)
- Sterile culture vessels (e.g. Petri dishes, tubes, jars, sizes see specific protocols)
- Sterilized instruments for specimen preparation (forceps, pipettes and tips, Pasteur pipettes, hypodermic needles, scalpels)
- LN

PROTOCOL 8.1 Controlled-Rate Cooling of Dormant Buds of Willow (*Salix* L. Species)[a] [27]

Equipment and Reagents

- Temperature controlled room (2–4 °C)
- Refrigerator
- Controlled rate freezer (e.g. Kryo 520; Planer)
- Heat sealer for the polyolefin tubes
- Heated mat
- Crisper container (26 × 32 cm)
- Polyolefin tubes (19–42 mm diam.; 3M Corp.)
- Rooting medium Dip-N-Grow (20×): 500 mg/l indole-3-butyric acid (IBA) and 250 mg/l naphthalene 2-acetic acid (NAA)
- Sterile substratum perlite, vermiculite, peat moss and sand, in equal proportions

Method

Pretreatment and cooling:

1 Collect branches in winter, when buds are dormant.

2 Place branches into plastic bags; store in a refrigerator at −3.5 °C until required.

3 Cut branches into 4–6 cm nodal segments each with two to three buds; place five to six segments into polyolefin tubes; heat-seal the tubes.

4 Place tubes into the controlled-rate freezer at −3.5 °C; cool to −35 °C in steps of 1 °C/h at 5 h/day, and hold the respective intermediate temperatures overnight.

5 Incubate tubes at −35 °C for 24 h; place the latter with contents into the LN vapour phase.

Recovery:

6 Warm tubes in air at 2–4 °C for 24 h.

7 Remove the lowermost buds, which would come under the surface of the culture medium with a razor blade, notch the basal end of the nodal segments with a razor blade, dip them into rooting medium for 3–5 s, and place the segments into sterilized compost soaked with sterile water in crisper containers for rooting[b]

8 Place containers on a heated mat, creating a temperature in the root zone of 13 °C and 4 °C above the sterile substratum and under low light (25 μmol/m^2/s) using Cool White fluorescent illumination, with a 10 h photoperiod.

9 Keep the lids of the crisper containers open by 1–2 cm; mist the nodal segments daily. Rooting should occur within 6 weeks after thawing the cryopreserved material.

Notes

[a]Other methods that can be used for explants from dormant buds include vitrification, encapsulation-dehydration and encapsulation–vitrification.

[b]In some cases, e.g. in apple, cold-hardened buds can be used directly after cryopreservation as scions for grafting onto rootstocks [28]. Dried cold-hardened nodal segments (30% moisture content) are cryopreserved in polyolefin tubes; for rehydration they are covered with moist peat moss in moisture-tight plastic crisper containers and held at 2 °C for 15 days. Scions are then excised from the nodal segments and grafted directly onto rootstocks.

PROTOCOL 8.2 Controlled-Rate Freezing of Jerusalem Artichoke (*Helianthus tuberosus* L.) Suspension Cultures [29][a]

Equipment and Reagents

- Controlled rate freezer
- Rotary shaker (Bioasset Technologies PVT. Ltd)
- Ice and water baths

- Büchner funnel with a nylon net, pore size 100 μm
- Cryotubes (1.8 ml; Nunc)
- Petri dishes (9 cm in diam.)
- Sterile filter paper discs (Whatman No. 1), 5.5 cm in diam.
- Liquid plant growth medium: Murashige and Skoog (30; MS, [30]) medium, with 0.22 mg/l dichlorophenoxyacetic acid (2,4-D), 0.09 M sucrose
- Preculture medium: Liquid plant growth medium with 0.75 M sucrose
- Cryoprotectant solution: Liquid plant growth medium with 0.5 M glycerol, 0.5 M DMSO, 1.0 M sucrose, 0.086 M proline
- Recovery medium: plant growth medium, semisolidified with 0.8% (w/v) agar (Bactoagar; Difco)

Method

Pretreatment and cooling:

1. Use suspension cultures in their logarithmic growth phase as basic material. Logarithmic growth can be attained by mixing 50 ml of a cell suspension with 100 ml of new medium every 14 days.
2. Transfer the cells into preculture medium. Transfer 50 ml of cell suspension into 100 ml of preculture medium, with a final sucrose concentration of 0.5 M; incubate on a rotary shaker in the dark for 1–6 days at 24 °C.
3. Harvest cultures by allowing the cells to settle, or by filtering or centrifugation; cool the concentrated suspensions on ice for 30 min (optional step).
4. Place aliquots of cells (0.75 ml) into cryotubes, add 0.5 ml of chilled (on iced water) cryoprotectant solution.
5. Incubate cells in the cryoprotectant solution[b] at 0 °C for 1 h.
6. Transfer cryotubes into a controlled-rate freezer; cool the tubes with contents at 0.5 °C/min to −35 °C.
7. Maintain cryotubes at −35 °C for 35 min.
8. Transfer the cryotubes to LN.

Recovery:

9. Plunge the cryotubes into a water bath (45 °C) for 2 min; agitate the tubes.
10. Remove the tubes from the water bath; disinfect them on the outside with 70% (v/v) ethanol.
11. Transfer suspensions derived from the 0.75 ml aliquots of the initially suspended cells onto axenic filter paper, placed on the surface of 25 ml semisolid plant growth medium in Petri dishes and incubate in the dark at 24 °C.

12 Two weeks later, transfer growing cells onto new plant growth medium by moving the filter paper with the attached cells to the surface of the new medium.

Notes

[a]Rapid cooling methods, namely vitrification and encapsulation–dehydration, are also used for suspension cultures.

[b]Various cryoprotectant solutions have been described, e.g. 1.0 M DMSO + 1.0 M glycerol + 2.0 M sucrose [31].

PROTOCOL 8.3 Dehydration and Cooling of Wild Cherry (*Prunus avium* L.) Embryogenic Callus [32]

Equipment and Reagents

- Water bath
- Cryotubes (1.8 ml)
- Petri dishes (40 × 12 mm)
- Petri dishes (100 × 20 mm) with air-vented lids (Greiner)
- Callus growth medium: MS salts, Morel's vitamins [33], 500 mg/l casein hydrolysate, 0.1 mg/l NAA, 0.1 mg/l kinetin, 0.1 mg/l benzylaminopurine (BAP), 0.09 M sucrose, 0.2% (w/v) Phytagel
- Preculture medium: callus growth medium with sucrose concentrations of 0.25 M, 0.5 M, 0.75 M or 1.0 M
- Rinsing solution: liquid MS-based medium with 1.2 M sucrose
- Recovery medium = callus growth medium

Method

Pretreatment and cooling:

1 Excise callus clumps, each 1–3 mm diam.; culture the tissues on MS-based callus growth medium with 0.25 M sucrose at 23 °C for 1 day (20 clumps/10 mm Petri dish).

2 Transfer stepwise onto preculture growth medium with 0.5 M (1 day), 0.75 M (2 days) and 1.0 M sucrose (3 days).

3 Determine the fresh weight of tissues; transfer, using forceps, the 20 tissue clumps into previously weighed empty Petri dishes and reweigh.

4 Transfer tissues to air-vented Petri dishes; desiccate in the air stream in a laminar flow cabinet until the tissues are 50–60% of their original fresh weight.

5 Place tissues in cryotubes (20 tissue clumps per tube) containing LN; plunge the tubes into LN.

Recovery:

6 Warm the rinsing solution to 40 °C in a water bath

7 Remove the cryotubes from LN, open the lids and place the tissues directly into Petri dishes with warm rinsing solution for 1 min.

8 Transfer the Petri dishes with their contents onto ice for 10 min.

9 Transfer tissues stepwise to MS-based medium with 1.0 M (for 12 h), 0.75 M (12 h), 0.5 M (12 h) and 0.25 M sucrose (48 h), in the dark at 25 °C.

10 Transfer tissues to callus growth medium in the dark at 25 °C.

11 Subculture the callus every 21 days.

12 Measure the callus growth after 6 weeks as the fresh weight ratio in comparison to the initial weight.

PROTOCOL 8.4 Cryopreservation of Pollen from Solanaceous Species – Tomato (*Lycopersicon esculentum* Mill.), Aubergine (*Solanum melongena* L.) and Bellpepper (*Capsicum annuum* L.) [34]

Equipment and Reagents

- Desiccator
- Silica gel
- Gelatin capsules: sizes 1, 0, or 00 (Value Healthcare)
- Butter or waxed paper
- Laminated aluminium pouches
- Alexander's stain [35]: 20 ml ethanol, 2 ml of 10.8 mM malachite green (Merck) in ethanol, 50 ml distilled water, 40 ml glycerol, 10 ml of 17.3 mM acid fuchsin (Merck) mixed with 1 g phenol, 2 ml lactic acid

Method

Pretreatment and cooling:

1 Collect healthy flowers at the time of anther dehiscence[a].

2 Place the flowers in Petri dishes (40 × 12 mm) in desiccators containing silica gel at ambient temperature for 30–45 min to release pollen.

3 Tap the flowers over butter or waxed paper to collect the pollen.

4 Transfer the pollen to gelatin capsules; enclose the capsules in aluminium pouches and seal the pouches.

5 Transfer the pouches into LN.

Recovery:

6 Warm samples at room temperature for 30–60 min.

7 Culture pollen in hanging drops [36].

8 Test pollen viability by staining with Alexander's stain.

Note

[a]It is also possible to collect flowers that have just opened and to keep them in an incubator at 25 °C in the light for 1 h. Remove the styles and cut the anther cones. Hold the flowers upside down and tap to release the pollen.

PROTOCOL 8.5 Vitrification of Garlic (*Allium sativum* L.) Shoot Tips from *In Vitro*-Derived Plants[a] [37]

Equipment and Reagents

- Illuminated incubator (Percival Scientific; Geneva Scientific LLC)
- Dissection microscope
- Water bath
- Shaker (Vortex Genie; Scientific Industries)
- Cryotubes (1.8 ml)
- Petri dishes (5 cm diam.)
- Growth medium: MS medium with 0.5 mg/l *N*6-(2-iso-pentenyl)adenine (2iP) + 0.1 mg/l NAA, 0.09 M sucrose, 1% (w/v) agar (Serva Kobe I)
- Preculture medium: MS medium with 0.5 mg/l 2iP, 0.1 mg/l NAA, 0.3 M sucrose, 1% (w/v) agar
- Loading solution: liquid growth medium with 0.4 M sucrose, 2.0 M glycerol
- PVS3 solution: liquid growth medium with 1.46 M sucrose, 5.4 M glycerol[b]
- Washing solution: liquid growth medium with 1.2 M sucrose

Method

Pretreatment and cooling:

1 Culture single, well developed *in vitro*-derived plants on 10 ml growth medium in culture tubes (3 cm diam.) at 25 °C/−1 °C (light/dark) in a light incubator with a 16 h photoperiod (60–80 µmol/m^2/s; Day Light fluorescent tubes; Philips) for 6–8 weeks for cold acclimation.

2 Prepare shoot explants consisting of basal plates (each 1–2 mm thick) and meristematic domes with three to four leaf bases (each 3–5 mm in length).

3 Preculture explants in Petri dishes on 5 ml aliquots of preculture medium[c] at 25 °C with a 16 h photoperiod (60–80 µmol/m^2/s; Day Light fluorescent tubes) for 20–24 h.

4 Transfer the explants into cryotubes (10 explants per tube), add 1 ml loading solution[c] and shake; incubate at room temperature for 20 min, before removing the loading solution.

5 Add 1 ml PVS3 solution[c,d] to the tubes, shake and incubate at room temperature for 2 h, before removing the PVS3 solution.

6 Add 0.5 ml PVS3 solution to each tube, shake, and plunge each tube immediately into a Dewar vessel containing LN.

Recovery:

7 Warm the tubes in a water bath (40 °C) for 2.0–2.5 min.

8 Remove the PVS3 solution.

9 Add 1 ml washing solution[c] to the tubes, shake, and maintain the tubes at room temperature for 10 min; remove the solution.

10 Transfer explants to preculture medium at 25 °C in the dark for 1 day. Transfer the explants onto growth medium at 25 °C in the dark for 7 days, followed by culture under a 16 h photoperiod at 25 °C for other 7 days.

11 Count the surviving explants and transfer them to culture tubes each with 10 ml growth medium.

12 Count the regenerating plants 8 weeks after warming.

Notes

[a]Other possible sources are basal plates from cloves or bulbs, and unripe or ripe bulbils.

[b]Cryoprotectant solution PVS2 is also used by some researchers. In this case, shorter dehydration times must be used.

[c]These solutions are autoclavable; adjust the pH to 5.8 before autoclaving.

[d]PVS3 solution requires stirring for an extended period and warming to dissolve.

PROTOCOL 8.6 Droplet–Vitrification of Mint (*Mentha* L.) Shoot Tips from *In Vitro*-Derived Plants [38]

Equipment and Reagents

- Illuminated incubator (Percival Scientific; Geneva Scientific LLC)
- Dissection microscope
- Water bath
- Petri dishes (6 cm in diam.)
- Filter paper disks (4.5 cm; Schleicher & Schüll)
- Strips of aluminium foil ($25 \times 5 \times 0.03$ mm)
- Cryotubes; 1.8 ml
- Growth medium: MS medium with 0.09 M sucrose, 1% (w/v) agar (SERVA, Kobe I), lacking growth regulators[a]
- Preculture solution: MS medium with 0.3 M sucrose[a]
- Loading solution: liquid growth medium with 0.4 M sucrose, 2.0 M glycerol[a]
- PVS2 solution: liquid growth medium with 0.4 M sucrose, 3.2 M glycerol, 2.4 M ethylene glycol, 1.9 M DMSO[b]
- Washing solution: liquid growth medium with 1.2 M sucrose[a]
- Recovery medium: MS medium with 0.5 mg/l 2iP, 0.1 mg/l NAA, 0.09 M sucrose, 1% (w/v) agar[a]

Method

Pretreatment and cooling:

1. Culture nodal segments on MS medium with 0.09 M sucrose at 25 °C/−1 °C with a 16 h photoperiod (60–80 µmol/m^2/s; Day Light fluorescent tubes) for 2–4 weeks for cold acclimation in a light incubator.
2. Excise axillary shoot tips (each 1–2 mm in length).
3. Preculture the explants onto the surface of two filter paper discs overlaying 2 ml of preculture solution in Petri dishes at 25 °C with a 16 h photoperiod for 20–24 h.
4. Transfer explants onto the surface of two filter paper discs overlaying 2 ml loading solution in Petri dishes at room temperature for 2 h.
5. Transfer explants 2 ml PVS2 solution in Petri dishes at room temperature[c] for 20 min.
6. Transfer explants into 2 µl droplets of PVS2 solution on aluminium foil strips, with one explant per droplet, and 10 explants per strip.

7 Place two strips with the adhering droplets into one cryotube, cap the tube and plunge the latter directly into LN.

Recovery:

8 Plunge the cryotubes into a water bath (40 °C) for 3–5 s.

9 Add 1 ml washing solution to each of the tubes, shake, and transfer the contents of each cryotube into a Petri dish with 2 ml of washing solution; remove the aluminium foil.

10 Maintain at room temperature for 20 min.

11 Transfer explants to recovery medium. Maintain the explants at 25 °C in the dark for 1 day; transfer to a 16 h photoperiod (50 µmol/m^2/s; Day Light fluorescent tubes).

Notes

[a]These media are autoclavable. Adjust the pH to 5.8 before autoclaving.

[b]All constituents are autoclavable, except DMSO, which must be filter-sterilized. Mix the components immediately prior to cryopreservation treatment.

[c]PVS2 solution is toxic to cells. Therefore, the incubation time must be minimal. Some researchers use a low temperature (0 °C) for pretreatment in PVS2 solution. Split samples when handling large numbers of explants.

PROTOCOL 8.7 DMSO–Droplet Freezing of Potato (*Solanum tuberosum* L.) Shoot Tips (Modified from Reference [9])

Equipment and Reagents

- Dissection microscope
- Water bath
- Styropor boxes (Eprak, Microtube Rack; 1.5 ml)
- Cryotubes (1.8 ml)
- Screw capped glass jars; 175 ml capacity
- Filter paper discs (45 mm; Schleicher & Schüll)
- Strips of aluminium foil (25 × 5 × 0.03 mm)
- Growth medium: MS medium lacking vitamins, 0.06 M sucrose, 1% (w/v) agar
- Preculture medium = washing solution: liquid MS medium with 0.09 M sucrose
- 1.28 M DMSO in preculture medium[a]

- Recovery medium: MS medium with 0.5 mg/l zeatin riboside, 0.5 mg/l indole-3-acetic acid (IAA), 0.2 mg/l gibberellic acid (GA_3), 0.09 M sucrose, 1% (w/v) agar [39]

Method

Pretreatment and cooling:

1. Excise nodal and shoot tip explants from source cultures (microtubers or shoots).
2. Propagate the explants in screw-capped glass jars, each with 50 ml of growth medium for 3–7 weeks depending on the genotype.
3. Preculture the explants at alternating temperatures of 22 °C/4 °C with a 8 h photoperiod (60–80 μmol/m^2/s; Day Light fluorescent tubes) for 1–2 weeks.
4. Excise the shoot tips, isolate the apical buds; incubate the latter in preculture medium overnight.
5. Transfer explants into 1.28 M DMSO in liquid preculture medium for 2 h.
6. Place explants into 2.5 μl droplets of 1.28 M DMSO in liquid preculture medium on aluminium foils.
7. Drop the foils with adhering DMSO droplets and explants directly into cryotubes containing LN.

Recovery:

8. Rewarm by plunging aluminium foils with explants into liquid preculture medium at room temperature.
9. Culture the explants on 3 ml recovery medium in 6 cm Petri dishes with a 16 h photoperiod (50 μmol/m^2/s; Day Light fluorescent tubes).

Note

[a]DMSO must be filter-sterilized and added immediately before use.

PROTOCOL 8.8 Encapsulation Dehydration of Hop (*Humulus lupulus* L.) Shoot Tips[a] [40]

Equipment and Reagents

- Dissection microscope
- Cryotubes (1.8 ml)
- Sterile filter papers (8 cm in diam.)
- Petri dishes (9 cm diam.)
- Silica gel (Absortech GmbH)

- Multiplication medium: MS salts, vitamin mixture after Wetmore and Sorokin [41], 0.17 M glucose, 1 mg/l BAP, 0.1 mg/l IBA, 0.7% (w/v) agar (Roko)
- Preculture medium: MS medium vitamin mixture after Wetmore and Sorokin, with 0.75 M sucrose, 1 mg/l BAP, 0.01 mg/l GA_3, 0.7% (w/v) agar (Roko)
- Alginate solution: modified liquid MS preculture medium with 3% (w/v) Na-alginate, 0.5 M sucrose lacking calcium
- Liquid MS medium with 100 mM $CaCl_2$, 0.09 M sucrose
- Regrowth medium: MS medium with 1.0 mg/l BAP, 0.1 mg/l GA_3 0.17 M glucose

Method

Pretreatment and cooling:

1. Culture donor plants on MS medium at 25 °C with a 16 h photoperiod (40 μmol/m^2/s) for 4 weeks.
2. Cold-acclimate shoot tips at 4 °C in the dark for 1–2 weeks.
3. Excise apical and axillary shoot tips (each 0.5–2.0 mm in length); suspend them in alginate solution.
4. Using a Pasteur pipette, pick up individual explants each with some alginate solution, and drop into liquid preculture MS medium with 100 mM $CaCl_2$ to make beads. Incubate for 30 min.
5. Transfer the beads onto preculture medium in 9 cm Petri dishes (10 beads/dish) containing 25 ml medium at 25 °C in the dark for 2 days.
6. Blot the beads briefly with sterile filter paper to absorb excess moisture.
7. Place the beads on filter papers in Petri dishes each containing 30 g silica gel and dry the beads in the air current of a laminar flow cabinet to a water content of 16% (on a fresh weight basis) according to a previously determined calibration curve.
8. Place the beads into cryotubes (five beads/tube) and plunge the latter into LN.

Recovery:

9. Warm the beads at room temperature for 15 min.
10. Transfer the beads to regrowth medium; incubate at 25 °C with a 16 h photoperiod (40 μmol/m^2/s) for 30 days.
11. Remove emerging shoots, and incubate on MS medium lacking growth regulators.

Note

[a]Other materials for encapsulation include embryo axes, embryos, embryogenic callus and suspension cells.

PROTOCOL 8.9 Encapsulation–Vitrification of Mint (*Mentha* L.) Shoot Tips [42]

Equipment and Reagents

- Dissection microscope
- Shaker
- Petri dishes (9 cm diam.)
- Cryotubes (1.8 ml)
- Alginate solution: liquid MS medium lacking calcium with 20 g/l Na alginate, 0.4 M sucrose
- Calcium chloride solution: liquid MS medium with 0.1 M calcium chloride, 0.4 M sucrose
- Osmoprotection solution: liquid MS medium with 2 M glycerol, 0.4 M sucrose
- PVS2 solution: liquid MS medium with 3.2 M glycerol, 2.4 M ethylene glycol, 1.9 M DMSO, 0.4 M sucrose
- Rinsing solution: liquid MS medium with 1.2 M sucrose
- Recovery medium: MS medium with 0.09 M sucrose, 1 g/l casamino acid, 2 g/l Gellan-gum (Gelrite; Duchefa)

Method

Pretreatment and cooling:

1 Culture the nodal segments, each consisting of a pair of leaves and 8–10 mm long stems on growth medium in Petri dishes at 25 °C for 1 day with a 16 h photoperiod at 96 μmol/m^2/s to induce axillary buds.

2 Cold-acclimate the explants at 4 °C with a 12 h photoperiod (20 μmol/m^2/s) for 3 weeks[a].

3 Dissect the shoot tips in alginate solution in Petri dishes (~0.1 ml alginate solution/explant); use 10 explants/treatment.

4 Take up the alginate solution containing explants with a sterile Pasteur pipette, and drop the explants (one explant/drop) into the calcium chloride solution to make beads; avoid air bubbles.

5 Leave the beads in the solution for 20 min to gel.

6 Place the beads in osmoprotection solution in 100 ml flasks on a shaker; agitate gently (20 rpm) at 25 °C for 1 h.

7 Drain, add PVS2 solution that was cooled previously in a refrigerator at 0–4 °C; agitate gently (15 rpm) at 0 °C for 3 h.

8 Place 10 beads and 1 ml PVS2 solution in each cryotube; place the cryotubes in LN.

Recovery:

9 Warm the cryotubes in a water bath (40 °C) for 2 min; drain the PVS2 solution immediately and replace twice at 10 min intervals with 1 ml rinsing solution.

10 Place beads on recovery medium for growth of the explants.

Note

[a]Cold acclimation depends on the mint species.

8.3 Troubleshooting

- LN is dangerous. Therefore, the rules of safety at work must be followed strictly. Direct contact with LN must be avoided by wearing appropriate gloves, face protection, aprons or laboratory coats and shoes. Nitrogen in the atmosphere cannot be measured directly. Therefore, oxygen sensors must be installed in rooms housing LN storage tanks. They should alarm emergency personnel as soon as the oxygen content of the atmosphere falls below the critical level of 17%. Sufficient aeration of the rooms must be ensured.

- Strict cleanness must be maintained in all steps of the work commencing with sterilization of explants. Solutions have to be autoclaved or, in the case of heat-unstable substances, filter-sterilized. Work must be performed in laminar flow cabinets; vessels must be flamed and preparation instruments must be sterilized by flaming or the use of hot-bead sterilizers. Culture of plant material must be performed in dedicated, clean rooms with accurate control of temperature, photoperiod and relative humidity.

- The quality of donor plant material is important. Only vigorous and healthy material should be used for cryopreservation. Since bacteria (endophytes) often colonize cells, appropriate bacterial media should be used to test for bacteria. Infected material must be discarded. Some unpredicted reduction in regeneration capacity may be caused by unrecognized endophytes. When using *in vitro*-derived plants as donor material, the quality of the plants may decline with time *in vitro* due to bacterial accumulation or other factors. Preculture of material *in vitro* should be as short as possible.

- Explant preparation is usually the most labour-consuming part of the protocol, which must be considered in planning work. Preparation must be carried out very carefully, so that the explants are of the correct developmental stage. Homogeneity of explants is essential to attain reproducible results.

- As 100% regeneration may be expected only very rarely after rewarming, stored samples must be sufficiently large, and duplicates should be taken for safety. A control set should be taken with each set of cryosamples, and should amount to ~40% of the total number of explants [43].

- Safe storage technology should be used. This includes a warning system on tanks which alarm when the level of LN becomes too low, with precise documentation and labeling (by nitrogen-resistant pens and, if possible, bar codes). For safety, duplicates should be placed in a different tank and, preferably, in a different location.

- Viability assessments should be as reproducible as possible. All records of regrowth, such as normal shoot and root formation, pollen tube growth and fertilization capacity, and callus production by cultured cells, should be favored over simple viability tests. The latter may give only an estimation of any procedure and not an assessment of its final success.

References

**1. Benson EE (1999) *Plant Conservation Biotechnology*. Taylor & Francis, London.

Monograph giving a good survey about the topic, from theory to practice.

***2. Reed BM (2008) *Plant Cryopreservation: A Practical Guide*. Springer, New York.

Most recent survey about general matters and specific protocols.

**3. Day JG, McLellan MR (2007) *Methods in Molecular Biology: Cryopreservation and Freeze–Drying Protocols*, 2nd edn. Humana Press Inc., Clifton, NJ, USA.

A comprehensive survey.

***4. Fuller BJ, Lane N, Benson EE (2004) *Life in the frozen state*. CRC Press, Boca Raton, FL, USA.

Monograph; survey and comprehensive research background.

5. Fahy GM, MacFarlane DR, Angell CA, *et al.* (1984) *Cryobiology* **21**, 407–426.

6. Withers L (1979) *Plant Physiol*. **63**, 460–467.

7. Sakai A, Kobayashi S, Oiyama I (1990) *Plant Cell Rep*. **9**, 30–33.

8. Dereuddre J, Scottez C, Arnaud Y, *et al.* (1990) *C. R. Acad. Sci. Paris* **310 Sér. III**, 317–323.

9. Schäfer-Menuhr A, Schumacher HM, Mix-Wagner G (1997) *Acta Hort*. **447**, 477–482.

10. Halmagyi A, Deliu C, Coste A (2005) *CryoLetters* **26**, 313–322.

11. Matsumoto T, Sakai A, Takahashi C, *et al.* (1995) *CryoLetters* **16**, 189–196.

12. Sakai A (1960) *Nature* **185**, 392–394.

13. Roberts EH (1973) *Seed Science Technol*. **1**, 499–514.

14. Stanwood PC, Bass LN (1978) In: *Plant Cold Hardiness and Freezing Stress*. Edited by PH Li and A Sakai. Academic Press, New York, pp. 361–371.

15. Hanna WW, Towill LE (1995) In: *Plant Breeding Reviews*. Edited by J Janick. Wiley, New York, pp. 179–207.

16. Agarwal DC, Pawar SS, Mascarenhas AF (1993) *J. Plant Physiol.* **142**, 124–126.

17. Engelmann F (2004) *In Vitro Cell Dev. Biol.–Plant* **40**, 427–433.

18. Reed BM (1988) *CryoLetters* **9**, 166–171.

19. Steponkus PL, Lamphear FO (1967) *Plant Physiol.* **42**, 1423–1426.

20. Norton JD (1966) *Proc. Am. Soc. Hort. Sci.* **89**, 132–134.

21. Vertucci CW (1990) *Biophys. J.* **58**, 1463–1471.

22. Carpentier SC, Witters E, Laukens K, *et al.* (2007) *Proteomics* **7**, 92–105.

23. Grout BWW, Henshaw GG (1980) *Ann. Bot.* **46**, 243–248.

24. Morisset C, Gazeau C, Hansz J, *et al.* (1993) *Protoplasma* **173**, 35–47.

25. Ishikawa M, Ide H, Price WS, *et al.* (2000) In: *Cryopreservation of Tropical Plant Germplasm. Current Research Progress and Application*. Edited by F Engelmann and H Takagi. IPGRI, Rome, Italy, pp. 22–35.

26. Fleck RA, Day JG, Rana KJ, *et al.* (1997) *CryoLetters* **18**, 343–354.

27. Towill LE, Widrlechner M (2004) *CryoLetters* **25**, 71–80.

28. Forsline, PL, Towill LE, Waddell JW, *et al.* (1998) *J. Am. Soc. Hort. Sci.* **123**, 365–370.

29. Swan TW, O'Hare D, Gill RA, *et al.* (1999) *CryoLetters* **20**, 325–336.

30. Murashige T, Skoog F (1962) *Physiol. Plant.* **51**, 473–479.

31. Withers LA, King PJ (1980) *CryoLetters* **1**, 213–220.

32. Grenier-de March G, de Boucaud MT, Chmielarz P (2005) *CryoLetters* **26**, 341–348.

33. Morel G, Wetmore RG (1951) *Am. J. Bot.* **38**, 138–140.

34. Rajasekharan PE, Ganeshan S (2003) *Capsicum Eggplant Newsl.* **22**, 87–90.

35. Alexander MP (1980) *Stain Technol.* **55**, 13–18.

36. Stanley RH, Linskens HF (1974) *Pollen: Biology, Biochemistry, Management*. Springer-Verlag, Berlin, Heidelberg.

37. Keller ERJ (2005) *CryoLetters* **26**, 357–366.

38. Senula A, Keller, ERJ, Sanduijav T, *et al.* (2007) *CryoLetters* **28**, 1–12.

39. Towill LE (1983) *Cryobiology* **20**, 567–573.

40. Revilla MA, Martínez D (2002) In: *Biotechnology in Agriculture and Forestry*. Vol. 50. *Cryopreservation of Plant Germplasm*. Edited by LE Towill and YPS Bajaj. Springer-Verlag, Berlin, Heidelberg, Germany, pp. 136–150.

41. Wetmore RH, Sorokin S (1955) *J. Arnold Arboretum* **36**, 305–317.

42. Hirai D, Sakai A (1999) *Plant Cell Rep.* **19**, 150–155.

43. Dussert S, Engelmann F, Noirot M (2003) *CryoLetters* **24**, 149–160.

9

Plant Protoplasts: Isolation, Culture and Plant Regeneration

Michael R. Davey, Paul Anthony, Deval Patel and J. Brian Power
Plant and Crop Sciences Division, School of Biosciences, University of Nottingham, Sutton Bonington Campus, Loughborough, UK

9.1 Introduction

Isolated protoplasts provide experimental material to genetically manipulate plants by somatic hybridization and cybridization, and some transformation procedures. Such experiments consist of three stages, namely, protoplast isolation, the genetic manipulation event involving protoplast fusion or gene uptake and, finally, protoplast culture and regeneration of fertile plants. Additionally, the tissue culture process *per se* may expose naturally occurring somaclonal variation, or in the case of protoplasts, protoclonal variation, which may also be considered as a simple form of genetic manipulation.

In theory, all living plant cells contain the genetic information essential for their development into fertile plants. However, this 'totipotency' is not always expressed since some plant cells lose this ability during culture. Some cells are morphogenically more competent than others. Generally, morphogenic competence is governed by three main factors, these being the plant genotype, the ontogenetic state of the explant source, and the culture environment in which the protoplasts or protoplast-derived cells are maintained. The latter includes the composition of the culture medium and the physical growth conditions.

Protoplasts may be isolated by mechanical disruption or by enzymatic degradation of their surrounding cell walls. Historically, mechanical disruption, involving the slicing of plant tissues, was the first procedure to be exploited. However, because

Plant Cell Culture Edited by Michael R. Davey and Paul Anthony

of the limited number of protoplasts released mechanically, this technique was superseded by enzymatic degradation once suitable cell wall degrading enzymes became commercially available. Enzymatic digestion is now employed routinely for protoplast isolation. Although any primary tissue of most plants is a potential source of protoplasts, the ability to isolate protoplasts capable of cell wall regeneration followed by sustained mitotic division and shoot regeneration, is still restricted to a relatively limited number of genera, species and varieties. In general, leaf tissues from seedlings are used extensively as source material for protoplast isolation. However, sustained mitotic division leading to protoplast-derived tissues from which plants can be regenerated is still not routine for mesophyll-derived protoplasts of many monocotyledons, with the exception of examples in rice and sorghum [1, 2]. Recent progress includes the sustained division of protoplasts isolated from leaves of date palm (*Phoenix dactylifera*) to produce callus [3]. Generally, embryogenic cell suspensions are a source of competent cells for cereals, grasses and other plants. It is frequently observed, when isolating protoplasts directly from leaves, that tissues of young leaves release protoplasts with the highest viability. In this respect, axenic cultured shoots and seedlings are often preferable to glasshouse-grown plants as source material, since it is easier to regulate the growth conditions of the donor plants. Axenic shoots also provide a continuous supply of juvenile tissues, which facilitates protoplast isolation, particularly in woody species.

Technologies that incorporate protoplast-based procedures have declined in the last two decades. Probably, this is because of emphasis on genetic manipulation involving the transfer of specific genes into totipotent target tissues using *Agrobacterium* or Biolistics-mediated gene transfer. However, protoplast isolation and culture remains fundamental to gene transfer by fusion and some aspects of transformation, particularly transient gene expression studies. Importantly, the genetic combinations that can be achieved at the nuclear and organelle levels through protoplast fusion are more extensive than those that result from conventional sexual hybridization. Consequently, breeders should be encouraged to pursue such approaches alongside conventional breeding techniques. The applications, merits and limitations of protoplast-based technologies are discussed in several review articles [4–9].

9.2 Methods and approaches

9.2.1 Protoplast isolation

Enzyme treatment of primary plant tissues

The optimum incubation conditions, and the concentration and combination of enzymes required to release viable protoplasts from any living tissue of a particular plant and the concentration of sugar alcohols used as osmoticum [e.g. 13% (w/v) mannitol] must be determined empirically. The concentration of the osmoticum must be adjusted to maintain the isolated protoplasts in a spherical condition. Insufficient concentration of osmoticum will result in protoplast lysis; an excess osmoticum will induce protoplasts to shrink through plasmolysis. Usually, protoplasts are isolated at 25–28 °C for either a short period of enzyme incubation (e.g. 2–6 h) or a longer period (12–20 h; overnight), generally in the dark. A short plasmolysis treatment, often involving incubation for 1 h in a salts solution (e.g. CPW salts; [10]) with the same osmoticum as the enzyme mixture, but lacking wall-degrading enzymes, is beneficial in maintaining protoplast viability and reducing the extent of spontaneous protoplast fusion during the enzyme treatment. Protoplasts of cells of some tissues are more prone to spontaneous fusion than others, this process involving expansion of plasmadesmata, resulting in coalescence of the cytoplasms of adjacent cells. Enzyme mixtures for protoplast isolation usually consist of pectinases and cellulases, often in complex cocktails, such as the mixture required to release protoplasts from cell suspensions and seedling hypocotyls of *Gentiana kurroo* [11]. Comparative studies may be essential to optimize the most effective combination of enzymes and their concentrations to maximize protoplast release, as in the case of protoplasts of *Ulmus minor* [12]. Pectinases digest the middle lamella between adjacent cells separating the latter, while cellulases remove the walls to release a population of osmotically fragile naked cells (protoplasts). The latter may range from ~20 μm in diameter (e.g. those of rice), to about 50 μm for protoplasts from leaf tissues of plants such as tobacco.

Protoplast purification

After enzyme treatment, a suspension is obtained consisting of released protoplasts, undigested tissue and cellular debris in the enzyme solution. Purification involves passing the suspension through nylon or metal sieves of decreasing pore size to remove the larger material, followed by gentle centrifugation to pellet the protoplasts. Finer debris remains in the supernatant. The protoplast pellets are resuspended in a solution containing an osmoticum of the same concentration as that used in the enzyme mixture [e.g. CPW salts solution containing 13% (w/v) mannitol as the osmoticum; designated CPW13M]. Centrifugation and resuspension of the pelleted protoplasts may need to be repeated several times until a pure suspension of protoplasts is obtained. Preparations of leaf-derived protoplasts are often contaminated with debris which can be removed by resuspending the protoplast-derived pellets, following the initial centrifugation in the enzyme mixture, in a solution of 21% (w/v) sucrose with CPW salts (CPW21S). Following slow-speed centrifugation (e.g. 100 g for 10 min), protoplasts form a band at the meniscus of the sucrose washing solution; other debris forms a pellet or remains suspended. The speed and duration of centrifugation may need to be determined empirically for different protoplast systems. The dense band of protoplasts is carefully removed from the top of the sucrose solution using a Pasteur pipette and, if necessary, the protoplasts are again suspended in the sucrose washing solution and the procedure repeated. Protoplasts are returned to a solution with a sugar alcohol as the osmoticum (e.g. CPW13M solution).

Visualization of the efficiency of cell wall removal and determination of protoplast viability

Isolated protoplasts should have a spherical shape when observed by light microscopy. The absence of birefringence indicates complete enzymatic removal of the cell wall. Staining protoplasts with Calcofluor White [13] or the fluorescent brightener Tinapol [14] will indicate the presence of any remaining wall material. Any remaining cell walls fluoresce yellow when stained with Tinapol, while those stained with Calcofluor White produce an intense blue fluorescence when examined under UV illumination.

Fluorescein diacetate (FDA; [15]) may be used to determine protoplast viability. FDA passes across the plasma membrane of cells but does not fluoresce until cleaved by esterases within the cytoplasm of living cells to release the fluorescent compound fluorescein. The latter remains in the cytoplasm as it is unable to pass out through the plasma membrane. Viable protoplasts fluoresce green/yellow, while non-viable protoplasts remain unstained. The number of viable protoplasts in a preparation can be counted using a haemocytometer.

9.2.2 Protoplast culture

Culture media

The nutritional requirements of protoplasts and cell suspension cultures are usually similar. Consequently, media used to culture protoplasts are often based on those employed for cell culture. Media prepared according to the formulations of Murashige and Skoog (1962; MS; [16]), Gamborg *et al.* (1968; B5; [17]), Kao and Michayluk (1975; [18]) and Kao [19] are used most extensively for protoplast culture, as in examples such as *Lupinus* [20], *Gossypium* [21], *Cucumis* [22] and *Solanum* [23]. However, in order to induce sustained mitotic division in protoplast-derived cells, modifications of the original formulations may be necessary. For example, ammonium ions are detrimental to protoplast survival and have been reduced, as in the culture of protoplasts from cell suspensions of ginger (*Zingiber officinale*) [24], or removed from many protoplast culture media. Microelements and organic components of published formulations may also need to be changed.

Since isolated protoplasts are osmotically fragile, the osmotic pressure of the culture medium is crucial to prevent lysis or plasmolysis of protoplasts, especially during the early stages of culture. The osmotic pressure of the culture medium is adjusted by the addition of sugars (e.g. sucrose, glucose) or sugar alcohols (e.g. mannitol, sorbitol) to the culture medium. Sucrose and glucose are generally employed as carbon sources and also act as osmotica; protoplasts synthesizing new cell walls rapidly remove sugars from the medium, especially during the early stages of their culture. Protoplasts undergo rapid cell wall synthesis immediately upon removal from the enzyme solution. Maltose as the carbon source may increase the frequency of plant regeneration from protoplast-derived tissues, compared to sucrose, especially in cereals such as rice [25, 26]. Rich culture media prepared to the formulations given by Kao and colleagues [18, 19] each contain a range of compounds (sugars) that act as carbon sources.

Procedures for culture of isolated protoplasts

Several procedures are available to culture isolated protoplasts, including their suspension in liquid medium, embedding in a semisolid medium, and suspension in liquid medium overlaying semisolid medium of the same composition. A filter paper or microbial membrane is sometimes included at the interface of the two phases. Liquid media permit more rapid diffusion of nutrients into, and waste products out, of protoplasts during culture, and facilitate reduction of the osmotic pressure to accompany protoplast growth. Media semisolidified with agar or agarose enhance support which encourages cell wall development. Pure, low gelling temperature agaroses, such as SeaPlaque (FMC BioProducts, Rockland, ME, USA) or Sigma types VII and IX are used extensively for protoplast culture. Techniques that are frequently exploited include those detailed below.

Culture in liquid medium Protoplasts are suspended in culture medium at the required plating density and dispensed into culture dishes (e.g. 3, 5 or 9 cm diam. Petri dishes). The latter are sealed with an expandable, gas permeable tape (e.g. Parafilm, Nescofilm). Cultures are incubated in a growth room (e.g. 25 °C) under low intensity illumination (e.g. 7 μmol/m^2/s from 'Daylight' fluorescent tubes) with a suitable photoperiod (e.g. 16 h).

Culture of protoplasts in hanging drops of liquid medium Isolated protoplasts may be cultured in drops of culture medium (each approx. 50 μl in size) hanging from the lids of Petri dishes. Protoplasts sink to the menisci of the droplets where they receive adequate aeration. This approach is useful when culturing protoplasts at low densities and for evaluating the composition of a range of culture media. However, the droplets are time consuming and tedious to prepare.

Embedding of isolated protoplasts in media semisolidified with agar, agarose or alginate Protoplasts are suspended at double the required plating density in liquid culture medium, prepared at twice its final strength, and mixed with the same volume of warm (40 °C) gelling agent prepared in water, also at twice the required final concentration. Protoplasts are suspended in the resulting medium immediately before the medium gels and dispensed into Petri dishes (e.g. 3 or 5 cm diam.), allowed to cool and to become semisolid. The dishes are sealed and incubated as described for liquid cultures. The type and concentration of gelling agent may influence protoplast development as in protoplasts of indica rice where protoplast-derived tissues required transfer from medium semisolidified with 1% (w/v) agarose to medium with 0.4% (w/v) agarose to maximize shoot regeneration [27]. The semisolid agar or agarose layer containing the embedded protoplasts may be cut into sectors and the latter transferred into liquid culture medium of the same composition in larger Petri dishes (e.g. 9 cm diam.). The liquid medium bathes the embedded protoplasts. The molten agarose medium containing the suspended protoplasts may also be dispensed as droplets or beads, each ~25–150 µl in size in the bottom of Petri dishes [12, 28]. After gelling of the medium, the droplets are bathed in liquid medium of the same composition.

Alginate is a useful gelling agent for protoplasts which are heat sensitive (e.g. protoplasts of *Arabidopsis thaliana*). Alginate is also employed if it is necessary to depolymerise the culture medium to release developing protoplast-derived cell colonies. Culture media containing alginate are gelled by exposure to Ca^{2+} ions. Embedded protoplasts are maintained in a medium with a concentration of Ca^{2+} which is just sufficient to keep the alginate semisolidified. Media containing alginate with the suspended protoplasts may be gelled as a thin layer (film) by pouring over an agar layer containing Ca^{2+} ions, as for protoplasts of *Cyclamen persicum* [29], or gelled as beads (each about 50 µl in volume) as in the culture of protoplasts of *Phalaenopsis* [30], by allowing droplets to fall into liquid culture medium containing Ca^{2+} ions. The thickness of the alginate layer influences the growth of embedded protoplasts [31, 32]. If depolymerization of the medium is required to release embedded protoplasts or protoplast-derived cells, the Ca^{2+} may be removed by a brief exposure of the cultures to sodium citrate. The released protoplast-derived cell colonies are washed free of alginate and citrate.

Liquid-over-semisolid medium A layer of semisolid medium is dispensed in the bottom of a Petri dish, allowed to gel and the same volume of liquid medium containing protoplasts at twice the required plating density is poured over the semisolid layer. A filter paper (e.g. Whatman No. 3) or a bacterial membrane at the interface of the two phases, may stimulate cell wall regeneration and sustained mitotic division.

Plating density and nurse cultures

Isolated protoplasts must be cultured at an optimum density, usually $1.0 \times 10^5 - 1.0 \times 10^6$/ml, to ensure cell wall regeneration and sustained mitotic division. A minimum plating (inoculum) density, which may be determined empirically, is essential to ensure protoplast division and sustained growth. Nurse cells may be used to promote protoplast division, particularly when the protoplasts are cultured at low density. For example, nurse cells were essential in promoting growth of protoplasts from cell suspensions of *Lilium japonicum* [33] and shoot regeneration from protoplast-derived tissues of banana [34]. Protoplasts or cells capable of rapid division, from the same genus, species, or cultivar can be used as a nurse culture, with protoplasts or cells from embryogenic cell suspensions being preferable to those of non-embryogenic cultures. Alternatively, nurse cells can be from a different genus or species. For example, protoplasts of red cabbage (*Brassica oleracea*) can be nursed by protoplasts of tuber mustard (*B. juncea* var. *tumida*; [35]). Protoplasts and dividing cells, if used as a nurse culture, must be separated physically from the experimental protoplasts unless they are phenotypically distinct. This can be achieved by spreading the isolated experimental protoplasts in a liquid layer on a membrane (e.g. 0.2–12 µm pore size) laid over a semisolid layer containing the nurse cells or protoplasts [36, 37], or by enclosing the test protoplasts in a cylinder made from a microbial membrane (e.g. 0.2 μm pore size), with the nurse cells in a surrounding liquid layer [38]. However, it is not essential for nurse cells to be capable of mitotic division, since X- or gamma-irradiated cells or protoplasts can also be used as a nurse. In this case, such protoplasts, because they are incapable of sustained growth and division, can be mixed with the protoplasts under investigation, or separated physically from the experimental material.

Whilst nurse cells utilize nutrients from the culture medium, dividing cells/protoplasts also release growth promoting factors, particularly amino acids, into the surrounding culture medium, contributing to the nurse effect. Protoplasts in culture may also be stimulated by 'conditioned' medium, the latter being prepared by culturing protoplasts or cells in liquid medium for a limited time. Subsequently, the protoplasts are removed, the medium filter-sterilized, and used to culture the protoplasts under investigation.

Additional approaches for maximizing protoplast yield and protoplast-derived cells in culture

The development of protoplast-to-plant systems demands optimum cell growth and differentiation. Several novel approaches have been described to maximize the regeneration of plants from protoplast-derived tissues, including electrical stimulation and manipulation of the gaseous environment during culture [7, 39].

The protocols described below provide details of the culture of protoplasts from embryogenic suspensions of a cereal (rice). Emphasis has been given to the Japonica type rice Taipei 309. In general, Japonica-type rice protoplasts/cells are more responsive to culture than those of Indica-type rices. The development of protocols for specific rice cultivars may require an empirical approach, using protocols for japonica rices as a guide. A protocol is also described for isolation and culture of protoplasts from a member of the Solanaceae, namely *Petunia parodii*. Similarly, this protocol can be adapted for protoplasts of other common members of this family. In general, because complex factors regulate plant cell division and growth, each parameter must be optimized to develop an efficient protocol for protoplasts from a target plant.

PROTOCOL 9.1 Initiation of Embryogenic Callus of Rice (*Oryza sativa* cv. Taipei 309)

Equipment and Reagents

- Rice seed of the cv. Taipei 309 (The International Rice Research Station IRRI, The Philippines)[a]
- Fine grain sand paper
- Laminar air flow cabinet
- Jeweller's forceps (No. 9 watchmaker; Arnold R. Horwell, UK)
- Ethanol or methylated spirits

- Spirit burner
- Commercial bleach solution containing about 5% available chlorine (e.g. 'Domestos'; Johnson Diversey UK)
- Heavy duty Duran-type screw capped glass bottles of 100, 200, 500 ml capacity (Schott Glass, UK)
- Sterile (autoclaved) reverse osmosis water[b]
- Sterile containers e.g. Screw-capped glass or plastic Universal bottles (Beatson Clark, UK; Bibby Sterilin, UK)
- 9 cm diameter Petri dishes (Bibby Sterilin, UK)
- Sealing tape e.g. Nescofilm (Bando Chemical Industries, Japan) or Parafilm M (Pechiney Plastic Packaging, USA)
- Linsmaier and Skoog liquid medium (designated LS2.5): Prepare according to the LS formulation [40], but with 1.0 mg/l thiamine HCl and 2.5 mg/l 2,4-dichlorophenoxyacetic acid (2,4-D), and at double strength (twice the required final concentration) [41][c]
- SeaKem Le agarose (FMC BioProducts, USA) in water at 0.8% (w/v)[c]

Method

1. Dehusk the rice seed by gently rolling the dry seed between sheets of fine grade sand paper; store the dehusked seed until required in a screw-capped glass or plastic Universal bottle.
2. Surface sterilise the seed by immersion in 30% (v/v) 'Domestos' bleach solution for 1 h in a suitable container (e.g. 50 ml Duran bottle); wash the seed at least three times with sterile reverse-osmosis water to remove the bleach solution[d].
3. Mix equal volumes of double strength LS2.5 liquid medium with an equal volume of 0.8% (w/v) SeaKem Le agarose at 40 °C[e].
4. Immediately dispense 25 ml aliquots of the diluted molten LS2.5 culture medium into 9 cm diameter Petri dishes and allow the medium to gel.
5. Place surface-sterilised seeds using sterile (flamed) jeweller's forceps on the LS2.5 medium, with eight seeds/9 cm diam. Petri dish. Seal the dishes with Nescofilm or Parafilm M and incubate in the dark at 28 ± 2 °C.
6. Excise and transfer aliquots (each approx. 1 g) of embryogenic callus[f] to new semisolid LS2.5 medium after 28 days from the initiation of cultures, and every 28 days thereafter.

Notes

[a]IRRI is a major resource of rice germplasm from which seeds are available on request.

[b]Dispense approx. 300 ml volumes into 500 ml bottles. Sterilize by autoclaving at 121 °C in saturated steam for 20 min at 100 kPa (1 bar). Use heavy duty Duran-type screw capped bottles to prevent breakage during autoclaving and handling.

[c]Dispense separately 200 ml volumes of double strength LS2.5 liquid medium and SeaKem Le agarose into 500 ml Duran bottles. Autoclave as in Note (b).

[d]The wash water will cease to foam when the bleach has been removed.

[e]LS2.5 culture medium and agarose (dissolved in water) are prepared at double strength and autoclaved separately, prior to being mixed in equal volumes when the agarose is still molten at 40 °C after autoclaving. Alternatively, the agarose can be liquefied by heating in a microwave oven and mixed with the double strength LS2.5 liquid medium before dispensing into Petri dishes. Local rules relating to the use of microwave ovens must be observed.

[f]Embryogenic callus is recognised by its compact and nodular appearance and a yellow/white colouration. Non-embryogenic callus is often mucilaginous. The careful selection of callus of the correct phenotype is essential to establish cultures that maintain their totipotency for the maximum time in both the callus and, subsequently, the cell suspension stages.

PROTOCOL 9.2 Initiation of Embryogenic Cell Suspension Cultures of Rice (*Oryza sativa* cv. Taipei 309)

Equipment and Reagents

- Laminar air-flow cabinet
- Jeweller's forceps (No. 9 watchmaker; Arnold R. Horwell, UK)
- Ethanol or methylated spirits
- Spirit burner
- Sterile 100 and 250 ml Erlenmeyer flasks with aluminium foil closures
- LS2.5 liquid medium: see Protocol 9.1
- AA2 liquid medium: Prepare according to the published formulation [42, 43]; filter sterilize[a]
- Orbital shaker for flasks containing cell suspensions
- Autoclaved nylon sieves (Wilson Sieves, UK) or metal sieves with a pore size of 500 μm

Method

1. Transfer using sterile (flamed) jeweller's forceps aliquots of 1–2 g fresh weight of embryogenic callus to 100 ml Erlenmeyer flasks each containing 25 ml of LS2.5 liquid medium.
2. Incubate the cultures on an orbital shaker at 120 r.p.m. in the dark at 27 ± 2 °C.
3. Replace 80% of the LS2.5 liquid medium in the 25 ml flasks every 5 d, avoiding any loss of cells[b] and reduction in cell density of the rice suspensions.

4 After 42 days, transfer the cell suspensions to 250 ml capacity Erlenmeyer flasks; add 15 ml of LS2.5 liquid medium to each suspension. Every 5 days, allow the cells to settle, remove 30 ml of spent medium and replace with new LS2.5 liquid medium.

5 At the third subculture, pass the cell suspensions through sieves of 500 μm pore size to remove the larger cell aggregates. Discard the large aggregates, but retain the suspensions.

6 After 90–120 days, transfer the cells into the same volume of AA2 liquid medium. Transfer the cultures every 7 days to new liquid medium by mixing 1 vol. of cell suspension with 3 vol. of new medium[c].

Notes

[a]Sterilize AA2 liquid medium by passage through a microbial filter (e.g. Minisart NML; Sartorius, UK) of pore size 0.2 μm.

[b]Remove the flasks from the shaker and allow the cells to settle. Carefully remove 80% of the spent medium by decanting the medium or using a sterile pipette.

[c]Each 250 ml Erlenmeyer flask should contain 10 ml of spent LS2.5 liquid medium with 2–3 ml settled cell volume (SCV) of cells, plus 30 ml of new AA2 medium. The settled cell volume can be determined using Erlenmeyer flasks with graduated side arms (made in a laboratory workshop/glass blowing facility). Gently swirl the cultures to suspend the cells and tilt the flask to fill each side arm with culture. Allow the cells to settle and record the SCV. It is crucial that the correct volume of cells is transferred to new medium at each subculture, otherwise the cultures will not attain their minimum inoculum density to ensure growth.

PROTOCOL 9.3 Isolation of Protoplasts from Embryogenic Cell Suspension Cultures of Rice (*Oryza sativa* cv. Taipei 309)

Equipment and Reagents

- Cell suspension cultures of rice cv. Taipei 309, 3–5 days after subculture, initiated and maintained as described in Protocols 9.1 and 9.2
- Autoclaved nylon or steel sieves with pore sizes of 30, 45, 64 and 500 μm
- CPW13M solution: CPW salts solution with 13% (w/v) mannitol
- Enzyme solution: 0.3% (w/v) Cellulase RS (Duchefa Biochemie BV, The Netherlands), 0.03% (w/v) Pectolyase Y23 (Duchefa), and 0.05 mM MES in CPW13M solution, pH 5.6[a]
- 9 cm diameter Petri dishes
- Sealing tape, e.g. Nescofilm or Parafilm M
- Sterile Pasteur pipettes with teats, 10 ml pipettes and 15 ml centrifuge tubes (Bibby Sterilin, UK)

- Orbital platform shaker
- Haemocytometer (modified Fuchs-Rosenthal; Scientific Laboratory Supplies, UK)
- 0.1% (w/v) Calcofluor White or Tinapol dissolved in CPW13M solution
- Aqueous solution of fluorescein diacetate (FDA) at 3 mg/ml

Method

1. Filter the cell suspension through a nylon sieve of pore size 500 μm into a preweighed 9 cm diam. Petri dish; remove the liquid culture medium with a sterile Pasteur pipette, leaving the cells in the dish.
2. Reweigh the Petri dish and add the appropriate volume of enzyme mixture (10 ml of enzyme solution/g fresh weight of cells).
3. Seal the Petri dish with Nescofilm or Parafilm M and incubate the enzyme/cell mixture on an orbital shaker at slow speed (30 rpm)[b], for 16 h in the dark at 27 ± 2 °C.
4. Filter the protoplast suspension through sieves of 64, 45 and 30 μm pore size to remove undigested cell clumps.
5. Transfer the protoplast suspension to sterile 15 ml centrifuge tubes and wash the protoplasts three times by gentle centrifugation (80 g, 10 min each) and resuspension in CPW13M solution.
6. Resuspend the protoplasts in a known volume (e.g. 10 ml) of CPW13M solution.
7. Count protoplasts using a haemocytometer[c].
8. Stain an aliquot of the protoplast suspension with Tinopal or Calcufluor White to confirm that the cell walls have been digested completely[d].
9. Check the viability of the isolated protoplasts[e].

Notes

[a]The enzyme solution should be pre-filtered, using a nitrocellulose membrane filter (47 mm diam., 0.2 μm pore size [Whatman, UK] to remove insoluble impurities). This prevents blockage of the filter during subsequent sterilization. Pass the enzyme solution through a microbial filter of pore size 0.2 μm (e.g. Minisart NML; Sartorius, UK) before use. Enzyme solutions may be stored at −20 °C until required, but should be frozen and thawed only once before use.

[b]The cells must be agitated on a rotary shaker at slow speed to avoid lysis of the protoplasts during their release. The enzyme/cell mixture should swirl gently in the Petri dish. It is essential that the mixture does not come into contact with the lid of the Petri dish or the space between the base and the lid of the dish as this will result in microbial contamination.

[c]It is crucial to know the number of protoplasts that are isolated to ensure that the protoplasts can be adjusted to the correct density during subsequent culture. Count the protoplasts using a modified Fuchs–Rosenthal haemocytometer. Prepare the haemocytometer by moistening the sides of the chamber and placing on the cover-slip (as

supplied with the haemocytometer), while pressing down towards the chamber. The correct distance between the counting area and the cover-slip is obtained when a diffraction pattern (Newton's rings) is observed where the cover slip makes contact with the body of the haemocytometer. Resuspend the protoplasts in a known volume (usually 10 ml) of solution (e.g. CPW13M solution). Remove a sample with a Pasteur pipette and immediately introduce the sample beneath the cover-slip to fill the counting area. Do not overfill the chamber. Examine the chamber under the light microscope to reveal a grid of small squares with a triple line every fourth line. Each triple lined square encloses 16 smaller squares. Count the number of protoplasts enclosed by a triple lined square (n), including those touching the top and left edges, but not the bottom or right edges. Calculate the number of protoplasts per ml as $5n \times 10^3$; the yield of protoplasts for a total volume of 10 ml is $5n \times 10^4$.

[d]Mix one drop of a 0.1% (w/v) solution of Calcofluor White or Tinapol in CPW salts solution containing 13% (w/v) mannitol (CPW13M) with an equal volume of the protoplast suspension on a microscope slide. Incubate for 5 min. at room temperature. Examine the protoplasts under UV illumination. Any remaining cell walls will fluoresce an intense blue colour with Calcofluor White, and yellow with Tinapol.

[e]Mix 100 µl of a 3 mg/ml stock solution of FDA with 10 ml of CPW13M solution to prepare a working dilution. Mix equal volumes of the working dilution of FDA and the protoplast suspension. Incubate for 5 min at room temperature. Examine the protoplasts using UV illumination. Viable protoplasts will fluoresce yellow-green.

PROTOCOL 9.4 Culture and Regeneration of Plants from Protoplasts Isolated from Embryogenic Cell Suspension Cultures of Rice (*Oryza sativa* cv. Taipei 309)

Equipment and Reagents

- KPR liquid medium (normal strength): Prepare K8P medium according to the published formulation [19] as modified [38], and supplemented with 3 mg/l 2,4-D[a]
- KPR liquid medium (double strength): As above but at twice the required final concentration[a]
- SeaPlaque agarose in water at 24 g/l[a]
- MSKN liquid medium (double strength): Prepare MS-based medium [16] with 2.0 mg/l α-naphthaleneacetic acid (NAA), 0.5 mg/l zeatin and 30g/l sucrose, at twice the required final concentration[a]
- SeaKem Le agarose in water at 8.0 g/l[a]
- Jeweller's forceps (No. 9 watchmaker; Arnold R. Horwell, UK)
- Ethanol or methylated spirits
- Spirit burner
- Water bath

- Ice bath
- 3.5 and 5 cm diam. Petri dishes
- Sealing tape, e.g. Nescofilm or Parafilm M
- 50 ml capacity screw-capped glass jars (Beatson Clark, UK); autoclaved
- 9 cm diameter plant pots
- Polythene bags (20 × 30 cm)
- Potting compost: Levington M3 (Fisons, UK), John Innes No. 3 (J. Bentley, UK) and perlite (Silvaperl; J. Bentley, UK)

Method

1 Isolate protoplasts as described in Protocol 9.3.

2 Resuspend the protoplasts in KPR liquid medium (normal strength) at a density of 5.0×10^5/ml in 15 ml screw-capped centrifuge tubes. Heat shock the protoplasts by placing the tubes in a water bath at 45 °C for 5 min; plunge the tubes into ice for 30 s[b].

3 Pellet the protoplasts by centrifugation at 80 g. Remove the supernatant and resuspend the pelleted protoplasts in new KPR liquid medium (normal strength). Repeat this procedure. Pellet the protoplasts.

4 Mix equal volumes of KPR liquid medium (double strength) with SeaPlaque agarose (24 g/l) at 40 °C. Carefully resuspend the protoplasts at a density of 3.5×10^5/ml in the resulting KPR agarose culture medium[c].

5 Immediately dispense 2 ml aliquots of the protoplast suspension in KPR agarose medium into 3.5 cm diam. Petri dishes. Allow the KPR agarose medium with the suspended protoplasts to gel for at least 1 h. Seal the dishes with Nescofilm or Parafilm M and incubate the cultures in the dark at 27 ± 2 °C.

6 After 14 days, divide the agarose layers from each dish into quarters with a sterile scalpel; transfer each quarter to a separate 5 cm diam. Petri dish. Add 3 ml of KPR liquid medium (normal strength) to each dish. Incubate the cultures in the dark at 27 ± 1 °C until cell colonies develop from the embedded protoplasts[d].

7 Mix MSKN liquid medium (double strength) with SeaPlaque agarose (24 g/l) at 40 °C. Immediately dispense 20 ml aliquots into 9 cm diam. Petri dishes. Allow the medium to gel for at least 1 h.

8 Transfer protoplast-derived cell colonies[e] using sterile (flamed) jeweller's forceps to the semisolid MSKN agarose medium. Seal and incubate the culture as in step 5.

9 Mix MSKN liquid medium (double strength) with SeaKem Le agarose (8 g/l) at 40 °C. Immediately dispense 20 ml aliquots into autoclaved screw-capped 50 ml glass jars[f].

10 After 7–14 days, transfer somatic embryo-derived shoots with coleoptiles and roots to MKN medium semisolidified with 4 g/l SeaKem Le agarose from step 9 (one shoot per jar). Incubate at 25 ± 1 °C in the light (50 μmol/m^2/s, 16 h photoperiod, 'Daylight' fluorescent tubes).

11 Remove rooted plants from the jars and gently wash their roots free of semisolid culture medium. Transfer the plants to compost[g] in 9 cm diam. pots, water the plants and cover with polythene bags. Stand the pots in trays containing water to a depth of approx. 10 cm in a controlled environment room (27 ± 2 °C with a 12 h photoperiod, 180 μmol/m^2/s, 'Daylight' fluorescent tubes).

12 After 7 days, remove one corner from each bag and a second corner 3 days later. Continue to open gradually the top of the bags during the next 10 days. Remove the bags after 21 days[h].

13 Maintain the protoplast-derived plants in a controlled environment room at 27 ± 1 °C with an 18 h photoperiod provided by mercury vapour lamps (310 μmol/m^2/s, Venture HiT 400 W/u/Euro/4K Kr85; Ventura Lighting International, USA). Transfer to glasshouse/field conditions as appropriate.

Notes

[a]See Protocol 9.1, Notes b, c.

[b]Heat shock increases the number of protoplast-derived cells forming cell colonies and, hence, the plating efficiency. The latter is defined as the number of protoplast-derived cell colonies that develop expressed as a percentage of the number of isolated protoplasts introduced into culture. This treatment probably synchronizes mitosis in some of the protoplast-derived cells.

[c]Adjust the volume of the molten medium to ensure that the final required plating density is achieved.

[d]Some protoplast-derived cell colonies will remain in the agarose medium, while others will become free floating in the liquid medium bathing the semisolid KPR medium.

[e]Transfer protoplast-derived cell colonies that are embedded/attached to the semisolid medium and those that are free floating to the surface of MSKN medium. Colonies may be selected with a pair of fine jeweller's forceps and transferred to new medium.

[f]The gelling agent may be changed from SeaPlaque agarose to SeaKem Le agarose at this stage. The purity of the agarose is not so critical at this stage, enabling less expensive SeaKem Le agarose to be used.

[g]Use a 6:1:1 by vol. mixture of Levington M3 compost, John Innes No. 3 compost and perlite to pot the rooted plants.

[h]Acclimation of protoplast-derived plants to *ex vitro* conditions is an exacting part of the schedule and is a stage when major plant losses may occur. Plants must to be checked twice daily to ensure that they do not desiccate, as they will have inadequately developed cuticles and poorly functioning stomata when transferred from culture.

PROTOCOL 9.5 Isolation of Protoplasts from Leaves of Glasshouse-Grown Seedlings of *Petunia parodii*

Equipment and Reagents

- Seeds and glasshouse-grown seedlings of *Petunia parodii*

- Glass casserole dish: wrapped in a heat-sealed nylon bag (Westfield Medical, UK); autoclaved
- Autoclaved reverse osmosis water: 300 ml aliquots in 500 ml Duran bottles
- Commercial bleach solution containing ca. 5% available chlorine (e.g. 'Domestos'; Johnson Diversey UK)
- White ceramic tiles wrapped in aluminium foil, placed in nylon bags and autoclaved
- Autoclaved reverse osmosis water[a]
- Jeweller's forceps (No. 9, watchmaker; Arnold R. Horwell, UK)
- Ethanol or methylated spirits
- Spirit burner
- CPW13M solution: CPW salts solution with 13% (w/v) mannitol
- CPW21S solution: CPW salts solution with 21% (w/v) sucrose
- Enzyme mixture: 1.5% (w/v) Meicelase (Meiji Seika Kaisha, Japan), 0.05% (w/v) Macerozyme R10 (Yakult Honsha, Japan) in CPW13M solution, pH 5.6
- 14 cm diameter Petri dishes (Bibby Sterilin, UK)
- Sealing tape, e.g. Nescofilm or Parafilm M
- Sterile Pasteur pipettes, 10 ml pipettes and 15 ml screw-capped centrifuge tubes (Bibby Sterilin, UK)
- Autoclaved nylon or steel sieves with pore sizes of 45 and 80 μm
- Bench top centrifuge, e.g. Centaur 2 (MSE, UK)

Method

1 Detach fully expanded leaves (approx. 30) from glasshouse-grown plants of *Petunia parodii*[a].

2 Place the leaves in a sterile casserole dish and surface sterilize the leaves by immersion in 8% (v/v) 'Domestos' bleach solution for 20 min. Wash the leaves thoroughly with at least three changes of sterile, reverse-osmosis water[b].

3 Transfer a leaf to the surface of a sterile ceramic tile and remove the lower epidermis by peeling with the aid of sterile (flamed) jeweller's forceps[c]. Excise leaf pieces (peeled areas only) with a sterile scalpel[d] and place the explants with their exposed mesophyll and palisade tissues on the surface of 30 ml of CPW13M solution in a 14 cm diameter Petri dish. Repeat the procedure until all the leaves have been used.

4 When the surface of the CPW13M solution is covered with leaf explants, remove the solution with a Pasteur pipette and replace with 25 ml of enzyme mixture[e]. Seal the Petri dish with Nescofilm and incubate at $25 \pm 2\,^{\circ}C$ for 16 h in the dark.

5 Following incubation, release the protoplasts by gently manipulating the leaf tissues with a pair of flamed jeweller's forceps[f].

6 Filter the enzyme-protoplast mixture through sterile (autoclaved) nylon sieves of 80 and 45 μm pore size into a 14 cm diameter Petri dish[g].

7 Transfer the protoplast suspension using a sterile Pasteur pipette to 15 ml screw-capped tubes[h] and centrifuge at 80 g for 10 min. Discard the supernatant and resuspend the protoplast pellets very gently in CPW21S solution. Repeat the centrifugation and collect the protoplasts from the surface of the solution using a Pasteur pipette[i].

8 Transfer the protoplasts to a measured volume of CPW13M solution in a centrifuge tube[j] and count the yield of protoplasts (see Protocol 9.2, step 7).

Notes

[a]Store seeds in a refrigerator at 5 °C. Seedlings are best grown in modules in good quality compost (e.g. Levington M3; Fisons, UK) in a controlled environment room or glasshouse. Plant material grown in a glasshouse under a natural photoperiod may require supplementary illumination (e.g. 180 μmol/m^2/s from 'Daylight' fluorescent tubes; 16 h photoperiod). Plants must be growing rapidly when used for experimentation, usually 5–6 weeks after sowing of the seed. Leaves of the plants must be free from diseases and pests.

[b]See Protocol 9.1, Note (b).

[c]Insert the points of the forceps at the junction of the midrib and the main veins on the underside of the leaf. Keep the tips of the forceps as near to the leaf surface as possible and gently pull away the epidermis. Repeat the procedure until the lower surface of each leaf has been removed to expose the underlying photosynthetic tissues.

[d]Change the scalpel blade frequently to ensure precise cutting rather than tearing and bruising of leaf material.

[e]Placing the peeled leaf explants on the surface of CPW13M solution prevents the explants from drying and also plasmolyses the cells, severing plasmodesmata connections between adjacent cells. This reduces spontaneous fusion of protoplasts from adjacent cells.

[f]Gently rolling and squeezing the leaf explants in the enzyme solution will release protoplasts.

[g]Very gently suck the suspension containing the released protoplasts into a Pasteur pipette. Holding the pipette at an angle of 45 ° will allow the suspension to 'flow' into the pipette, reducing the formation of air bubbles and minimizing protoplast lysis. Leaf-derived protoplasts must be handled with care; they burst easily because of the chloroplasts in the cytoplasm.

[h]When dispensing the protoplast suspension from the Pasteur pipette, hold the receiving tube at an angle of 45 ° and gently run the protoplast suspension down the inner wall of the tube.

[i]Very gently suck the protoplasts into the Pasteur pipette as described in Note (g).

[j]Very gently and slowly, resuspend the protoplasts in CPW13M solution to enable the protoplasts to accommodate the change from a sucrose-based to a mannitol-based osmoticum, limiting protoplast lysis.

PROTOCOL 9.6 Culture and Plant Regeneration from Leaf Protoplasts of *Petunia parodii*

Equipment and Reagents

- Freshly isolated leaf protoplasts of *Petunia parodii*
- 9 cm diam. Petri dishes
- Sealing tape, e.g. Nescofilm or Parafilm M
- MSP1 liquid medium: MS-based medium [16] with 2.0 mg/l NAA and 0.5 mg/l benzylaminopurine (BAP)
- MSP19M liquid medium: MS-based medium with 2.0 mg/l NAA, 0.5 mg/l BAP and 9% (w/v) mannitol
- MSP19M agar medium: As above with the addition of 1.2% (w/v) agar (Sigma)
- MSZ medium: MS-based culture medium with 1.0 mg/l zeatin
- Jeweller's forceps: No. 9, Watchmaker (Arnold R. Horwell, UK)
- Ethanol or methylated spirits
- Spirit burner
- Sterile Pasteur pipettes and 15 ml screw-capped centrifuge tubes
- Bench top centrifuge, e.g. Centaur 2 (MSE, UK).
- Haemocytometer: see Protocol 9.3
- Pots and potting composts: see Protocol 9.4

Method

1. Transfer the protoplast suspension from Protocol 9.5, step 8, using a Pasteur pipette, to 15 ml screw-capped tubes and centrifuge at 80 g for 10 min. Discard the CPW13M solution and resuspend the protoplast pellets very gently in MSP19M medium at a final density of 1×10^5 protoplasts/ml (see Protocol 9.3, step 7).
2. Dispense 8 ml aliquots of molten (40 °C) MSP19M medium with 1.2% (w/v) agar into 9 cm Petri dishes and allow the medium to gel.
3. Dispense 8 ml aliquots of protoplast suspension in MSP19M liquid medium over the surface of MSP19M agar medium in 9 cm Petri dishes to give a final plating density[a] of 5×10^4 protoplasts/ml. Seal the dishes with Nescofilm or Parafilm M.
4. Incubate the cultures at 25 ± 2 °C under low intensity continuous illumination of 20 μmol/m^2/s, 'Daylight' fluorescent tubes.
5. After approx. 50 days of culture, transfer protoplast-derived colonies using sterile (flamed) jeweller's forceps to 20 ml aliquots of 0.8% (w/v) agar-solidified MSP1

medium in 9 cm Petri dishes (30 colonies per dish). Maintain the cultures at 25 ± 2 °C in the light (50 μmol/m^2/s; 16 h photoperiod, 'Daylight' fluorescent tubes).

6 Transfer protoplast-derived callus to 50 ml aliquots of MSZ medium semisolidified with 0.8% (w/v) agar (Sigma) in 175 ml capacity glass jars. Incubate under the same conditions as in step 5.

7 After 21–28 days of culture, excise the regenerated shoots and transfer to 50 ml aliquots of MS-based agar medium, lacking growth regulators, in 175 ml glass jars to induce roots on the regenerated shoots. Incubate as in step 5.

8 Transfer the regenerated plants to *ex vitro* conditions as in Protocol 9.4, steps 1–13, but do not stand the potted plants in trays of water.

Note

[a]The final plating density must be calculated on the total volume of the liquid and semisolid layers of medium in each Petri dish.

9.3 Troubleshooting

- Laboratory working areas and culture rooms must be clean and tidy at all times to minimize the possibility of microbial contamination. All experiments must be performed with reference to local guidelines of safety and good laboratory practice.

- Adequate supplies of materials must always be available prior to the commencement of experiments. In particular, culture media and solutions that must be sterile should be prepared in advance of experiments (often several days), incubated at room temperature for 7–14 days before use in order to check for microbial contamination. Incubation of samples of culture media in Luria broth [44] at 37 °C, should reveal the presence of any contaminating microorganisms.

- Isolated plant protoplasts are 'naked' cells, each bounded only by the plasma membrane. Consequently, they are extremely fragile and all preparations must be handled with care, for example, during pipetting of suspensions and embedding in semisolid culture media. Chloroplast-containing protoplasts isolated from leaves are especially prone to lysis; those from cell suspensions are more robust.

- Source material must be in excellent condition. Seedlings must be actively growing and free from contamination. Those of plants such as Petunia, must not be flowering when used for protoplast isolation. Cultured cells should be actively dividing and in exponential growth when harvested as a source of protoplasts.

- The age of the cell suspensions used as a source of protoplasts is crucial to successful isolation of totipotent protoplasts. In the case of rice, cell suspensions may remain totipotent for about 10 months. Totipotency declines rapidly after this time. It may be essential to initiate new callus and cell suspensions at frequent intervals (e.g. every 6 months) in order to ensure a totipotent source of protoplasts. Alternatively, totipotent cells may be harvested from suspension, cryopreserved [45] and subsequently reinstated in suspension as required.

- It may be necessary to develop empirically enzyme mixtures to isolate protoplasts from a specific target plant, especially if an enzyme mixture has not been described previously in the literature. Enzyme mixtures used for protoplast isolation may be decanted into small volumes (e.g. 10–20 ml) following filter sterilization and stored at −20 °C. Powdered enzymes purchased from suppliers should also be stored at this temperature until required.

- Culture media containing a gelling agent (e.g. agarose) must be at 35–40 °C when used to resuspend and dispense protoplasts into culture vessels. Following addition of the culture medium to pellets of protoplasts, the latter can be gently resuspended by gentle inversion of tubes containing the mixture of protoplasts and the molten culture medium. These procedures must be performed rapidly but carefully before the medium becomes semisolid. Temperatures above 40 °C must be avoided.

- Inexpensive nylon sieves, of various pore sizes, for filtration of cell and protoplast suspensions, may be obtained from Wilson Sieves, Nottingham, UK. These sieves are useful for removing cellular debris. Excessive debris in the protoplasts cultures results in phenolic oxidation and a subsequent reduced plating efficiency of the cultured protoplasts.

References

1. Gupta HS, Pattanayak A (1993) *Bio/Technology* **11, 90–94.

The first report of plant regeneration from leaf protoplasts of a monocotyledon, namely rice.

2. Sairam RV, Seetharama N, Devi PS, Verma A, Murthy UR, Potrykus I (1999) *Plant Cell Rep*. **18, 972–977.

The first report of plant regeneration from leaf protoplasts of sorghum.

3. Chabane D, Assani A, Bouguedoura N, Haicour R, Ducreux G (2007) *Comptes Rend. Biol*. **330**, 392–401.

4. Davey MR, Power JB, Lowe KC (2000) In: *Encyclopedia of Cell Technology*. Edited by RE Spier. John Wiley and Sons, New York, pp. 1034–1042.

5. Davey MR, Anthony P, Power JB, Lowe KC (2005) *In Vitro Cell Dev. Biol.-Plant* **41**, 202–212.

6. Davey MR, Anthony P, Power JB, Lowe KC (2005) In: *Journey of a Single Cell to a Plant*. Edited by SJ Murch SJ and PK Saxena. Science Publishers, Inc., Enfield, NH, USA, pp. 37–57.

7. Davey MR, Anthony P, Power JB, Lowe KC (2005) *Biotechnol. Adv*. **23, 131–171.

A comprehensive review of the isolation, culture and applications of protoplast technology.

**8. Fehér A, Pasternak TP, Ötuüs K, Dudits D (2005) In: *Journey of a Single Cell to a Plant*. Edited by SJ Murch SJ and PK Saxena. Science Publishers, Inc., Enfield (NH), USA, pp. 59–89.

Excellent background information on the consequences of cell wall removal to release viable protoplasts.

**9. Pauls KP (2005) In: *Journey of a Single Cell to a Plant*. Edited by SJ Murch and PK Saxena. Science Publishers, Inc., Enfield, NH, USA, pp. 91–132.

A summary of some of the biotechnological applications of isolated plant protoplasts.

10. Frearson EM, Power JB, Cocking EC (1973) *Dev. Biol*. **33, 130–137.

A key paper describing the use of CPW salts solution in protoplast isolation.

11. Fink A, Rybczynski JJ (2007) *Plant Cell Tissue Organ Cult*. **91**, 263–271.

12. Conde P, Santos C (2006) *Plant Cell Tissue Organ Cult*. **86**, 359–366.

*13. Galbraith DW (1981) *Bio/Technology* **3**, 1104–106.

The use of Calcofluor White to visualize cell walls.

*14. Cocking EC (1985) *Physiol. Plant*. **53**, 111–116.

The use of Tinapol to visualize cell walls.

*15. Widholm J (1972) *Stain Technol*. **47**, 186–194.

Fluorescein diacetate to estimate the viability of isolated protoplasts.

***16. Murashige T, Skoog F (1962) *Physiol. Plant*. **56**, 473–497.

The formulation of the classic MS-based culture medium for plant tissue culture.

17. Gamborg OL, Miller RA, Ojima K (1968) *Exp. Cell Res*. **50, 151–158.

Gamborg's B5 culture medium for isolated protoplasts.

18. Kao KN, Michayluk MR (1975) *Planta* **126, 105–110.

Formulation of rich media for protoplast culture.

19. Kao KN (1977) *Mol. Gen. Genet*. **150, 225–230.

Rich media for protoplast culture.

20. Sonntag K, Ruge-Wehling B, Wehling P (2009) *Plant Cell Tissue Org. Cult*. **96**, 297–305.

21. Wang J, Sun Y, Yan S, Daud MK, Zhu S (2008) *Biol. Plant*. **52**, 616–620.

22. Gajdova J, Navratilova B, Smolna J, Lebeda A (2007) *J. Appl. Bot. Food Qual*. **81**, 1–6.

23. Oda N, Isshiki S, Sadohara T, Ozaki Y, Okubo H (2006) *J. Fac. Agric. Kyushu Univ*. **51**, 63–66.

24. Guo YH, Bai JH, Zhang Z (2007) *Plant Cell Tissue Org. Cult*. **89**, 151–157.

25. Biswas GCG, Zapata FJ (1992) *J. Plant Physiol.* **141**, 470–475.

26. Jain RK, Davey MR, Cocking EC, Wu R (1997) *J. Exp. Bot.* **308**, 751–26.

27. Tang KX, Zhao EP, Hu QN, Yao JH, Wu AZ (2000) *In Vitro Cell. Dev. Biol.-Plant* **36**, 362–365

28. Sinha A, Caligari PDS (2005) *Ann. Appl. Biol.* **146**, 441–448.

29. Winkelmann T, Orange ANS, Specht J, Serek M (2008) *Prop. Ornamental Plants* **8**, 9–12.

30. Shrestha BR, Tokuhara K, Mii M (2007) *Plant Cell Rep.* **26**, 719–725.

31. Pati PK, Sharma M, Ahuja PS (2005) *Protoplasma* **226**, 217–221.

32. Pati PK, Sharma M, Ahuja PS (2008) *Protoplasma* **233**, 165–171.

33. Komai F, Morohashi H, Horita M (2006) *In Vitro Cell. Dev. Biol.-Plant* **42**, 252–255.

34. Assani A, Chabane D, Foroughi-Wehr B, Wenzel G (2006) *Plant Cell Tissue Organ Cult.* **85**, 257–264.

35. Chen L-P, Zhang M-F, Xiao Q-B, Wu J-G, Hirata Y (2004) *Plant Cell Tissue Organ Cult.* **77**, 133–138.

36. Lee SH, Shon YG, Kim CY, *et al.* (1999) *Plant Cell Tissue Organ Cult.* **57**, 179–187.

37. Azhakanandam K, Lowe KC, Power JB, Davey MR (1997) *Enzyme Microb. Technol.* **21**, 572–577.

*38. Gilmour DM, Golds TJ, Davey MR (1989) In: *Biotechnology in Agriculture and Forestry*. Vol. 8. *Plant Protoplasts and Genetic Engineering*. Edited by YPS Bajaj. Springer-Verlag, Heidelberg, pp. 370–388.

The use of microbial membranes to separate protoplasts from nurse cell in culture.

39. Rakosy-Tican E, Aurori A, Vesa S, Kovacs KM (2007) *Plant Cell Tissue Organ Cult.* **90**, 55–62.

40. Linsmaier EM, Skoog F (1965) *Physiol. Plant.* **18, 100–127.

Formulation of a well-established medium for plant cell cultures.

41. Thompson JA, Abdullah R, Cocking EC (1986) *Plant Sci.* **49**, 123–133.

42. Müller AJ, Grafe R (1978) *Mol. Gen. Genet.* **161**, 67–76.

43. Abdullah R, Thompson JA, Cocking EC (1986) *Bio/Technology* **4**, 1087–1090.

44. Sambrook J, Russell D (2001) *Molecular Cloning: A Laboratory Manual*, 3rd edn. Cold Spring Harbor Laboratory Press, Cold Spring Harbor, NY, USA.

45. Lynch PT, Benson EE, Jones J, Cocking EC, Power JB, Davey MR (1994) *Plant Sci.* **98**, 185–192.

10

Protoplast Fusion Technology – Somatic Hybridization and Cybridization

Jude W. Grosser[1], Milica Ćalović[1] and Eliezer S. Louzada[2]

[1] *University of Florida IFAS, Citrus Research and Education Center, Lake Alfred, FL, USA*
[2] *Texas A&M University, Weslaco, TX, USA*

10.1 Introduction

Plant somatic hybridization via protoplast fusion has become an important tool in plant improvement, allowing researchers to combine somatic cells (whole or partial) from different cultivars, species or genera resulting in novel genetic combinations including symmetric allotetraploid somatic hybrids, asymmetric somatic hybrids or somatic cybrids. This technique can facilitate breeding and gene transfer, bypassing problems sometimes associated with conventional sexual crossing, including sexual incompatibility, polyembryony and male or female sterility. The pioneer of plant protoplasts, Edward C. Cocking, initiated this technology with his landmark paper on plant protoplast isolation published in *Nature* [1]. Since the first successful report on somatic hybridization with tobacco in 1972 [2], hundreds of reports have been published during the past three decades which extend the procedures to additional plant genera and evaluate the utilization potential of somatic hybrids in many crops, including rice, rapeseed, tomato, potato and citrus. Some key papers published during the evolution of this technology include those listed [3–13]. Plant somatic hybridization has been reviewed several times in general [3, 13–15] and, specifically, for citrus [16, 17] and potato [18]. Key reviews are also available that focus on somatic cybridization and organelle inheritance [8, 19], with the latter

Plant Cell Culture Edited by Michael R. Davey and Paul Anthony

reference also featuring current methodologies for molecular characterization of somatic hybrid and cybrid plants.

10.2 General applications of somatic hybridization

Applications of somatic hybridization in crop improvement are constantly evolving. Most original experiments targeted gene transfer from wild accessions to cultivated selections that were either difficult or impossible by conventional methods. The most common target using somatic hybridization is the generation of symmetric hybrids that contain the complete nuclear genomes of both parents (see Figure 10.1). Somatic hybrid recovery following protoplast fusion is often facilitated by hybrid vigour [20]. In rare cases, a new somatic hybrid may have direct utility as an improved cultivar [21]. However, the most important application of somatic hybridization is the building of novel germplasm as a source of elite breeding parents for various types of conventional crosses. This is especially true in citrus where somatic hybridization is generating key allotetraploid breeding parents for use in interploid crosses to generate seedless triploids [22]. Successful somatic hybridization in citrus rootstock improvement has allowed the creation of a rootstock breeding programme at the tetraploid level that achieves maximum genetic diversity in zygotic progeny and has great potential for controlling tree size [23]. Much of the excitement generated from somatic hybridization, has been the expanded opportunities for wide hybridization especially the production of intergeneric combinations that maximize genetic diversity [5, 24–29]. Many somatic hybrids have been produced to access genes that confer disease resistance [30, 31].

Somatic cybridization is the process of combining the nuclear genome of one parent with the mitochondrial and/or chloroplast genome of a second parent [19, 32]. Cybrids can be produced by the donor-recipient method [33, 34] (see Figure 10.2) or by cytoplast–protoplast fusion [35] but can also occur spontaneously from intraspecific, interspecific or intergeneric symmetric hybridization [36]. This is a common phenomenon in some species especially, tobacco and citrus. In interspecific asymmetric somatic hybridization in *Nicotiana*, half of all regenerated plants were confirmed to be cybrids [14]. Citrus cybrids frequently occur as a by-product from the application of standard symmetric somatic hybridization procedures [17, 36, 37]. A primary target of somatic cybridization experiments has been the transfer of cytoplasmic male sterility (CMS) to facilitate conventional breeding [34, 38–45] or to produce seedless fruit [37].

In addition to somatic cybridization discussed above, incomplete asymmetric somatic hybridization also provides opportunities for transfer of fragments of the nuclear genome, including one or more intact chromosomes from one parent (donor) into the intact genome of a second parent (recipient) [46, 47]. The evolution of techniques to facilitate partial genome transfer also has a long history [48]. This approach advanced with the development of microprotoplast mediated chromosome transfer (MMCT), first established in plants by Ramulu *et al.* [49] by fusing protoplasts of *Lycopersicon peruvianum* (L.) Mill. with microprotoplasts of transgenic potato (*Solanum tuberosum* L.). From this fusion, a hybrid of *Lycopersicon*

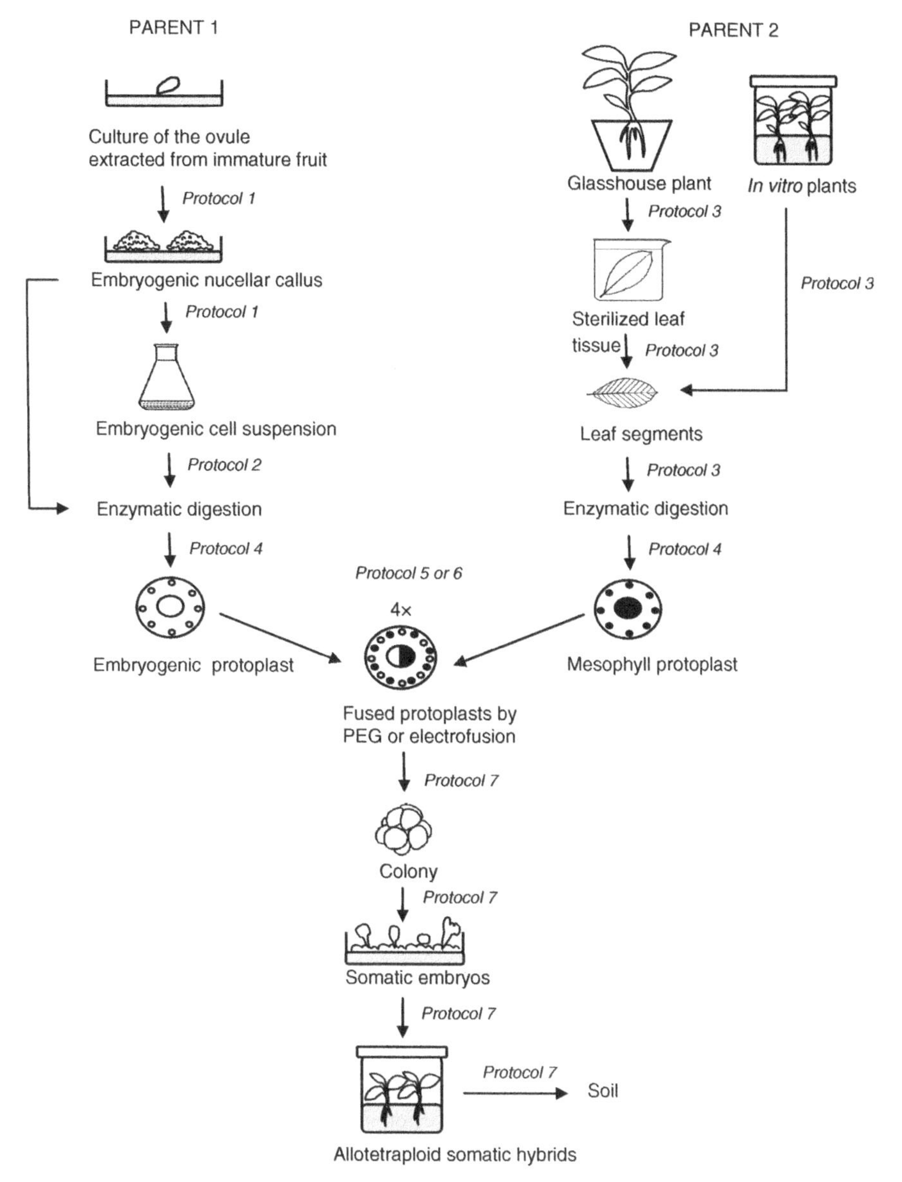

Figure 10.1 Steps in symmetric protoplast fusion in citrus. Parent 1 is introduced into culture as embryogenic nucellar callus derived from the ovule (Protocol 10.1) and as cell suspensions (Protocol 10.1) from which embryogenic protoplasts are obtained through processes of enzymatic digestion (Protocol 10.2) and protoplast isolation (Protocol 10.4). Similarly, leaves from Parent 2 (glasshouse and/or in vitro plants) serve as a source for obtaining mesophyll protoplasts through enzymatic digestion (Protocol 10.3) and protoplast isolation (Protocol 10.4). Protoplasts from both parents are fused using PEG (Protocol 10.5) or electrofusion (Protocol 10.6) and the main products are heterokaryons that combine cytoplasms and nuclei from both parents. Creation of symmetric allotetraploid somatic hybrids from these heterokaryons goes through phases involving the formation of cell colonies, calli, somatic embryos and plantlets that can eventually be transferred to soil (Protocol 10.7).

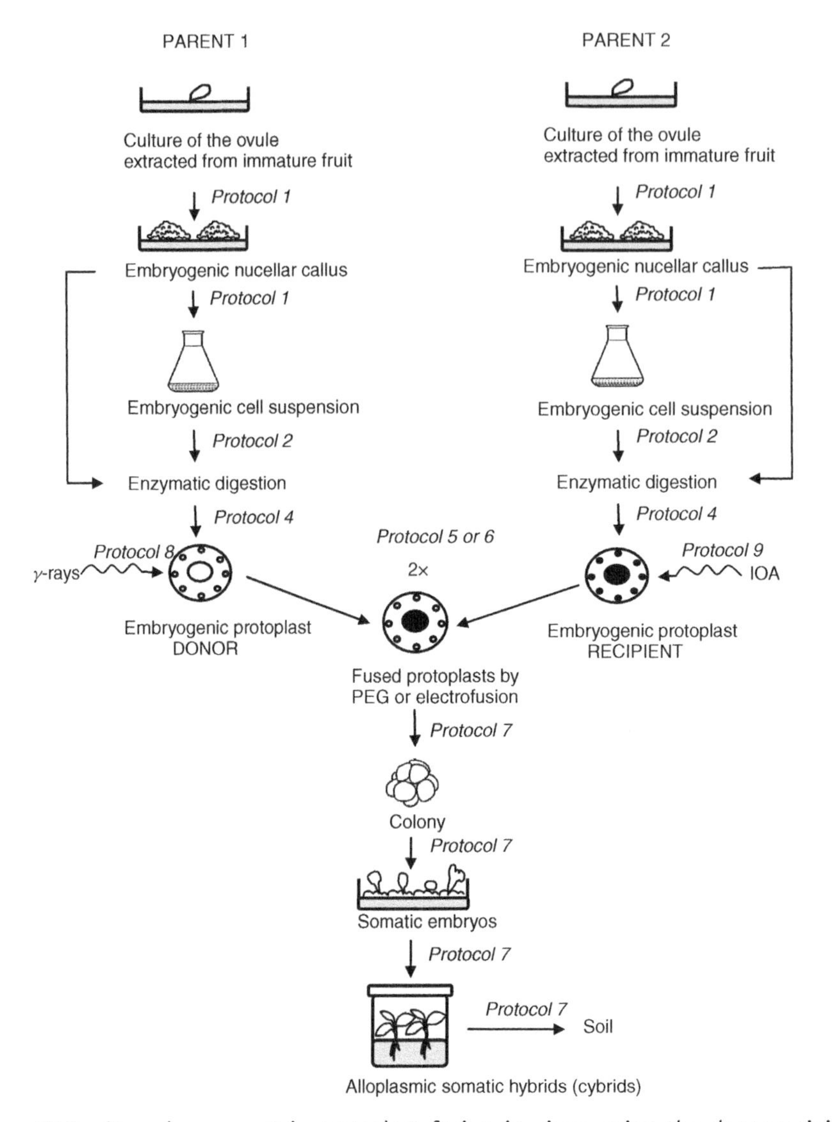

Figure 10.2 Steps in asymmetric protoplast fusion in citrus using the donor-recipient method. Parents 1 and 2 are introduced separately into culture as embryogenic nucellar calli derived from the ovule (Protocol 10.1) and as cell suspensions (Protocol 10.1) from which embryogenic protoplasts are obtained through processes of enzymatic digestion (Protocol 10.2) and protoplast isolation (Protocol 10.4). Embryogenic protoplasts from Parent 1 are irradiated with gamma-rays (Protocol 10.8) for nuclei destruction and cytoplasm donation, while recipient embryogenic protoplasts (Parent 2) are treated with iodoacetic acid (IOA) for metabolical inhibition of organelle genomes (Protocol 10.9). Treated protoplasts from both parents are fused using PEG (Protocol 10.5) or electrofusion (Protocol 10.6) and the main products are heterokaryons that combine cytoplasm from the donor parent with the intact nucleus from the recipient parent. The creation of alloplasmic somatic hybrids (cybrids) from these heterokaryons goes through multiple phases involving formation of colonies, calli, somatic embryos and plantlets that can eventually be transferred to compost (Protocol 10.7).

peruvianum containing one chromosome of potato was obtained. Later Binsfeld *et al.* [50] obtained hybrid plants of common sunflower (*Helianthus annuus* L.) containing from two to eight chromosomes of Maximilian sunflower (*Helianthus maximiliani* L.) or giant sunflower (*Helianthus giganteus* L.) using MMCT without any selection pressure. This proved that there is no requirement of designed pressure to maintain donor chromosomes in the recipient background. In citrus, microprotoplast isolation was first accomplished by Louzada *et al.* [51] and embryos of sweet orange containing a few additional chromosomes from sour orange were obtained. The presence of a high concentration of cytochalasin B was later determined to be the cause of non-regeneration of embryos (unpublished data). Recently, Zhang *et al.* [52] isolated microprotoplasts of satsuma mandarin (*Citrus unshiu*), containing one or a few chromosomes, further expanding the possibilities of using this technique for gene transfer and the creation of novel genetic diversity.

The following protocols were developed for citrus and have been very successful. They were developed with the goal of minimizing genetic specificity. The application of these protocols has resulted in the regeneration of somatic hybrid plants from more than 500 parental combinations and somatic cybrids from more than 50 combinations. The protocols can be easily fine-tuned and adapted to other plant genera and species [53], as evidenced by successes in avocado [54], and grape [55]. Successful protoplast culture media for a specific plant species (for citrus protoplast culture medium is 0.6 M BH3 liquid medium; see Protocol 10.2) can be developed by combining the previously successful tissue culture basal medium for the given species with appropriate osmoticum and the 8P multivitamin and sugar alcohol additives of Kao and Michayluk [56]. Subsequent plant regeneration schemes should be dependent on growth regulator combinations, already developed for any given species.

10.3 Methods and approaches

PROTOCOL 10.1 Initiation and Maintenance of Embryogenic (Callus and Cell Suspension) Cultures [16]

Equipment and Reagents

- Sterilization solution: 20% (v/v) commercial bleach solution
- Rotary shaker in plant growth chamber at 28 ± 2 °C
- Laminar flow cabinet
- Autoclave
- BH3 macronutrient stock: 150 g/l KCl, 37 g/l $MgSO_4.7H_2O$, 15 g/l KH_2PO_4, 2 g/l K_2HPO_4; dissolve in H_2O and store at 4 °C
- Murashige and Tucker (MT) macronutrient stock [57]: 95 g/l KNO_3, 82.5 g/l NH_4NO_3, 18.5 g/l $MgSO_4.7H_2O$, 7.5 g/l KH_2PO_4, 1 g/l K_2HPO_4; dissolve in H_2O and store at 4 °C

- MT micronutrient stock: 0.62 g/l H_3BO_3, 1.68 g/l $MnSO_4.H_2O$, 0.86 g/l $ZnSO_4.7H_2O$, 0.083 g/l KI, 0.025 g/l $Na_2MoO_4.2H_2O$, 0.0025 g/l $CuSO_4.5H_2O$, 0.0025 g/l $CoCl_2.6H_2O$; dissolve in H_2O and store at 4 °C
- MT vitamin stock: 10 g/l myoinositol, 1 g/l thiamine-HCl, 1 g/l pyridoxine-HCl, 0.5 g/l nicotinic acid, 0.2 g/l glycine; dissolve in H_2O and store at 4 °C
- MT calcium stock: 29.33 g/l $CaCl_2.2H_2O$; dissolve in H_2O and store at 4 °C
- MT iron stock: 7.45 g/l Na_2EDTA, 5.57 g/l $FeSO_4.7H_2O$; dissolve in H_2O and store at 4 °C
- Kinetin (KIN) (Sigma) stock solution: 1 mg/ml; dissolve the powder in a few drops of 1 N HCl; bring to final volume with H_2O and store at 4 °C
- Callus-induction media:
  - 0.15 M EME semisolid medium: 20 ml/l MT macronutrient stock, 10 ml/l MT micronutrient stock, 10 ml/l MT vitamin stock, 15 ml/l MT calcium stock, 5 ml/l MT iron stock, 50 g/l sucrose, 0.5 g/l malt extract, 8 g/l agar, pH 5.8; autoclave medium and pour into 100 × 20 mm Petri dishes; 35 ml/dish
  - DOG semisolid medium: same as 0.15 M EME semisolid medium plus 5 mg/l kinetin (5 ml kinetin stock solution); autoclave medium and pour into 100 × 20 mm Petri dishes; 35 ml/dish
  - H+H semisolid medium: 10 ml/l MT macronutrient stock, 5 ml/l BH3 macronutrient stock, 10 ml/l MT micronutrient stock, 10 ml/l MT vitamin stock, 15 ml/l MT calcium stock, 5 ml/l MT iron stock, 50 g/l sucrose, 0.5 g/l malt extract, 1.55 g/l glutamine, 8 g/l agar, pH 5.8; autoclave medium and pour into 100 × 20 mm Petri dishes; 35 ml/dish
- Cell suspension maintenance H+H liquid medium: 10 ml/l MT macronutrient stock, 5 ml/l BH3 macronutrient stock, 10 ml/l MT micronutrient stock, 10 ml/l MT vitamin stock, 15 ml/l MT calcium stock, 5 ml/l MT iron stock, 35 g/l sucrose, 0.5 g/l malt extract, 1.55 g/l glutamine, pH 5.8; pour 500 ml aliquots into 1000 ml glass Erlenmeyer flasks, autoclave and store at room temperature

Method

1. Immerse harvested immature fruit in sterilization solution in a beaker for 30 min.
2. Using sterile tongs, place fruit on sterilized paper plates in a laminar flow hood.
3. Using a sterile surgical blade, make an equatorial cut 1–2 cm deep and break open the fruit.
4. With sterile forceps extract ovules and place them onto callus-induction medium (0.15 M EME, H+H or DOG) (see Figures 10.1 and 10.2).
5. Incubate extracted ovules in the dark at 28 ± 2 °C and transfer them every 2–3 weeks to new callus-induction medium until embryogenic (yellow and friable) callus emerges from the ovules.
6. To maintain long-term cultures, transfer embryogenic undifferentiated calli[a] onto new medium every 4 weeks and incubate under the same conditions.

7 To initiate cell suspensions from embryogenic undifferentiated nucellus-derived callus, take approx. 2 g of calli from callus-induction medium and transfer to 125 ml Erlenmeyer flasks each containing 20 ml of H+H liquid medium.

8 Shake the cell suspension cultures on a rotary shaker at 125 rpm under a 16 h photoperiod (70 μmol/m^2/s) at 28 ± 2 °C. After 2 weeks, add 20 ml of new H+H liquid medium to Erlenmeyer flasks.

9 Maintain established embryogenic cell suspension cultures by subculture every 2 weeks to 40 ml aliquots of H+H liquid medium shaking at 125 rpm and incubating under the same conditions.

Note

[a]Since the nucellar callus has high embryogenic capacity, the best way to maintain the long-term callus in an undifferentiated state is to visually select and subculture only white/yellow friable callus. Differentiated callus types and organized tissues should be discarded.

PROTOCOL 10.2 Preparation and Enzymatic Incubation of Cultures from Embryogenic Parent [16]

Equipment and Reagents

- Rotary shaker in incubator at 28 °C
- Laminar flow cabinet
- Autoclave
- BH3 macronutrient stock: see Protocol 10.1
- MT micronutrient, vitamin, calcium and iron stocks: see Protocol 10.1
- BH3 multivitamin stock A: 1 g/l ascorbic acid, 0.5 g/l calcium pantothenate, 0.5 g/l choline chloride, 0.2 g/l folic acid, 0.1 g/l riboflavin, 0.01 g/l *p*-aminobenzoic acid, 0.01 g/l biotin; dissolve in H_2O and store at −20 °C
- BH3 multivitamin stock B: 0.01 g/l retinol dissolved in a few drops of alcohol, 0.01 g/l cholecalciferol dissolved in a few drops of ethanol, 0.02 g/l vitamin B12; dissolve in H_2O and store at −20 °C
- BH3 KI stock: 0.83 g/l KI; dissolve in H_2O and store at 4 °C
- BH3 sugar + sugar alcohol stock: 25 g/l fructose, 25 g/l ribose, 25 g/l xylose, 25 g/l mannose, 25 g/l rhamnose, 25 g/l cellobiose, 25 g/l galactose, 25 g/l mannitol; dissolve in H_2O and store at −20 °C
- BH3 organic acid stock: 2 g/l fumaric acid, 2 g/l citric acid, 2 g/l malic acid, 1 g/l pyruvic acid; dissolve in H_2O and store at −20 °C
- 0.6 M BH3 liquid medium: 10 ml/l BH3 macronutrient stock, 10 ml/l MT micronutrient stock, 10 ml/l MT vitamin stock, 15 ml/l MT calcium stock, 5 ml/l MT iron stock, 2 ml/l

BH3 multivitamin stock A, 1 ml/l BH3 multivitamin stock B, 1 ml/l BH3 KI stock, 10 ml/l BH3 sugar + sugar alcohol stock, 20 ml/l BH3 organic acid stock, 20 ml/l coconut water, 82 g/l mannitol, 51.3 g/l sucrose, 3.1 g/l glutamine, 1 g/l malt extract, 0.25 g/l casein enzyme hydrolysate, pH 5.8; filter-sterilize and store at room temperature

- Stock solutions for preparation of enzyme solution:
  - Calcium chloride ($CaCl_2.2H_2O$ stock solution, 0.98 M): dissolve 14.4 g in 100 ml H_2O and store at −20 °C
  - Monosodium phosphate (NaH_2PO_4 stock solution, 37 mM): dissolve 0.44 g in 100 ml H_2O and store at −20 °C
  - 2 (*N*-morpholino) ethanesulfonic acid (MES stock solution, 0.246 M): dissolve 4.8 g in 100 ml H_2O and store at −20 °C
- Enzyme solution: 0.7 M mannitol, 24 mM $CaCl_2$, 6.15 mM MES buffer, 0.92 mM NaH_2PO_4, 2% (w/v) Cellulase Onozuka RS (Yakult Honsha), 2% (w/v) Macerozyme R-10 (Yakult Honsha), pH 5.6. To prepare 40 ml of enzyme solution, dissolve 0.8 g Cellulase Onozuka RS, 0.8 g Macerozyme R-10 and 5.12 g mannitol in 20 ml H_2O and add 1 ml of $CaCl_2.2H_2O$, NaH_2PO_4 and MES stock solutions; bring volume to 40 ml with H_2O, pH to 5.6 using KOH, filter-sterilize; store at 4 °C for up to 3 weeks

Method

1. Transfer 1–2 g of friable callus into a 60 × 15 mm Petri dish. If using a suspension as a source for embryogenic cells[a] (see Figures 10.1 and 10.2) transfer approx. 2 ml of suspension[b] with a wide-mouth pipette and drain off the liquid using a Pasteur pipette.
2. Resuspend the cells in a mixture of 2.5 ml 0.6 M BH3 liquid medium and 1.5 ml enzyme solution.
3. Seal Petri dishes with Parafilm and incubate overnight (15–20 h) at 28 °C on a rotary shaker at 50 rpm in the dark.

Notes

[a]Cultured embryogenic cells used for protoplast isolation should be in the log phase of growth. Use 5–12-day-old suspensions from a 2 week subculture cycle, or 7–21-day-old callus from a 4 week subculture cycle.

[b]Correlates to approx. 1 g fresh weight.

PROTOCOL 10.3 Preparation and Enzymatic Incubation of Cultures from Leaf Parent [16]

Equipment and Reagents

- Sterilization solution: see Protocol 10.1

- Rotary shaker in incubator at 28 °C
- Vacuum pump
- Laminar flow hood
- Autoclave
- MT macronutrient, micronutrient, vitamin, calcium and iron stock: see Protocol 10.1
- α-naphthalene acetic acid (NAA; Sigma; stock solution, 1 mg/10 ml): dissolve the powder in a few drops of 5 M NaOH bring to final volume with H_2O and store at 4 °C
- Root induction and propagation RMAN medium: 10 ml/l MT macronutrient stock, 5 ml/l MT micronutrient stock, 5 ml/l MT vitamin stock, 15 ml/l MT calcium stock, 5 ml/l MT iron stock, 25 g/l sucrose, 0.5 g/l activated charcoal, 8 g/l agar, 0.02 mg/l NAA (200 µl NAA stock solution); pH 5.8). Autoclave medium and pour into sterile Magenta GA-7 boxes (Sigma); 80 ml/box
- 0.6 M BH3 liquid medium: see Protocol 10.2
- Enzyme solution: see Protocol 10.2

Method

1 Excise 10–15 leaves from 2-month-old plants growing *in vitro* on propagation RMAN medium[a] (see Figure 10.1). If using glasshouse-grown plants as an explant source, before incubation immerse one to three leaves[b] in sterilization solution for 25 min and rinse three times in sterile H_2O.

2 Remove damaged vascular tissue and midvein region with a sterile surgical blade; remaining leaf material cut or feather into 1–2 mm wide segments.

3 Incubate plant material[c] in a 125 ml Erlenmeyer flask containing a mixture of 8 ml 0.6 M BH3 liquid medium and 3 ml enzyme solution.

4 Evacuate leaf material in the medium/enzyme solution for 15 min at 50 kPa to facilitate enzyme infiltration.

5 Incubate this preparation under the same conditions as in Protocol 10.2, step 3.

Notes

[a]Propagate plants aseptically by shoot tip or nodal cuttings and transfer to new RMAN medium every 8–9 weeks.

[b]Leaves should be young, but fully expanded, and taken only from new flushes that have not fully hardened. Best results are generally obtained when leaf explant sources come from seedlings or recently budded plants maintained in a heavily shaded glasshouse.

[c]To ensure an adequate yield of protoplasts ($5-10 \times 10^6$ protoplasts/flask) cut enough leaf segments to cover the surface of the medium/enzyme solution.

PROTOCOL 10.4 Protoplast Isolation and Purification [16]

Equipment and Reagents

- Centrifuge with 500–4000 rpm capability
- Laminar flow cabinet
- Autoclave
- CPW salts stock solution 1: 25 g/l $MgSO_4.7H_2O$, 10 g/l KNO_3, 2.72 g/l KH_2PO_4, 0.016 g/l KI, 0.025 ng/l $CuSO_4.5H_2O$; dissolve in H_2O and store at 4 °C
- CPW salts stock solution 2: 15 g/l $CaCl_2.2H_2O$; dissolve in H_2O and store at 4 °C
- 13% (w/v) mannitol solution with CPW salts (CPW 13M): dissolve 13 g mannitol in 80 ml H_2O, add 1 ml each of CPW salts stock solutions 1 and 2; bring volume to 100 ml with H_2O, pH to 5.8, filter-sterilize; store at room temperature
- 25% (w/v) sucrose solution with CPW salts (CPW 25S): dissolve 25 g sucrose in 80 ml H_2O, add 1 ml each of CPW salts stock solutions 1 and 2; bring to 100 ml with H_2O, pH to 5.8, filter-sterilize and store at room temperature
- 0.6 M BH3 liquid medium: see Protocol 10.2
- 40 ml Pyrex tubes (Fisher Scientific)

Method

1 Following overnight incubation, pass enzymatic preparations from the two parental sources (see Protocols 10.2 and/or 10.3; see Figures 10.1 and 10.2) through a sterile 45 µm nylon mesh sieve[a] to remove undigested tissues and other cellular debris; collect the filtrate in 40 ml Pyrex tubes.

2 Transfer the protoplast-containing filtrate to a 15 ml calibrated screw-cap centrifuge tube and centrifuge at 900 rpm for 5–8 min.

3 Remove the supernatant with a Pasteur pipette and gently resuspend the protoplast pellet in 5 ml of CPW 25S solution.

4 Slowly pipette 2 ml of CPW 13M solution directly on top of the sucrose layer. Avoid mixing the layers.

5 Centrifuge at 900 rpm for 8–10 min.

6 Only viable protoplasts form a band at the interface between the sucrose and the mannitol layers. Remove the protoplasts from this interface with a Pasteur pipette and resuspend them in 10–13 ml of 0.6 M BH3 liquid medium (using a new screw-cap centrifuge tube).

7 Centrifuge at 900 rpm for 5–8 min.

8 Remove the supernatant and gently resuspend the pellet in 10–13 ml of 0.6 M BH3 medium.

9 Centrifuge at 900 rpm for 5–8 min.

10 Remove the supernatant and resuspend the pellet with 0.6 M BH3 medium in a volume that is approx. 10 × the size of the pellet.

Note

[a]Nylon mesh is sealed to a 4 cm long plastic cylindrical tube made from a syringe. In order to make a similar piece of equipment, take a 30 ml syringe, cut it at the 25 ml mark and keep the upper part with wings. Place a nylon membrane on a pre-heated hot plate beneath the cylindrical tube and seal the two parts.

PROTOCOL 10.5 Polyethylene Glycol (PEG)-Induced Protoplast Fusion [16]

Equipment and Reagents

- Centrifuge with 500–4000 rpm capabilities
- Laminar flow hood
- Autoclave
- PEG 1500 MW (Sigma; stock solution, 50%): Place the bottle of PEG in a water bath at 80 °C until it melts, take 250 ml and mix it with 250 ml H_2O, add 4 g of resin AG501-X8 (Bio-Rad), stir for 30 min, filter out the resin through a layer of cotton and allow to stand for several hours before use; store at room temperature
- Polyethylene glycol (PEG) working solution: 40% (wv) PEG, 0.3 M glucose, 66 mM $CaCl_2.2H_2O$, pH 6.0. To prepare 100 ml of PEG solution, dissolve 0.97 g $CaCl_2.2H_2O$ and 5.41 g glucose in 10 ml H_2O, add 80 ml of PEG stock solution and adjust the volume to 100 ml with H_2O, pH; filter-sterilize and store at 4 °C. Check the pH every 2–3 weeks, since this solution acidifies with time
- Elution solutions for PEG removal. Solution A: 0.4 M glucose, 66 mM $CaCl_2.2H_2O$, 10% dimethyl sulfoxide (DMSO), pH 6.0. Solution B: 0.3 M glycine adjusted with NaOH pellets to pH 10.5. Filter-sterilize both solutions; store at room temperature and mix together (9:1, v:v) prior to use to avoid precipitation
- 0.6 M BH3 liquid medium: see Protocol 10.2
- 0.6 M EME liquid medium: ingredients and pH are the same as in 0.15 M EME semisolid medium (see Protocol 10.1) with two modifications; instead of 50 g/l sucrose, add 205.4 g/l sucrose and omit agar; filter-sterilize and store at room temperature

Method

1 Mix isolated protoplasts from the two parental sources (see Figures 10.1 and 10.2) at a ratio of 1:1 and centrifuge at 900 rpm for 5–8 min. If fusing protoplasts from one parent with microprotoplasts from the other parent, then mix them at a ratio of 1:2 or 1:3.

2. Remove the supernatant with a Pasteur pipette and resuspend the pellet of mixed protoplasts with 0.6 M BH3 medium in the volume that is 4–10 times the size of the pellet.
3. Pipette two to four drops of the resuspended mixture into 60 × 15 mm Petri dishes[a] for fusion.
4. Immediately add two to four drops of PEG solution directly into the centre of the protoplast mixture, allowing the PEG to mix with the protoplast droplet.
5. After 10–15 min, add two to four drops of A + B solution (9:1 v:v) into each fusion dish but this time gently on the periphery of the protoplast mixture trying not to disturb fusing protoplasts.
6. Following another incubation period of 10–15 min, gently add 15–20 drops of 0.6 M BH3 medium around the periphery of the fusing protoplasts, again trying not to disturb them.
7. After incubating for an additional 5 min, gently remove all of the fluid from the dish with a Pasteur pipette and replace it with 15–20 drops of 0.6 M BH3 medium.
8. Repeat the washing procedure (step 7) twice carefully avoiding the loss of protoplasts.
9. Finally, add 1.5–2.0 ml of a 1:1 (v:v) mixture of 0.6 M BH3 and 0.6 M EME liquid media. Spread into a thin layer by gently swirling the Petri dishes[b].
10. Seal the dishes with Parafilm and culture in the dark at 28 ± 2 °C for 4–6 weeks.

Notes

[a]The number of dishes is determined by the total volume of mixed protoplasts.

[b]In each fusion dish, protoplasts are plated at a density of approx. 1–5 × 10^6 protoplasts/ml of culture medium. If necessary, determine and adjust protoplast density using a haemocytometer.

PROTOCOL 10.6 Protoplast Electrofusion [58]

Equipment and Reagents

- Somatic hybridizer SSH-2 equipped with a 1.6 ml FTC-04 electrofusion chamber (Shimadzu Corporation)
- Light microscope (standard type, 400 × magnification)
- Centrifuge with 500–4000 rpm capabilities
- Laminar flow cabinet
- Autoclave
- Electrofusion solution: 0.7 M mannitol, 0.25 mM $CaCl_2.2H_2O$, pH 5.8; filter-sterilize; store at room temperature

- 0.6 M BH3 liquid medium: see Protocol 10.2
- 0.6 M EME liquid medium: see Protocol 10.5

Method

1. Recover separately viable protoplasts of each parent (see Figures 10.1 and 10.2) with a Pasteur pipette from the sucrose-mannitol gradient (see Protocol 10.4, step 6). If fusing protoplasts that have been treated with gamma-irradiation (see Protocol 10.8) and iodoacetic acid (see Protocol 10.9) delete this step and proceed to step 2.
2. Resuspend protoplasts in 10–13 ml of electrofusion solution using a new 15 ml screw-cap centrifuge tube.
3. Centrifuge at 900 rpm for 5–8 min.
4. Remove the supernatant with a Pasteur pipette and gently resuspend the pellet containing the protoplasts in electrofusion solution at a concentration $0.5–1.5 \times 10^6$ protoplasts/ml.
5. Mix isolated protoplasts from the two parental sources in equal numbers. If fusing protoplasts from one parent with microprotoplasts from the other, mix them at a ratio of 1:2 or 1:3.
6. Sterilize the fusion chamber of the somatic hybridizer by immersion in 70% (v/v) ethanol (5 min) and allow to dry in the laminar air flow cabinet.
7. Immediately prior to use, rinse the fusion chamber with Electrofusion solution and add 2 ml of this solution to the inner compartment.
8. Load the outer compartment of the fusion chamber with 1.6 ml of protoplast mixture.
9. Seal the chamber with Parafilm and leave undisturbed for 5 min.
10. Apply an AC-alignment field (1 MHz, 95 V/cm) with a duration of 1 min, followed by five pulses of DC field (1250 V/cm, 40 μs each) at 0.5 s intervals[a]; gradually reduce the AC field to 0 V/cm during the next 1–2 min.
11. Leave protoplasts undisturbed for 20 min to enable fusion products to regain a spherical shape.
12. After the recovery period, gently transfer the treated protoplasts with a Pasteur pipette into a centrifuge tube and centrifuge at 900 rpm for 5–8 min.
13. Remove the supernatant and gently resuspend the pellet of fusion-treated protoplasts in 1.5–2.0 ml of a 1:1 (v:v) mixture of 0.6 M BH3 and 0.6 M EME liquid media; transfer resuspended protoplasts into a 60×15 mm Petri dish and spread the suspension as a thin layer by gently swirling the dish.
14. Seal the dish with Parafilm and culture in the dark at $28 \pm 2\,^{\circ}C$ for 4–6 weeks.

Note

[a]Before applying the AC/DC voltage, place the fusion chamber on the stage of an inverted microscope to observe alignment and fusion of protoplasts.

PROTOCOL 10.7 Protoplast Culture and Plant Regeneration [16]

Equipment and Reagents

- Laminar flow cabinet
- Autoclave
- MT macronutrient, micronutrient, vitamin, calcium and iron stock: see Protocol 10.1
- Coumarin (Sigma; stock solution, 1.46 mg/ml): dissolve the powder in warm H_2O; store at 4 °C
- NAA stock solution (1 mg/10 ml): see Protocol 10.3
- 2,4-dichlorophenoxyacetic acid (2,4-D; Sigma; stock solution, 1 mg/10 ml): dissolve the powder in a few drops of 95% (v/v) ethanol, bring to final volume with H_2O; store at 4 °C
- 6-benzylaminopurine (BAP; Sigma; stock solution, 1 mg/ml): dissolve the powder in a few drops of 5 M NaOH, bring to final volume with H_2O; store at 4 °C
- Gibberellic acid (GA3; Sigma; stock solution, 1 mg/ml): dissolve the powder in a few drops of 95% (v/v) ethanol bring to final volume with H_2O, filter-sterilize; store in small aliquots at 4 °C; add to the medium after autoclaving and cooling the medium to 55 °C in a water bath
- 0.6 M BH3 liquid medium: see Protocol 10.2
- 0.15 M EME liquid medium: ingredients and pH are the same as in 0.15 M EME semisolid medium (see Protocol 10.1) only omit agar, filter-sterilize; store at room temperature
- 0.15 M EME–malt semisolid medium: same as 0.15 M EME semisolid medium (see Protocol 10.1), but substitute 50 g/l maltose for the 50 g/l sucrose; autoclave medium and pour into 100 × 20 mm Petri dishes, 35 ml/dish
- 0.15 M EME–malt liquid medium: ingredients and pH are the same as in 0.15 M EME–malt semisolid medium, but omit the agar; filter-sterilize and store at room temperature
- 0.6 M EME liquid medium: see Protocol 10.5
- EME 1500 semisolid medium: 20 ml/l MT macronutrient stock, 10 ml/l MT micronutrient stock, 10 ml/l MT vitamin stock, 15 ml/l MT calcium stock, 5 ml/l MT iron stock, 50 g/l sucrose, 1.5 g/l malt extract, 8 g/l agar, pH 5.8; autoclave medium and pour into 100 × 20 mm Petri dishes, 35 ml/dish
- B+ semisolid medium: 20 ml/l MT macronutrient stock, 10 ml/l MT micronutrient stock, 10 ml/l MT vitamin stock, 15 ml/l MT calcium stock, 5 ml/l MT iron stock, 25 g/l sucrose, 20 ml/l coconut water, 14.6 mg/l coumarin (10 ml coumarin stock), 0.02 mg/l NAA (200 µl NAA stock), 1 mg/l GA3 (add 1 ml GA3 stock after medium is autoclaved and cooled to 55 °C in water bath), 8 g/l agar, pH 5.8); autoclave medium and pour into 100 × 20 mm Petri dishes, 35 ml/dish
- DBA3 semisolid medium: 20 ml/l MT macronutrient stock, 10 ml/l MT micronutrient stock, 10 ml/l MT vitamin stock, 15 ml/l MT calcium stock, 5 ml/l MT iron stock, 25 g/l

sucrose, 1.5 g/l malt extract, 20 ml/l coconut water, 0.01 mg/l 2,4-D (100 μl 2,4-D stock), 3 mg/l BAP (3 ml BAP stock); 8 g/l agar, pH 5.8; autoclave medium and pour into 100 × 20 mm Petri dishes, 35 ml/dish

- Root induction and propagation RMAN medium: see Protocol 10.3

Method

1. After 4–6 weeks of incubation, supplement cultures of fused protoplasts, or protoplasts with microprotoplasts, with new medium containing reduced osmoticum. Accomplish this by adding 10–12 drops of 1:1:1 (by vol) mixture of 0.6 M BH3, 0.6 M EME and 0.15 M EME liquid media.
2. Incubate cultures for another 2 weeks in low light (20 $\mu E/m^2/s$ intensity) with a 16 h photoperiod at 28 ± 2 °C.
3. Accomplish another reduction of osmoticum in the cultures by the following steps:
   - add 2 ml of 1:2 (v:v) mixture of 0.6 M BH3 and 0.15 M EME–malt liquid media to each dish of fusion-treated protoplasts
   - immediately pour the entire contents onto Petri dishes with agar-solidified 0.15 M EME-malt medium and swirl gently each dish in order to spread the liquid containing protoplast-derived colonies evenly over the entire semisolid agar surface.
4. Incubate cultures with a 16 h photoperiod (70 $\mu mol/m^2/s$ intensity) at 28 ± 2 °C and, from this point until somatic hybrids are planted in compost, keep the cultures under the same growth conditions.
5. Transfer regenerated somatic embryos as soon as they appear from callus colonies to new agar-solidified 0.15 M EME-malt medium (see Figures 10.1 and 10.2).
6. After 3–4 weeks, move small somatic embryos to semisolid EME 1500 medium for enlargement and germination and further to semisolid B+ medium for axis elongation.
7. Dissect abnormal embryos that fail to germinate into large sections and place on DBA3 medium for shoot induction.
8. Transfer all resulting shoots into RMAN medium to induce rooting.
9. Transfer rooted plants into compost in the glasshouse and cover with rigid clear plastic for 3–4 weeks maintaining high humidity. Remove the plastic covers following this period of acclimatization.

PROTOCOL 10.8 Gamma Irradiation

Equipment and Reagents

- Centrifuge with 500–4000 rpm capabilities
- Light microscope
- Laminar flow cabinet

- Cobalt-60 gamma-ray source
- 0.6 M BH3 liquid medium: see Protocol 10.2
- Haemocytometer

Method

1. Separately recover viable embryogenic protoplasts (from Parent 1) (see Figure 10.2) with a Pasteur pipette from the sucrose-mannitol gradient (see Protocol 10.4, step 6).
2. Resuspend protoplasts in 10–13 ml of 0.6 M BH3 liquid medium in a new 15 ml screw capped centrifuge tube.
3. Centrifuge at 900 rpm for 5–8 min.
4. Remove the supernatant with a Pasteur pipette and gently resuspend the pellet in 0.6 M BH3 medium at a concentration of $1-5 \times 10^6$ protoplasts/ml using a haemocytometer.
5. Transfer 5–8 ml of protoplast suspension to a 60 × 15 mm Petri dish.
6. Expose protoplasts to 500–600 Gy from gamma-ray source.
7. Transfer protoplasts to a 15 ml calibrated screw-capped centrifuge tube and wash twice in 10–13 ml of 0.6 M BH3 medium; after each wash, pellet the protoplasts by centrifugation (900 rpm, 5–8 min).
8. Remove the supernatant with a Pasteur pipette and gently resuspend the pellet in 0.6 M BH3 medium in a volume that is 10× the size of the pellet.

PROTOCOL 10.9 Iodoacetic Acid Treatment

Equipment and Reagents

- Centrifuge with 500–4000 rpm capabilities
- Light microscope
- Laminar flow hood
- 0.6 M BH3 liquid medium: see Protocol 10.2
- Iodoacetic acid (IOA; Sigma; stock solution, 25 mM): dissolve 0.232 g IOA in 50 ml H_2O and store at 4 °C for up to 3–4 months
- IOA working solution (0.25 mM): add 1 ml IOA stock solution to 99 ml 0.6 M BH3 medium, pH to 5.8, filter-sterilize; use promptly

Method

1. Separately recover viable embryogenic protoplasts (from Parent 2; see Figure 10.2) with a Pasteur pipette from the sucrose-mannitol gradient (see Protocol 10.4, step 6).

2 Resuspend protoplasts in 10–13 ml of 0.6 M BH3 liquid medium in a new 15 ml screw-capped centrifuge tube.

3 Centrifuge at 900 rpm for 5–8 min.

4 Remove the supernatant with a Pasteur pipette and gently resuspend the pellet containing the protoplasts in 0.25 mM IOA solution at a concentration of $1–5 \times 10^6$ protoplasts/ml; use a haemocytometer.

5 Incubate for 15–20 min at room temperature.

6 Wash protoplasts twice in 10–13 ml of 0.6 M BH3 medium; after each wash, pellet protoplasts by centrifugation (900 rpm 5–8 min).

7 Remove the supernatant with a Pasteur pipette and gently resuspend the pellet with 0.6 M BH3 medium in the volume that is 10 × the size of the pellet.

PROTOCOL 10.10 Pretreatment of Embryogenic Cell Suspensions and Enzymatic Incubation for Microprotoplast Isolation [51]

Equipment and Reagents

- Rotary shaker in plant growth chamber at $28 \pm 2\,^{\circ}C$
- Rotary shaker in incubator at $28\,^{\circ}C$
- Laminar flow cabinet
- Autoclave
- BH3 macronutrient stock: see Protocol 10.1
- MT micronutrient, vitamin, calcium, and iron stock: see Protocol 10.1
- Cell suspension maintenance H + H-MP liquid medium: 5 ml/l MT macronutrient stock, 2.5 ml/l BH3 macronutrient stock, 10 ml/l MT micronutrient stock, 10 ml/l MT vitamin stock, 10 ml/l MT calcium stock, 2.5 ml/l MT iron stock, 50 g/l sucrose, 0.5 g/l malt extract, 1.55 g/l glutamine, pH 5.8; pour 500 ml aliquots into 1000 ml glass Erlenmeyer flasks, autoclave and store at room temperature
- 0.6 M BH3 liquid medium: see Protocol 10.2
- Enzyme solution: see Protocol 10.2
- Hydroxyurea (HU; Sigma; stock solution, 75 mg/ml): dissolve the powder in water-free DMSO; use promptly
- Amiprophos-methyl (APM; Bayer Corp; stock solution, 10 mg/ml): dissolve the powder in water-free DMSO; store at $-20\,^{\circ}C$ for up to 6 months
- Cytochalasin B (CB; Sigma; stock solution, 2 mg/ml): dissolve the powder in water-free DMSO; store at $-20\,^{\circ}C$ for up to 1 year

- 4′6-diamidino-2 phenylindole dihydrochloride (DAPI; Sigma; stock solution): dissolve 1 mg DAPI in 4 ml H_2O; store at −20 °C in a foil-wrapped tube
- DAPI working solution (0.4 μg/ml): add 1.6 μl DAPI stock solution to 1 ml H_2O; use promptly

Method

1 Thirty days prior to microprotoplast isolation, start subculturing established embryogenic cell suspension cultures (see Protocol 10.1) every 3–4 days in 125 ml Erlenmeyer flasks, each containing 40 ml of H+H-MP liquid medium[a]. For growth chamber conditions and speed of rotary shaker, see Protocol 10.1.

2 Add HU to suspension cultures to a final concentration of 10 mM (1 ml HU stock solution/100 ml medium) and shake cultures at 125 rpm for 24 h. This step is aimed to arrest cells at the S phase and synchronize cell growth.

3 Wash cell suspension cultures three times by adding to Erlenmeyer flasks H+H-MP medium (30 ml/flask) shaking them (100 rpm) for 15 min each time.

4 Add APM and CB to the synchronized suspension cultures to a final concentration of 32 μM (0.1 ml APM stock solution/100 ml medium) and 10 μM (0.25 ml CB stock solution/100 ml medium).

5 Return suspension cultures to the shaker (125 rpm) for 24 h.

6 Transfer into a 60 × 15 mm Petri dish with a wide-mouth pipette approx. 2 ml of suspension[bc] and drain off the liquid using a Pasteur pipette.

7 Resuspend the cells in a mixture of 2.5 ml 0.6 M BH3 liquid medium and 1.5 ml enzyme solution containing 32 μM APM and 10 μM CB.

8 Seal Petri dishes with Parafilm and incubate overnight (15–20 h) at 28 °C on a rotary shaker at 50 rpm in the dark.

Notes

[a]Use of full strength H+H medium (see Protocol 10.1) with a 3–4 day subculture cycle may induce browning.

[b]At this point scattered chromosomes can be visualized with DAPI. Transfer a few drops of the suspension culture into a microcentrifuge tube, add 1 ml 1 N HCl and incubate for 10 min at 60°C. Using a Pasteur pipette, remove the acid and wash the suspension culture with phosphate-buffered saline (PBS). Transfer a small number of washed cells to a glass microscope slide, add one drop of DAPI working solution, cover with a glass cover slip and squash the cells by gentle pressure on the cover slip. Observe the nuclei using a fluorescence microscope with excitation filter 360/40 nm and emission barrier filter 460/50 nm (Nikon EF-4UV-2E/C).

[c]Correlates to approx. 1 g fresh weight.

PROTOCOL 10.11 Microprotoplast Isolation and Purification [51]

At this stage the majority of protoplasts should contain multiple nuclei.

Equipment and Reagents

- Centrifuge with 500–4000 rpm capabilities
- Ultracentrifuge with 28 000 rpm capabilities
- Swinging bucket rotor (SW 41TI; Beckman)
- Laminar flow cabinet
- Autoclave
- 0.6 M BH3 liquid medium: see Protocol 10.2
- Sterile Percoll solution (GE Healthcare)
- Mannitol/Percoll solution (7.2% w:v): dissolve 7.2 g mannitol in 100 ml sterile Percoll solution; filter-sterilize; use promptly
- Amiprophos-methyl (APM; Bayer Corp.; stock solution): see Protocol 10.10
- Cytochalasin B (CB; Sigma) stock solution: see Protocol 10.10
- Acridine orange (AO; Sigma; stock solution): dissolve 2 mg of powder in 1 ml H_2O; store at $-20\,^{\circ}C$ in a foil-wrapped tube
- AO working solution (10 μg/ml): add 5 μl AO stock solution to 1 ml H_2O; use promptly

Method

1 Following overnight incubation, isolate and purify the protoplasts from APM and CB pre-treated enzymatic preparations, as described in Protocol 10.4, steps 1–6[a].

2 After recovering viable protoplasts of each parent with a Pasteur pipette from the sucrose–mannitol gradient (see Protocol 10.4, step 6), resuspend the protoplasts in 10–13 ml of BH3 liquid medium containing 32 μM APM (0.1 ml APM stock solution/100 ml medium) and 10 μM CB (0.25 ml CB stock solution/100 ml medium).

3 Centrifuge at 900 rpm for 5–8 min.

4 Remove the supernatant and gently resuspend the pellet in 10–13 ml of 0.6 M BH3 medium containing 32 μM APM and 20 μM CB (0.5 ml CB stock solution/100 ml medium).

5 Centrifuge at 900 rpm for 5–8 min.

6 Remove most of the supernatant and leave just enough medium to maintain the protoplasts as a dense suspension.

7 Pipette 8 ml of mannitol/Percoll solution into a sterile Polyallomer centrifuge tube (14 × 89 mm) and centrifuge at 28 000 rpm at 20 °C for 30 min in a swinging bucket rotor[b] to form an iso-osmotic gradient.

8 Remove the top 5 mm from the preformed iso-osmotic gradient and add the dense solution of protoplasts[c] (from step 6) to the iso-osmotic gradient.

9 Centrifuge at 28 000 rpm for 2 h at 20 °C in a swinging bucket rotor.

10 Carefully remove the tubes from the ultracentrifuge. The tube with microprotoplasts should contain up to nine bands depending on the amount of protoplast suspension loaded in the gradient.

11 Starting from the surface of the formed gradient do not collect with a Pasteur pipette the first band composed of sticky material. Instead, use a sterile glass rod to take the first band out of the way of the Pasteur pipette used to collect the other bands.

12 The second band is usually very thick and contains a large amount of microprotoplasts of various size. Collect this band alone or with the third[d] band.

13 Dilute the microprotoplast suspension from the second, or second plus third bands, in 13–15 ml of 0.6 M BH3 medium[e].

14 Filter the collected bands sequentially through sterile 20 and 10 μm nylon mesh sieves[f,g] using only gravity force. To facilitate the flow, apply gentle pressure using a sterile syringe plunger and/or add an additional amount of 0.6 M BH3 medium.

15 Transfer the microprotoplast-containing filtrate to a 15 ml calibrated screw-cap centrifuge tube and centrifuge at 700 rpm for 10 min.

16 Remove the supernatant (retain for the next step) and resuspend the pellet with 0.6 M BH3 medium in a volume that is 10 times the size of the pellet.

17 Centrifuge saved supernatant at 1400 rpm but this time discard supernatant after centrifugation and resuspend the pellet in the same way as in step 16.

Notes

[a]All media and solutions used for protoplast isolation and purification must contain 32 μM APM and 10 μM CB.

[b]Balance the tubes in the centrifuge using the same Percoll solution which is denser than water.

[c]Up to 1 ml of the dense protoplast solution can be load per tube of mannitol/Percoll gradient.

[d]Do not collect bands 4–9 since they contain the smallest microprotoplasts that are difficult to pellet.

[e]There is no need for APM or CB in the 0.6 M BH3 medium used during the purification of microprotoplasts.

[f]To make these sieves, see Protocol 10.4, note (a).

[g]At this point, microprotoplasts can be visualized with acridine orange (AO). Add 3 μl of AO working solution to one drop of microprotoplast suspension on a glass microscope slide and observe using a fluorescence microscope with excitation filter 450–490/40 nm and emission barrier filter 515 nm (Nikon EF-4 B-2A).

10.4 Troubleshooting

- Protoplasts are fragile. Take extra care when filtering the protoplast/enzyme solution and later when centrifuging and resuspending protoplasts. When being transferred from one tube to another it is important that the protoplasts are drawn gently into the Pasteur pipette and dispensed slowly down the inside wall of the receiving centrifuge tube. Also, when resuspending pellets of protoplasts with different solutions, ensure a gentle technique of breaking clumps by introducing small bubbles of air with a Pasteur pipette instead of sucking suspensions in and out of the pipette. Mishandling of the protoplasts can affect their integrity and thereby affect the efficiency of the procedure.

- If, after isolation and purification a good yield of protoplasts ($5-10 \times 10^6$ protoplasts/incubation plate or flask) is not obtained, it may be necessary to vary both the enzyme concentration and length of incubation time to optimize digestion efficiency. Adding Pectolyase Y23 to the enzyme solution is no longer recommended as the fungal source of this enzyme provided by commercial sources has changed, negatively affecting its performance in citrus protoplast isolation.

- When recovering protoplasts from the sucrose–mannitol gradient (see Protocol 10.4, step 6) take as little of the sucrose as possible with the protoplasts. Retention of too much sucrose makes it difficult to pellet the protoplasts at later steps. It is recommended to repeat steps 8 and 9 in Protocol 10.4 until a tight clean pellet is obtained. Similarly in Protocol 10.11, step 12, when collecting second and third bands with microprotoplasts, take as little of the mannitol/Percoll solution as possible.

- In order to retain viability and induce cell division, fused protoplasts have to be plated in thin-layer culture at high cell density. In the case of citrus protoplasts, the best results are obtained when the cell density exceeds 1×10^6 protoplasts/ml of medium.

- Perform fusion (PEG-induced and/or electrofusion method) within 1–2 h (preferably immediately) after protoplast isolation since protoplasts start to regenerate cell walls as soon as they are rinsed from the enzyme solution. Cell wall regeneration may hinder fusion. If irradiating donor parent protoplasts prior to fusion takes more than 1–2 h (due to the immediate unavailability of irradiation equipment) the problem can be circumvented by directly irradiating the leaves (instead of isolated protoplasts) prior to protoplast isolation.

- Obtain high-quality protoplast preparations for electrofusion. In poor quality preparations, protoplast lysis leads to salt accumulation in the electrofusion solution resulting in changed electrical conductivity. This, in turn, can cause improper alignment of protoplasts in the AC field.

- Optimization of electrical conditions is critical for efficient electrofusion. Under the electrical settings suggested good alignment ('pearl chain' formation of protoplasts) and efficient fusion is expected. Efficient fusion is recognized by the relevant changes seen under the microscope, namely increased contact between plasma membranes, membrane mixing and fused protoplasts becoming spherical. If these changes are not observed, it may be necessary to vary the strength, duration and number of DC pulses to maximize protoplast fusion. Although increasing the AC alignment voltage increases fusion efficiency, increasing the AC field greater than 125 V/cm is not recommended since it causes protoplast lysis when DC pulses are given. However, reducing the AC field below 95 V/cm reduces protoplast adhesion.

- If isolated microprotoplasts in Protocol 10.11 contain high chromosome numbers, it is advised that after passing the microplasts through the 20 and 10 μm sieves, to pass the preparation through a 5 μm sieve. This assures recovery of only small microprotoplasts with low chromosome numbers that are excellent for fusion experiments.

Acknowledgements

Dedicated to Edward C. Cocking and Oluf L. Gamborg – two pioneers of *in vitro* plant science whose research and enthusiasm inspired a whole generation of scientists. Our appreciation goes to Barbara Thompson for her assistance in formatting this chapter.

References

1. Cocking EC (1960) *Nature* **187, 927–929.

Pioneering paper on the isolation of plant protoplasts.

2. Carlson PS, Smith H, Dearing RD (1972) *Proc. Natl. Acad. Sci. USA* **69**, 2292–2294.

3. Bravo JE, Evans DA (1985) *Plant Breed. Rev.* **3**, 193–218.

*4. Davey MR, Kumar A (1983) *Int. Rev. Cytol. Suppl.* **16**, 219–299.

Early review on protoplast-based technologies.

5. Dudits D, Fejer O, Hadlaczky G, Koncz C, Lazar GB, Horvath G (1980) *Mol. Gen. Genet.* **179**, 283–288.

6. Fowke LC, Gamborg OL (1980) *Int. Rev. Cytol.* **68**, 9–51.

7. Kao KN, Constabel F, Michayluk MR, Gamborg OL (1974) *Planta* **120, 215–227.

Key paper on protoplast-based technologies.

8. Kumar A, Cocking EC (1987) *Am. J. Bot.* **74**, 1289–1303.

*9. Melchers G, Sacristan MD, Holder AA (1974) *Mol. Gen. Genet.* **135**, 277–294.

10. Saito W, Ohgawara T, Shimizu J, Kobayashi S (1994) *Plant Sci*. **99**, 89–95.

11. Schieder O, Vasil IK (1980) *Int. Rev. Cytol. Suppl*. **11B**, 21–46.

12. Shepard JF, Bidney D, Barsby T, Kemble R (1983) *Science* **219**, 683–688.

*13. Waara S, Glimelius K (1995) *Euphytica* **85**, 217–233.

Review of plant somatic hybridization.

*14. Gleba YY, Sytnik KM (1984) *Protoplast Fusion: Genetic Engineering in Higher Plants*. Edited by R Shoeman. Springer-Verlag, Berlin, Heidelberg.

Review of plant somatic hybridization.

*15. Johnson AAT, Veilleux RE (2001) *Plant Breed. Rev*. **20**, 167–225.

Review of plant somatic hybridization.

16. Grosser JW, Gmitter FG (1990) *Plant Breed. Rev*. **8, 339–374.

Review relating to somatic hybridization in citrus.

*17. Grosser JW, Ollitrault P, Olivares-Fuster O (2000) *In Vitro* Cell. Dev. Biol.-Plant **36**, 434–449.

Review of somatic hybridization in citrus.

18. Orczyk W, Przetakiewicz J, Nadolska-Orczyk A (2003) *Plant Cell Tissue Org. Cult*. **73**, 245–256.

*19. Guo WW, Cai XD, Grosser JW (2004) In: *Molecular Biology and Biotechnology of Plant Organelles*: *Somatic Cell Cybrids and Hybrids in Plant Improvement*. Edited by H Daniell and CD Chase. Kluwer Academic Publisher Dordrecht, The Netherlands, pp. 635–659.

Review of somatic hybridization and cybridization in citrus.

20. Guo WW, Grosser JW (2005) *Plant Sci*. **168**, 1541–1545.

21. Guo WW, Prasad D, Serrano P, Gmitter FG Jr, Grosser JW (2004) *J. Hort. Sci. Biotechnol*. **79**, 400–405.

*22. Grosser JW, Gmitter FG Jr (2005) *In Vitro Cell. Dev. Biol.-Plant* **41**, 220–225.

Supporting of somatic hybridization manuscripts.

23. Grosser JW, Graham JH, McCoy CW, *et al.* (2003) *Proc. Fla. State Hort. Soc*. **116**, 262–267.

24. Grosser JW, Mourao FAA, Gmitter FG Jr, *et al.* (1996) *Theor. Appl. Genet*. **92**, 577–582.

25. Skarzhinskaya M, Landgren M, Glimelius K (1996) *Theor. Appl. Genet*. **93**, 1242–1250.

26. Escalante A, Imanishi S, Hossain M, Ohmido N, Fukui K (1998) *Theor. Appl. Genet*. **96**, 719–726.

27. Bastia T, Scotti N, Cardi T (2001) *Theor. Appl. Genet*. **102**, 1265–1272.

28. Wang YP, Sonntag K, Rudloff E (2003) *Theor. Appl. Genet*. **106**, 1147–1155.

29. Xu CH, Xia GM, Zhi DY, Xiang FN, Chen HM (2003) *Plant Sci*. **165**, 1001–1008.

30. Fock I, Collonnier C, Purwito A, *et al.* (2000) *Plant Sci*. **160** 165–176.

31. Collonnier C, Fock I, Daunay M-C, *et al.* (2003) *Plant Sci*. **164**, 849–861.

32. Smith MV, Pay A, Dudits D (1989) *Theor. Appl. Genet.* **77**, 641–644.

*33. Vardi A, Breiman A, Galun E (1987) *Theor. Appl. Genet.* **75**, 51–58.

34. Melchers G, Mohri Y, Watanabe K, Wakabayashi S, Harada K (1992) *Proc. Natl Acad. Sci. USA* **89**, 6832–6836.

35. Xu XY, Liu JH, Deng XX (2006) *Plant Cell Rep.* **25**, 533–539.

36. Grosser JW, Gmitter FG, Tusa N, Recupero GR, Cucinotta P (1996) *Plant Cell Rep.* **15**, 672–676.

37. Guo WW, Prasad D, Cheng YJ, Serrano P, Deng XX, Grosser JW (2004) *Plant Cell Rep.* **22**, 752–758.

38. Bhattacharjee B, Sane AP, Gupta HS (1999) *Mol. Breed.* **5**, 319–327.

39. Cardi T, Earle ED (1997) *Theor. Appl. Genet.* **94**, 204–212.

40. Creemers-Molenaar J, Hall RD, Krens FA (1992) *Theor. Appl. Genet.* **84**, 763–770.

41. Grelon M, Budar F, Bonhomme S, Pelletier G (1994) *Mol. Gen. Genet.* **243**, 540–547.

42. Kirti PB, Banga SS, Prakash S, Chopra VL (1995) *Theor. Appl. Genet.* **91**, 517–521.

43. Leino M, Teixeira R, Landgren M, Glimelius K (2003) *Theor. Appl. Genet.* **106**, 1156–1163.

44. Perl A, Aviv D, Galun E (1990) *J. Heredity* **81**, 438–442.

45. Sigareva MA, Earle ED (1997) *Theor. Appl. Genet.* **94**, 213–220.

46. Varotto S, Nenz E, Lucchin M, Parrini P (2001) *Theor. Appl. Genet.* **102**, 950–956.

47. Yamagishi H, Landgren M, Forsberg J, Glimelius K (2002) *Theor. Appl. Genet.* **104**, 959–964.

*48. Lörz H, Paszkowshi J, Dierks-Ventling C, Potrykus I (1981) *Physiol. Plant.* **53**, 385–391.

49. Ramulu KS, Diijkhuis P, Rutgers E, *et al.* (1995) *Euphytica* **85** 255–268.

50. Binsfeld PC, Wingender R, Schnabl H (2000) *Theor. Appl. Genet.* **101**, 1250–1258.

51. Louzada ES, Del Rio HS, Xia D, Moran-Mirabal JM (2002) *J. Am. Soc. Hort. Sci.* **127**, 484–488.

52. Zhang Q, Liu J, Deng XX (2006) *J. Plant Physiol.* **163**, 1185–1192.

53. Grosser JW (1994) *HortScience* **29**, 1241–1243.

54. Witjaksono, Litz RE, Grosser JW (1998) *Plant Cell Rep.* **18**, 235–242.

55. Xu X, Lu J, Grosser JW, Dalling D, Jittayasothorn Y (2007) *Acta Hort.* **738**, 787–790.

56. Kao KN, Michayluk MR (1974) *Planta* **115, 355–367.

57. Murashige T, Tucker DPH (1969) *Proc First Intl Citrus Symp.* **3**, 1155–1161.

58. Liu J, Deng XX (2002) *Euphytica* **125**, 13–20.

11 Genetic Transformation – *Agrobacterium*

Ian S. Curtis
Texas A&M AgriLife Research, Weslaco, TX, USA

11.1 Introduction

For more than a century, it has been known that the pathogen responsible for inducing crown gall disease in plants is the Gram-negative soil bacterium, *Agrobacterium tumefaciens* [1]. However, it was not until improvements in molecular analyses during the 1970s revealed that such a symptom was the result of genetic material being transferred from the bacterium into the host plant genome [2]. *Agrobacterium* is the only natural vector for inter-kingdom gene transfer [3]. This discovery formed the platform for plant researchers to develop an important tool for understanding plant development and improving crop performance through the transfer of agronomically useful traits.

The production of transgenic plants by *Agrobacterium*-mediated transformation has now become the method of choice compared to biolistic or electroporation procedures. *Agrobacterium*-based gene transfer methods result in the transfer of low numbers of copies of genes into the plant nucleus, giving a reduced frequency of gene silencing events caused by gene dosage. Despite the earlier difficulties in using *Agrobacterium* for the transformation of monocotyledons, significant improvements in plant tissue culture, the discovery of supervirulent strains of *Agrobacterium* and the engineering of novel vectors, has enabled this natural vector system to be used for the production of transgenic plants from a wide range of species. This chapter describes the way in which *Agrobacterium* can be used to genetically transform plants using methods which will assist researchers to understand more about plant

Plant Cell Culture Edited by Michael R. Davey and Paul Anthony

development and allow designer crops to be created through the expression of specific gene traits.

11.2 Methods and approaches

11.2.1 *Agrobacterium* as a natural genetic engineer

A. tumefaciens and *A. rhizogenes* infect a wide range of dicotyledonous plants at wound sites to incite the development of galls or hairy roots [4, 5]. Such symptoms are a result of the transfer and integration of a segment of DNA (T-DNA) from the bacterium Ti (tumour-inducing) and Ri (root-inducing) plasmids into the nucleus of plant cells, followed by their expression. The T-DNA encodes genes responsible for the synthesis of opines which support the growth of the *Agrobacterium* strain inciting the disease, thus creating a metabolic advantage over unrelated strains. In addition, the T-DNA also contains genes responsible for cell growth and development, such genes being involved in the production of auxins and cytokinins. Galls induced by *Agrobacterium* carrying Ti plasmids usually remain undifferentiated, even when transferred to culture. Excised transformed roots from plants infected by *A. rhizogenes* carrying the Ri plasmid can develop into shoots spontaneously or through supplementation of the culture medium with growth regulators. However, such Ri transformed plants usually exhibit wrinkled leaves, dwarfism and are often sterile.

During the last 10 years, advances in molecular biology have improved considerably our understanding of the interaction between *Agrobacterium* with plants enabling researchers to transform a diverse range of species. *Agrobacterium* is attracted chemically towards wounded plant cells and binds to them by a polar attachment mechanism [6]. The genes involved in the production of the transferred DNA intermediate, and the membrane-bound DNA transfer, are located on the virulence (*vir*) regulon (operons *virB, virC, virD, virE* and *virG*) sited on the Ti plasmid. These operons are co-ordinately regulated by a VirA/VirG 'two-component' system common to many bacteria to mediate responses to environmental stimuli [7]. The presence of appropriate chemicals at the infection site causes, either directly or indirectly, the autophosphorylation of the VirA membrane-bound histidine kinase transmitter, which in turn phosphorylates the cytoplasmic transcriptional factor VirG. However, in the case of some cereals, especially maize seedlings, this component system is blocked by the roots exuding a VirA-mediated induction inhibitor, 2-hydroxy-4, 7-dimethoxybenzoxazin-3-one (MDIBOA) which is a resistance mechanism against the transformation process [8]. For transformation to be successful, the phosphorylated VirG binds to a specific region of the *vir* promoters or 'vir box', resulting in the stimulation of transcription of all *vir* genes [9]. VirD1 and VirD2 proteins are responsible for cleavage of the T-strand at the T-DNA left and right border [10]. The VirD2 protein binds to the 5′-end of the T-strand prior to being coated with the single strand binding protein VirE2 to form a T-complex [11]. The T-complex is then exported by a bacterial secretion system involving the *virB* operon and VirD4 protein. Once the T-complex is inside the plant cell, the VirD2 and VirE2 proteins interact with plant components to aid targeting to the plant nucleus [6]. The T-strand finally integrates into the nuclear genome via plant encoded proteins involved in recombination and/or repair processes [12].

Most cereals, especially maize, are considered resistant to transformation by *Agrobacterium* due to limitations in the signal-induced expression of genes involved in the T-DNA transfer process [13]. However, overexpression of *vir* genes in the presence of high concentrations of acetosyringone has greatly improved the transformation efficiency of maize when embryogenic cultures are inoculated with *Agrobacterium* [14]. The isolation of mutant strains of *Agrobacterium* resistant to the VirA induction process, MDIBOA, could help to further improve the transformation of maize and other cereals [8].

11.2.2 Vector systems for transformation

For *Agrobacterium*-mediated transformation to occur, the T-DNA and the *vir* region must be present in the bacterium. One of the first vectors developed for transformation of plants involved the removal of wild type T-DNA, or oncogenes, to create a disarmed strain [15]. The introduction of engineered T-DNA into *A. tumefaciens* involved the insertion of genes into an *Escherichia coli* vector, such as pBR322, that could be integrated into the disarmed Ti plasmid to create a cointegrative vector [16]. Although the system was successful, the resulting vector of ~150 kb was difficult to handle in the laboratory because of its size and instability. The discovery that the T-DNA and the *vir* region could operate on separate plasmids, or in *trans*, to allow transformation, resulted in the evolution of the most important tool in gene transfer, the binary vector [15]. In this case, the T-DNA was integrated into a plasmid which could replicate in both *Agrobacterium* and *E. coli*. Following construction, the vector was transferred into *Agrobacterium* to produce a strain suitable for introducing genes into target plants. The mid-1990s saw the development of the 'super-binary' system which enabled researchers to enhance the transformation efficiency by employing additional virulence genes [17, 18]. This was achieved by inserting a DNA fragment containing *virB, virC* and *virG* genes from pTiB0542 into a small T-DNA carrying plasmid. The final step of making a super-binary vector was to integrate the intermediate vector with an acceptor vector in *Agrobacterium*. Details on the vectors available, their construction, the marker genes which they carry and their limitations to transform plants have been reviewed recently [19].

The ideal vector for plant transformation is one that has several gene cloning sites, a high efficiency for transforming plant cells, wide compatibility with *Agrobacterium* strains, several plant selectable markers and is readily available and robust for handling gene constructs. Unfortunately, no one vector is suitable for all plant transformation studies. However, many of the protocols used for transforming plants use derivatives of pBIN19 [20] because of its convenient cloning sites, stability and availability. Nevertheless, one of the main obstacles is the cloning of DNA fragments of 15 kb or more in size, since these can result in a low efficiency of transformation in bacteria and with associated DNA rearrangements in the bacteria [21]. Hence, careful management of vector construction is critical to any plant transformation project.

Following the construction of the vector, the next step is to transfer the vector into *Agrobacterium*, the most efficient method involving electroporation. This procedure exploits an electric current to create pores in bacterial membranes to allow the DNA to enter the cells. Survival of the bacteria depends on the membranes to reassemble and their ability to tolerate the electrical shock. In the absence of an electroporator, the simplest method of transferring macromolecules into *Agrobacterium* is the freeze/thaw procedure. It is thought that the rapid change in temperature alters the fluidity of the cell membranes of the bacteria, allowing DNA to enter the cells. Finally, another method of transferring plasmid DNA into *Agrobacterium* involves the mating of two strains of *E. coli* (one helper the other donor) with a recipient *Agrobacterium* strain by a technique known as triparental mating. These three methods have been described in a detailed review [22].

11.2.3 Inoculation procedures

Plants can be inoculated by *Agrobacterium* using several methods. One of the first procedures employed for transforming dicotyledonous plants in tissue culture used leaf discs (Protocol 11.1). This system relies on wounded cells at the edges of the explant being transformed by *Agrobacterium* and then developing shoot buds. The protocol essentially describes a method of transforming tobacco and, with slight modifications, can also be applied to other members of the Solanaceae, including tomato and petunia. Refinements to the leaf disc transformation system allowed seedling explants, such as cotyledons and hypocotyls, to be used as target tissues. Floating explants on a culture of *Agrobacterium* enabled other dicotyledonous plants to be transformed. In optimizing the transformation efficiency of *Brassica napus*, the presence of an intact petiole greatly improves transgenic shoot production due to the high regenerative potential of petiole tissues [23]. In the case of petunia, explant size is a critical factor in establishing a highly efficient transformation system using leaf discs inoculated with *Agrobacterium* [24]. In lettuce, the number of bacteria within the inoculum is a critical factor in transformation [25] as large numbers of *Agrobacterium* cells cause the plant cells to become stressed, reducing the transformation efficiency. The preculturing of explants on a shoot regeneration medium prior to *Agrobacterium*-inoculation has also been shown to be beneficial in terms of improving the number of transformed tobacco shoots. In addition, prolonging the time that agrobacteria are in contact with plant tissues in the absence of antibiotics (the cocultivation period) can also be of benefit in some cases. Most plant species require a cocultivation period of 2 days. In the case of the ornamental, *Kalanchoë laciniata*, a period of 7 days can greatly increase the number of transformed shoots compared to shorter co-culture times [26]. Other factors may be used to improve leaf disc transformation for a specific plant species, including the use of a nurse culture, the inclusion of a supervirulent plasmid within the *Agrobacterium* [27], the presence of *vir*-inducing phenolic compounds in the inoculation medium [28] and optimizing the pH of the cocultivation medium [29]. Many dicotyledonous plants can be transformed by floating explants on a suspension of *Agrobacterium*, but each individual plant species may require specific modifications to achieve optimal transformation efficiency.

The discovery that somatic embryogenic calli can be used for transformation enabled several of the cereals and other recalcitrant crops to be transformed by *Agrobacterium*. The ability of rice scutellum tissue to be induced to develop somatic embryos on medium containing 2,4-dichlorophenoxyacetic acid (2,4-D), and the transformation of somatic embryos by *Agrobacterium*, resulted in the first model system for gene targeting in monocotyledons (Protocol 11.2). This milestone in plant genetic transformation accelerated *Agrobacterium*-mediated transformation of maize and wheat. The use of embryogenic calli to produce transgenic plants has extended to the transformation of crops such as banana, grapevine, coffee, tea, cotton and sugarcane [30]. Although the routine production of embryogenic calli is restricted to a few genotypes, further improvement in tissue culture will enhance the importance of such an explant system to increase genetic diversity of our crops.

So far, the procedures used for *Agrobacterium*-mediated transformation have relied on efficient tissue culture systems being available for the regeneration of transgenic shoots from explants. If the latter are not amenable to shoot regeneration in culture, then alternative strategies need to be employed to generate transgenic plants. Radish (*Raphanus sativus*) is a classic example in which the culture of explants from *in vitro*-derived seedling infected with *Agrobacterium* fail to generate transformed shoots. Although improvements in shoot regeneration from hypocotyl and cotyledon explants have been achieved by the addition of ethylene-inhibitors to culture media, there are no reports on the production of transgenic plants through tissue culture. However, if a flowering radish plant is submerged in a suspension of *Agrobacterium* for approx. 5 sec, a small proportion (1.2–1.4%) of the developing seeds from the inoculated plant will become transformed (Protocol 11.3). This simple approach of producing transgenic plants is commonly referred to as the 'floral-dip' procedure. The technique was first applied to the production of transgenic plants of *Arabidopsis thaliana* [31] and later used to transform plants such as pakchoi [32] and *Medicago truncatula* [33]. The floral-dip technique can be performed by researchers without any previous experience of plant tissue culture and serves as a valuable tool in extending the pool of plants that are normally difficult to transform in culture.

PROTOCOL 11.1 Leaf Disc Transformation of Tobacco [34]

Equipment and Reagents

- *Agrobacterium tumefaciens* strain GV3 Ti11SE carrying pTiB6SESE::pMON200 (Monsanto Company). The cointegrative vector pMON200 carries the neomycin phosphotransferase II gene (*npt*II) as a plant selectable marker and the nopaline synthase gene (*nos*), both under the control of the *nos* promoter and terminator sequences
- Seeds of *Nicotiana tabacum* cvs. Samson, Xanthi or SR1
- Surface sterilant: 10% (v/v) 'Domestos' bleach (Johnson Diversey)
- Dissecting instruments, cork borer, sterile tiles, bacterial loops, nylon sieves (64 μm pore size)

- Sterile 7 cm diam. Whatman filter papers
- 9 cm diam. Petri dishes (Bibby Sterilin)
- GA-7 Magenta boxes (Sigma-Aldrich)

Culture medium for *A. tumefaciens*:

- Semisolidified Luria broth (LB): 10 g/l Oxoid Bacto tryptone, 5 g/l Oxoid Bacto-yeast extract, 10 g/l NaCl, 18 g/l agar (Sigma-Aldrich), pH 7.2
- Kanamycin sulfate (Sigma-Aldrich): 10 mg/ml stock in water. Filter-sterilize by passage through a 0.2 μm membrane (Minisart); store at −20 °C
- Streptomycin sulfate (Sigma-Aldrich): 25 mg/ml stock in water. Filter-sterilize; store at −20 °C
- Chloramphenicol (Sigma-Aldrich): 10 mg/ml stock in absolute ethanol. Filter-sterilize; store at −20 °C
- Liquid culture medium for bacteria: LB medium with 50 mg/l kanamycin sulfate, 25 mg/l streptomycin sulfate and 25 mg/l chloramphenicol

Plant tissue culture:

- B5 medium vitamin stock [35]: 100 mg/ml myoinositol, 10 mg/ml thiamine-HCl, 1 mg/ml nicotinic acid, 1 mg/ml pyridoxine-HCl; store at 4 °C
- MSB medium: 4.3 g/l Murashige and Skoog (MS) salts [36], 1 ml/l B5 vitamin stock solution, 30 g/l sucrose, 8 g/l Difco-Bacto agar, pH 5.7
- α-naphthaleneacetic acid (NAA; Sigma-Aldrich): 1 mg/ml stock in 70% (v/v) ethanol; store at 4 °C
- 6-benzylaminopurine (BAP; Sigma-Aldrich): 50 mg/100 ml stock solution. Dissolve BAP in a few drops of 5 N HCl with agitation before making up to volume with water; store at 4 °C
- Cefotaxime (Claforan; Roussel Laboratories): 10 mg/ml stock in water. Filter-sterilize; store at −20 °C
- Carbenicillin (Pyopen; Beechams Research Laboratories): 100 mg/ml stock in water. Filter-sterilize; store at −20 °C
- MS104 medium: MSB medium with 1 mg/l BAP, 0.1 mg/l NAA
- MS104 selection medium: MS104 medium with 500 mg/l carbenicillin, 300 mg/l kanamycin sulfate
- MS rooting medium: MSB medium with 6 g/l agar (Sigma-Aldrich), 500 mg/l carbenicillin, 100 mg/l kanamycin sulfate
- Kanamycin sulfate (Sigma-Aldrich): see culture medium for *A. tumefaciens*.

Sterilize all media by autoclaving at 121 °C for 20 min. Add antibiotics to the media after allowing the media to cool to 40 °C.

Method

1 Immerse tobacco seeds in 10% (v/v) 'Domestos' bleach solution for 20–30 min and rinse three times in sterile water. Place the seeds on a 64 μm nylon mesh during surface sterilization to ease handling.

2 Sow seeds onto 20 ml aliquots of MSB agar medium contained in 9 cm diam. Petri dishes. Incubate the cultures at 24–28 °C, and a 16 h photoperiod with a light intensity of 48 μmol/m^2/s (Daylight fluorescent tubes) for 7 days when the cotyledons will be fully expanded.

3 Transfer individual seedlings (one seedling per box) to Magenta boxes each containing 40 ml of MSB agar medium. Incubate cultures for 21–28 days under the same conditions as used for seed germination until plants have developed four to five true leaves suitable for preparing leaf discs.

4 Excise leaves from parent plants using a scalpel and transfer to a sterile tile for dissection[a]. Alternatively, excise leaves from plants grown in the glasshouse for 28 d (Figure 11.1a). Using either a cork borer or scalpel, excise 1 cm diam. discs and transfer, abaxial surface down, onto MS104 agar medium (seven to eight discs per plate; Figure 11.1b)[b]. Place a single sterile filter paper over the discs to help keep the explants flat on the surface of the medium (Figure 11.1c). Incubate for 2 days as for seed germination (step 2).

5 Take an overnight liquid culture of *A. tumefaciens* and dilute 1:0 (v:v) with MSB liquid medium (2 ml of bacterial culture/20 ml MSB medium in a 9 cm diam. Petri dish).

6 Float the discs in the bacterial suspension, making sure the wounded surface of each explant is immersed in the suspension. After 5 min, blot dry the explants on sterile filter paper. Transfer the explants back to MS104 medium and incubate at 24–26 °C at a low light intensity (24–48 μmol/m^2/s) for 2 days.

7 Transfer the explants to MS104 selection medium and incubate as in step 2.

8 Shoots should be visible from the wounded edges of the disc 18–21 days post-inoculation (Figure 11.1d). Putatively transformed shoots will emerge from kanamycin-resistant calli (Figure 11.1e). After 28 days from *Agrobacterium*-inoculation, regenerated shoots should be large enough (approx. 1 cm in height) to be excised and transferred to rooting medium (Figure 11.1f).

Notes

[a]Leaf discs can also be prepared from plants grown in a glasshouse. Sow seeds (four to five) directly onto the surface of Levington M3 compost (Fisons) contained in a 9 cm diam. pot at 26 ± 2 °C, under natural light supplemented with 16 h photoperiod (150 μmol/m^2/s). Once germinated, select the strongest growing seedling from each pot and allow to grow until four to five true leaves have developed (Figure 11.1a). Excise the fully expanded leaves and surface sterilize by immersion in a 10% (v/v) 'Domestos' bleach solution contained in a sterile casserole dish for 10 min. Discard the bleach solution and rinse the leaves five times with sterile water.

[b]When preparing the discs from surface-sterilized leaves, it is important to remove any bleached tissues prior to culture. Leaves which appear darker green or 'water-soaked' due

to penetration of the surface sterilant should be avoided for preparing leaf discs, since they will not regenerate shoots in culture. Explants should be excised from the leaf laminae and midribs removed to maximize transformation efficiency.

Figure 11.1 Production of putatively transformed shoots of *Nicotiana tabacum* cv. Xanthi by *Agrobacterium*-mediated transformation of leaf discs. (a) Twenty-one-day-old tobacco plants with four to five mature leaves ready for preparing leaf discs. (b) Leaf explants excised from mature leaves. (c) Explants covered with a single, sterile filter paper to keep the wounded edges of the leaves in contact with the culture medium. (d) Leaf discs regenerating shoots from their wounded edges. (e) Transformed shoots emerging from leaf explant-derived tissue. (f) A regenerated shoot ready to be excised from the selected parent tissue prior to rooting. Bars = 5.5 cm (a), 3.3 cm (b–d), 1 cm (e), 3.8 cm (f).

PROTOCOL 11.2 Transformation of Somatic Embryogenic Calli of Rice [37]

Materials

- *A. tumefaciens* strain EHA101 or LBA4404 carrying a CAMBIA vector such as pCAMBIA1201 (CAMBIA, Canberra, Australia). The binary vector pCAMBIA1201 carries the hygromycin phosphotransferase gene (*hpt*) as a plant selectable marker under the control of the cauliflower mosaic virus (CaMV) 35S promoter and terminator sequences. A reporter β-glucuronidase (*gus)*-intron gene is also located between the T-DNA borders and is also under the control of the CaMV35S promoter and *nos* terminator
- Seeds of *Oryza sativa* indica-type rice cultivars, such as BR29 and IR64
- Surface sterilant: 50% (v/v) 'Domestos' bleach solution
- 70% (v/v) ethanol
- 9 cm diam. Petri dishes (Bibby-Sterilin)
- Dissection instruments
- 50 ml capacity disposable screw-capped tubes (BD Biosciences)
- 1.5 ml capacity Cryotubes (Anachem)
- Avanti J-E Centrifuge (Beckman Coulter)
- 50 ml capacity BD Falcon tubes (BD Biosciences)
- Beckman Spectrophotometer DU-65 (GenTech Scientific Inc.)

Stock Solutions

- 2,4-dichlorophenoxyacetic acid (2,4-D; Sigma-Aldrich): 20 mg/20 ml stock solution. Dissolve powder in a few drops of absolute ethanol and add double distilled water to volume; store at 4 °C for 2–3 months
- Cefotaxime: 100 mg/l aqueous stock (see Protocol 11.1)
- Acetosyringone (3′,5′, dimethoxy-4′hydroxy-acetophenone; Merck Chemicals Ltd): 40 mg/ml stock solution. Dissolve the powder in a few drops of dimethyl sulfoxide (DMSO; Sigma-Aldrich); make to volume with double distilled water. Filter-sterilize; store in the dark at 4 °C
- Kinetin (Sigma-Aldrich): 20 mg/20 ml stock solution. Dissolve the powder in a few drops of concentrated HCl and add double distilled water to volume; store at 4 °C
- NAA stock solution: See Protocol 11.1
- Hygromycin: 50 mg/ml aqueous stock. Filter-sterilize; store at −20 °C

Culture Media for *A. tumefaciens*

- AB medium: 3 g/l K_2HPO_4, 1 g/l $NaH_2PO_4.H_2O$, 1 g/l NH_4Cl, 300 mg/l $MgSO_4.7H_2O$, 150 mg/l KCl, 10 mg/l $CaCl_2.2H_2O$, 2.5 mg/l $FeSO_4.7H_2O$, 10 g/l glucose, 30 g/l agar
- AAM medium: See Table 11.1

Table 11.1 Composition of AAM medium[a].

Components	Concentration (mg/l)
Macronutrients	
$CaCl_2.2H_2O$	150
$MgSO_4.7H_2O$	250
$NaH_2PO_4.H_2O$	150
KCl	2950
Micronutrients	
KI	0.75
H_3BO_3	3
$MnSO_4.H_2O$	10
$ZnSO_4.7H_2O$	2
$Na_2MoO_4.2H_2O$	0.25
$CuSO_4.5H_2O$	0.025
$CoCl_2.6H_2O$	0.025
Iron composition	
Na_2EDTA	37.3
$FeSO_4.7H_2O$	27.8
Vitamins	
Nicotinic acid	0.5
Pyridoxine-HCl	0.5
Thiamine-HCl	1
Glycine	2
Myoinositol	100
Others	
L-glutamine	876
Aspartic acid	266
Arginine	174
Casamino acid	500
Sucrose	68 500
Glucose	36 000

[a]Add 200 μM to the medium. Mix components thoroughly prior to adjusting to pH 5.2. Filter-sterilize (0.2 μm pore size).

Plant Tissue Culture Media

- MS 2,4-D medium: MS salts and vitamins, 300 mg/l casamino acid, 2 mg/l 2,4-D, 8 g/l agar, pH 5.8
- N6-AS liquid medium: N6 salts, MS vitamins, 300 mg/l casamino acid, 2 mg/l 2,4-D, 30 g/l sucrose, 10 g/l glucose, pH 5.2
- N6-AS medium: N6-AS liquid medium, 9 g/l agar
- Selection medium: MS salts and vitamins, 300 mg/l casamino acid, 2 mg/l 2,4-D, 500 mg/l cefotaxime, 50 mg/l hygromycin, 30 g/l sucrose, 8 g/l agar, pH 5.8
- MSKN medium: MS salts and vitamins, 2 mg/l kinetin, 1 mg/l NAA, 300 mg/l casamino acid, 50 mg/l cefotaxime, 30 g/l sucrose, 10 g/l sorbitol, 2.5 g/l Gelrite (Sigma-Aldrich), pH 5.8
- MS0 medium: MS salts and vitamins, 30 g/l sucrose, 2.5 g/l Gelrite, pH 5.8

Sterilize all media by autoclaving at 121 °C for 20 min (18 kPa nominal steam pressure). Add antibiotics to the agar media after the latter have cooled to 40 °C.

Method

Induction of embryogenic calli:

1 Remove the hulls from immature or mature seeds by hand.

2 Transfer approx. 100 seeds to a 50 ml capacity disposable screw-capped tube containing 70% ethanol and shake briefly for 1 min. Remove the ethanol and replace with 50% (v/v) 'Domestos' bleach solution and incubate on a shaker at 60 rpm for 30 min. Discard the bleach solution and wash seeds three times with sterile water.

3 Place the seeds on a sterile tile and isolate the embryos using a scapel and forceps. Transfer the isolated embryos to the surface of 20 ml aliquots of MS 2,4-D medium in 9 cm Petri dishes (25 embryos/dish). Embryos should be partially submerged (1–2 mm) in the medium with the scutellar tissue uppermost. Incubate cultures in the dark at 25 °C[a].

4 After 3–4 days, shoots and roots will begin to develop from the embryo. These should be excised to allow embryogenic calli to develop.

5 Yellowish white soft embryogenic calli should be visible developing from the scutellum tissue after 14–21 days of culture.

6 After 28–35 days of culture, the embryogenic calli (each approx. 1.5–3.0 mm in diam.) should be transferred to new MS 2,4-D medium (100 calli/dish) and incubated at 28 °C in the dark prior to inoculation 3 days later.

Preparation of a glycerol stock of *Agrobacterium*:

1 Using a sterile loop, collect a single colony of *Agrobacterium* strain EHA101 or LBA4404 carrying pCAMBIA1201 and inoculate 10 ml of AAM medium containing

50 mg/l rifampicin and 50 mg/l kanamycin in a 50 ml disposable tube. Place the tube on a shaker at 250 rpm in the dark at 28 °C overnight.

2 Transfer 12 ml of AAM liquid medium to a 50 ml capacity disposable tube containing 8 ml of glycerol solution (Sigma-Aldrich). Add stock solutions of rifampicin and kanamycin to give the same concentrations as for growing the bacteria in culture. Mix the solution and filter-sterilize through a 0.2 μm filter into a new 50 ml capacity sterile tube.

3 Transfer 0.5 ml of the overnight *Agrobacterium* suspension to a 1.5 ml capacity screw-capped cryotube containing 0.5 ml of the glycerol/AAM mixture. Invert the tube to mix and store at −70 °C.

Culture and pretreatment of *Agrobacterium*:

1 Transfer a loopful of a glycerol stock of *Agrobacterium* strain EHA101 or LBA4404 containing pCAMBIA1201 and streak on semisolid AB medium containing 50 mg/l rifampicin and 50 mg/l kanamycin; incubate in the dark for 2 days at 28 °C[b].

2 Using a sterile loop, collect three to four colonies of *Agrobacterium* and inoculate 50 ml of AAM medium containing the same concentration of antibiotics as in Step 1 in a 250 ml capacity Erlenmeyer flask. Place the flask on a shaker at 250 rpm in the dark at 28 °C overnight.

3 Add 200 μM of acetosyringone to the *Agrobacterium* suspension and continue shaking for 2 h.

4 Pellet the bacteria by centrifugation (3500 g, 30 min, 10 °C) and discard the supernatant. Resuspend the bacteria in 20 ml $MgSO_4$ (10 mM) contained in a 50 ml capacity BD Falcon tube[a].

5 Repeat the centrifugation and resuspend the pellet in a small volume of liquid N6-AS medium. Add medium to give an optical density (OD) of the *Agrobacterium* suspension of approx. 1.0 at 600 nm.

Inoculation of embryogenic calli, transformation and selection:

1 Transfer the embryogenic calli to a Petri dish (50 × 18 mm) containing the bacterial suspension and leave for 20 min[c]. Remove the suspension using a pipette and then remove the excess bacterial medium from the tissues by blotting with sterile filter papers. Carefully transfer the calli to semisolid N6-AS medium and incubate in the dark for 3 days at 28 °C[d].

2 Subculture the calli to MS 2,4-D medium containing 50 mg/l hygromycin and 250 mg/l cefotaxime. Incubate the cultures in the dark for 10 days at 25 °C.

3 Remove the calli and transfer to Selection medium containing 50 mg/l hygromycin and 250 mg/l cefotaxime. Incubate in the dark for 14 days at 25 °C. Continue to subculture the surviving calli to new selection medium every 14 days for two more passages[e].

4 Transfer surviving calli to MSKN regeneration medium and incubate in the dark for 20 days.

5 Transfer calli with emerging shoots to new MSKN medium and culture in the light (110 µmol/m^2/s) at 27 °C with a 16 h photoperiod for 10–20 days.

6 Excise healthy shoots and root the excised shoots on MS0 medium[f].

Notes

[a]Incubating cultures under continuous light at 32 °C for 5 days can improve the transformation efficiencies of recalcitrant rice varieties by increasing the proliferation of embryogenic calli from the scutellum.

[b]The use of *A. tumefaciens* strains EHA101 and EHA105 in transformation studies usually yields more primary transformants compared to the strain LBA4404. In terms of the number of transgenes integrated into the genome of such transformed plants, molecular studies have revealed that using LBA4404 produces a greater frequency (30–40%) of single copy inserts compared to strains EHA101 and EHA105 (10%).

[c]The transformation of rice calli can be improved by placing the dish into a vacuum desiccator for 10 min.

[d]The presence of acetosyringone in the culture medium for *Agrobacterium* and in the medium used after inoculating the calli is critical in generating transformed plants.

[e]It is important to transfer the fast growing healthy calli to MSKN medium as soon as possible to minimize the possibility of aberrant phenotypes *e.g.* plants exhibiting low seed yield.

[f]Choose only one shoot per callus to avoid the generation of sibling transformants (plants with the same T-DNA insertion pattern).

PROTOCOL 11.3 Transformation of Radish by the Floral-Dip Procedure [38]

Materials

- *A. tumefaciens* strain AGL1 carrying pCAMBIA3301. The binary vector pCAMBIA3301 carries the *gus*-intron and bialaphos resistance (*bar*) genes both under the control of the CaMV 35S promoter located between T-DNA border fragments
- Seeds of the Korean radish cv. 'Jin Ju Dae Pyong' (Kyoungshin Seeds Co.)
- Petri dishes (9 cm diam.; Bibby-Sterilin)
- Pointed dissecting scissors
- Fine paint brush
- Measuring cylinders and beakers (1–2 l capacity)
- 50 ml capacity BD Falcon tubes (BD Biosciences)
- Certomat IS UHK Orbital Shaker (DJB Labcare)
- Beckman Spectrophotometer DU-65 (GenTech Scientific)

Culture media for *A. tumefaciens*:

- YEP medium: 10 g/l tryptone, 10 g/l yeast extract, 5 g/l NaCl.
- Kanamycin sulfate: (see Protocol 11.1)
- Rifampicin (Sigma-Aldrich): 4 mg/ml stock. Dissolve powder in methanol. Filter-sterilize; store at −20 °C
- Bacterial culture medium: YEP medium, 50 mg/l kanamycin sulfate, 50 mg/l rifampicin
- Agar-solidified medium: YEP medium, 14 g/l agar, 50 mg/l kanamycin sulfate, 100 mg/l rifampicin

Sterilize all media by autoclaving at 121 °C for 20 min. Add antibiotics to the agar media after the latter have cooled to 40 °C.

Solutions

- Inoculation medium: 50 g/l sucrose, 0.05% (v/v) Silwet L-77 (Setre Chemical Co.), pH 5.2

Method

1. Sow seeds (one seed/3 cm^2) in a deep seed tray (12 cm depth) containing a peat-based compost and maintain in a glasshouse under natural daylight supplemented with 61 μmol/m^2/s Daylight fluorescent illumination (16 h photoperiod) at 26 °C (day) and 18 °C (night).
2. After approx. 21–28 days, carefully transfer individual plants to deep pots (20 cm diam., 30 cm depth) containing new compost to encourage plants to develop long taproots. Maintain the plants under the same glasshouse conditions for 10 days to aid recovery.
3. Transfer plants at the six-leaf stage of development to a cold chamber set at 4 ± 2 °C (16 h photoperiod, 45 μmol/m^2/s, daylight fluorescent tubes) for 10 days to promote bolting. Return the plants to the glasshouse under conditions described previously in step 1.
4. Plants with single thick stems with numerous immature floral buds are ideal for the floral dip technique[a].
5. Four days before the floral dip treatment, take a loop-full of a glycerol stock of *Agrobacterium* strain AGL1 carrying pCAMBIA3301 and streak onto agar-solidified medium containing 50 mg/l kanamycin and 100 mg/l rifampicin. Incubate the culture in the dark at 28 °C for 2 days[b] .
6. Using a sterile loop, transfer a loop-full of bacteria to a 50 ml capacity Falcon tube containing 10 ml of bacterial culture medium. Transfer the culture to an orbital shaker at 180 rpm and maintain in the dark at 28 °C overnight.
7. Transfer the 10 ml liquid bacterial culture to a 1 l capacity flask containing 500 ml of bacterial culture medium and incubate for 12–16 h as described earlier in step 6 until the OD reaches 1.0 at 600 nm.

8 Pellet the bacterial culture by centrifugation (3500 g, 20 min, 4 °C). Resuspend the culture in 500 ml of inoculation medium.

9 Remove any floral buds which show petal colour prior to the floral dip treatment. Carefully submerge the inflorescence into the inoculation medium and gently swirl for 5 s[c]. Transfer the plant to an upright position and cover the inflorescence with a polythene bag[d]. Place all floral-dipped plants under the staging of the glasshouse and leave overnight.

10 Transfer the dipped plants to the glasshouse staging and remove the bag. Allow the plants to grow under conditions described earlier. Hand-pollinate all flowers using a fine paint brush to aid seed set[e].

Notes

[a]The developmental stage of the inflorescence is a critical factor in the transformation of radish. Plants with a single primary bolting stem (1.4% of all harvested seeds) produce more transformed seeds compared to plants with secondary (0.2%) and tertiary (0%) bolting stems.

[b]*A. tumefaciens* strain AGL1 is known to transform a wide range of seedling explants of the Korean radish cv. 'Jin Ju Dae Pyong'. It is not known whether other strains of *Agrobacterium* are virulent on this or other cvs. of radish.

[c]Silwet L-77 is more efficient than Pluronic F-68 (Sigma-Aldrich) and Tween 20 (Sigma-Aldrich) as a surfactant in terms of the yield of transformed seeds from floral-dipped plants.

[d]It is important to remove any air pockets in the bag to prevent the inflorescence drying out. Keeping the inflorescence in contact with the inoculation medium containing agrobacteria aids the movement of bacteria to inoculation sites, such as the ovule.

[e]Radish will not form seeds in the absence of an effective pollinator, such as insects and wind. Hand pollinating open-flowered flowers daily for 3 d improves seed set and, importantly the production of transformed seeds.

11.3 Troubleshooting

- When designing transformation experiments, it is critical to employ a negative control to allow chemical, molecular and phenotypic studies to be compared between putative transformants and non-transformed plants. In terms of using leaf discs and calli as explants for *Agrobacterium* inoculation, there should also be uninoculated explants. Shoots that regenerate from uninoculated explants are ideal control material to determine whether somaclonal variation is a factor in the phenotypic characterisation of plants through tissue culture. In the case of the floral-dip approach, some plants should be dipped into inoculation medium lacking *Agrobacterium*, but hand-pollinated with a fine paint brush not used for inoculated plants.

- A positive control treatment is also important, especially when *Agrobacterium* is being used to deliver an agronomic trait into a target plant. Such a treatment will confirm whether the marker gene(s) alone influences the phenotype of transformants carrying the agronomic trait. In the case of the floral-dip technique on radish, a positive control treatment consists of dipping plants into a suspension of *Agrobacterium* carrying only marker genes (e.g. *bar* and *gus*-intron genes from pCAMBIA3301). Transformed seeds from these plants can be compared with seeds derived from the treatment in which the *Agrobacterium* carries both marker genes and gene(s) of agronomic interest.

- Optimum results are obtained when commencing with a rapidly growing cell line of *Agrobacterium*. This can be achieved using a glycerol stock of bacteria derived from a rapidly growing single colony. The use of an overnight liquid culture with an optical density less than 0.6 should be avoided in inoculation studies, as transformation rates are generally poor.

References

1. Smith EF, Townsend CO (1907) *Science* **25, 671–673.

The original publication describing *Agrobacterium*.

2. Chilton MD, Drummond MH, Merlo DJ, *et al.* (1977) *Cell* **11, 263–271.

The original publication confirming gene transfer from *Agrobacterium* to plants.

3. Hansen G, Chilton MD (1999) *Curr. Top. Microbiol. Immunol.* **240**, 21–57.

4. Erwin DC, Stuteville DL (1990) *Compendium of Alfalfa Diseases*, 2nd edn. APS Press, Minnesota.

5. Christey MC (2001) *In Vitro Cell Dev. Biol.-Plant* **37**, 687–700.

6. Tzfira T, Citovsky V (2002) *Trends Cell Biol.* **12**, 121–128.

7. Pirrung MC (1999) *Chem. Biol.* **6**, 167–175.

8. Zhang J, Boone L, Kocz R, Zhang C, Binns AN, Lynn DG (2000) *Chem. Biol.* **7**, 611–621.

9. Zupan J, Muth TR, Draper O, Zambryski P (2000) *Plant J.* **23**, 11–28.

10. Stachel SE, Zambryski PC (1985) *Cell* **46**, 325–333.

11. Vergunst AC, Schrammeijer B, der Dulk-Ras A, de Vlaam CMT, Regensburg-Tuink, TJG, Hooykaas PJJ (2000) *Science* **290**, 979–982.

12. van Attikum H, Bundock P, Hooykaas PJJ (2001) *EMBO J.* **20**, 6550–6558.

13. Raineri DM, Boulton MI, Davies JW, Nester EW (1993) *Proc. Natl Acad. Sci. USA* **90**, 3549–3553.

14. Ishida Y, Saito H, Ohta S, Hiei Y, Komari T, Kumashiro T (1996) *Nat. Biotechnol.* **14**, 745–750.

15. Hoekama A, Hirsch PR, Hooykaas PJJ, Schilperoort RA (1983) *Nature* **303, 179–180.

The original publication describing the use of a vector for *Agrobacterium*-mediated gene transfer to produce transgenic plants.

16. Fraley RT, Rogers SG, Horsch RB (1986) *Crit. Rev. Plant Sci.* **4**, 1–46.

17. Srivatanakul M, Park SH, Salas MG, Smith RH (2000) *J. Plant Physiol*. **157**, 685–690.

18. Vain P, Harvey A, Worland B, Ross S, Snape JW, Lonsdale D (2004) *Trans. Res*. **13**, 593–603.

19. Komari T, Takakura Y, Ueki J, Kato N, Ishida Y, Hiei Y (2006) In: *Methods in Molecular Biology*. Vol. 343. *Agrobacterium Protocols*. Edited by K Wang. Humana Press Inc., Totowa, NJ, USA, pp. 15–41.

20. Bevan M (1984) *Nucl. Acids Res*. **12**, 8711–8721.

21. Sambrook J, Russell DW (2001) *Molecular Cloning, A Laboratory Manual*, 3rd Edn. Cold Spring Harbor Laboratory Press, Cold Spring Harbor, NY, USA.

22. Wise AA, Liu Z, Binns A (2006) In *Methods in Molecular Biology*. Vol. 343. *Agrobacterium Protocols*. Edited by K. Wang. Humana Press Inc., Totowa, NJ, USA, pp. 43–53.

23. Moloney MM, Walker JM, Sharma KK (1989) *Plant Cell Rep*. **8**, 238–242.

24. Beck MJ, Camper ND (1991) *Plant Cell Tissue Organ Cult*. **26**, 101–106.

25. Michelmore R, Marsh E, Seely S, Landry B (1987) *Plant Cell Rep*. **6**, 439–442.

26. Jia SR, Yang MZ, Ott R, Chua N-H (1989) *Plant Cell Rep*. **8**, 336–340.

27. Curtis IS, Power JB, Blackhall NW, de Laat AMM, Davey MR (1994) *J. Exp. Bot*. **45**, 1441–1449.

28. Sheikholeslam SN, Weeks DP (1987) *Plant Mol. Biol*. **8**, 291–298.

29. Goodwin I, Todd G, Ford-Lloyd B, Newbury HJ (1991) *Plant Cell Rep*. **9**, 671–675.

30. Curtis IS (2004) *Transgenic Crops of the World – Essential Protocols*. Kluwer Academic Publishers, Dordrecht, The Netherlands.

31. Bechtold N, Ellis J, Pelletier G (1993) *C.R. Acad. Sci. Paris, Life Sci*. **316**, 1194–1199.

32. Qing CM, Fan L, Lei Y, *et al.* (2000) *Mol. Breed*. **6**, 67–72.

33. Trieu AT, Burleigh SH, Kardailsky IV, *et al.* (2000) *Plant J*. **22**, 531–541.

34. Horsch RB, Fry JE, Hoffmann NL, Eichholtz D, Rogers SG, Fraley RT (1985) *Science* **227, 1229–1231.

The first report of leaf disc transformation.

35. Gamborg OL, Miller RA, Ojima K (1968) *Exp. Cell Res*. **50**, 151–158.

36. Murashige T, Skoog F (1962) *Physiol. Plant*. **15**, 473–497.

37. Hiei Y, Ohta S, Komari T, Kumashiro T (1994) *Plant J*. **6**, 271–282.

38. Curtis IS, Nam HG (2001) *Trans. Res*. **10**, 363–371.

12
Genetic Transformation – Biolistics

Fredy Altpeter and Sukhpreet Sandhu

Agronomy Department, Plant Molecular Biology Programme, Genetics Institute, University of Florida – IFAS, Gainesville, FL, USA

12.1 Introduction

The term 'biolistic' is derived from biological + ballistics and is often used interchangeably with terms such as 'microprojectile bombardment', 'particle bombardment', or 'the particle gun method'. Biolistic gene transfer employs high-velocity metal particles to deliver biologically active DNA into plant cells. This concept has been described in detail by Sanford [1]. A comprehensive review on microparticle bombardment technology and its applications has been provided by Altpeter *et al.* [2] and Taylor and Fauquet [3]. Biolistic gene transfer has become the most commonly used direct gene transfer method in plants. Its versatility, ease of adaptability to a wide range of cells and tissues, and high transformation efficiency, makes it a popular system of choice for many crop species. It supports gene stacking and pathway engineering by transfer of multiple unlinked transgene expression cassettes. In contrast to *Agrobacterium*-mediated gene transfer, biolistic transfer of minimal expression cassettes effectively avoids integration of prokaryotic vector backbone sequences into the recipient genome. Alternative direct gene transfer systems, including electroporation [4], polyethylene glycol (PEG)-mediated DNA uptake [5], silicon carbide fibres [6] and microtargeting [7], are typically less efficient or versatile than biolistic gene transfer. Biolistic gene transfer is also the most efficient system for gene transfer to the chloroplast genome (for review see [2]).

Plant Cell Culture Edited by Michael R. Davey and Paul Anthony

12.2 Methods and approaches

12.2.1 Biolistic technology

Direct gene transfer through particle bombardment was developed in the 1980s in an attempt to overcome both the host limitations of *Agrobacterium*-mediated transformation, and the technical difficulty of protoplast-mediated gene transfer. Particle bombardment employs the use of accelerated DNA coated particles directly into cells, a concept first reported in 1987 [8]. The first particle delivery device utilized DNA-coated tungsten metal particles as microcarriers adhered to a plastic macroprojectile. A gunpowder cartridge was used to propel the macroprojectile into a stopping screen. The stopping screen arrested the macroprojectile and the tungsten particles were launched through the openings in the stopping screen to penetrate the plant tissue situated below the screen [1, 8]. This device was marketed by Biolistics Inc., and sold as the Biolistic device, Model BPG. DuPont Inc. later developed the PDS-1000 device, which was further modified to form the PDS-1000/He instrument by BioRad Laboratories. This is the most commonly used apparatus for particle bombardment and utilizes inert helium gas as the accelerating force. A simple cost-effective alternative to the PDS-1000/He was developed by Finer *et al.* in 1992 [9], known as the particle inflow gun (PIG). Other instruments used for direct gene transfer include those based on ACCELL technology [10], the microtargeting bombardment device [7] and the Helios gene gun (Biorad, 2002).

PDS-1000/He biolistic particle delivery system

This instrument was introduced in the 1990s and is the most widely used system for transient gene expression studies and the generation of transgenic plants by direct gene transfer. The Biolistic PDS-1000/He system uses high-pressure helium, released by a rupture disc and a partial vacuum, to propel a macrocarrier plastic sheet, loaded with DNA-coated tungsten or gold microcarriers, towards target tissues at high velocity [11]. A stopping screen arrests the macrocarrier after a short distance. The DNA-coated microcarriers continue traveling and penetrate the target tissues to affect gene transfer. The launch velocity of microcarriers for each bombardment is dependent upon the helium pressure which is typically adjusted to 6.2–8.9 MPa (900–1300 psi) by selection of the appropriate rupture disc, the vacuum in the bombardment chamber, the distance from the rupture disc to the macrocarrier, the macrocarrier travel distance to the stopping screen and the distance between the stopping screen and target cells.

Particle inflow gun (PIG)

The particle inflow gun is a low-cost alternative system for gene transfer which does not employ the use of rupture discs [9]. The target tissues are placed in a vacuum chamber and the gold particles travel with helium inflow generated by a solenoid. This assembly has been used to generate transgenic plants from several crops including soybean [9], barley [12], cassava [13] and bahiagrass [14].

PIG technology is an efficient transformation system for tissues that are highly susceptible to bursts of gas and acoustic shock [12].

Electrical discharge particle acceleration: ACCELL technology

ACCELL technology employs high-voltage electrical discharge for particle acceleration [10, 15]. By varying the intensity of an electric discharge through a water droplet a shock wave is created which accelerates the DNA-coated gold particles. A major advantage of this technology is that penetration of the target tissue can be controlled very accurately. It has been used for the genetic transformation of important crops such as soybean [16], rice [17], poplar [18] and cotton [19].

Microtargeting bombardment device

DNA is targeted to actively dividing totipotent cells in the shoot meristematic region. Pressurized gas, such as nitrogen, is applied to the droplet carrying the DNA-microcarrier mixture, which is forced through a small aperture and delivered to the target tissue under vacuum [7]. Since this device enables DNA targeting to regions as small as 150 μm, it can facilitate the use of shoot meristems for gene transfer in a genotype-independent manner [20, 21]. However, this technology does not support the same throughput and transformation efficiencies as the devices described earlier.

Helios gene gun

The Helios gene gun manufactured by BioRad is a semiportable particle bombardment instrument. Helium accelerates microcarriers (DNA coated gold particles) down the barrel to strike the tissue. A vacuum is not utilized, which makes the instrument portable and effective for field or glasshouse applications, unlike all other systems (BioRad 2002). This device has been used for transient expression studies in plants, but not to generate stably transformed plants [22, 23].

12.2.2 Optimization of gene delivery parameters

Gene transfer parameters, such as vacuum [24], size and density of microparticles [25, 26], distance between rupture disc, macrocarrier and the target tissue [12], helium pressure [27], osmoticum treatment [28] and time of preculture of the target tissue [26], are critical components to maximize transformation efficiency. The ability to adjust these parameters makes particle bombardment versatile. The transformation efficiency also depends on cell survival post-bombardment and maintenance of plant regeneration capacity. Hence, it is critical to determine gene transfer parameters that support the introduction of DNA with minimal tissue damage and to identify highly regenerable plant genotypes [26]. This article discusses these important factors for efficient gene transfer with respect to the PDS-1000/He device. A detailed analysis of factors affecting biolistic transformation is also described by Southgate *et al.* [29].

Bombardment conditions – acceleration pressure, vacuum, distance

Physical parameters like pressure, vacuum and distance of target tissues from the macrocarrier disc must be optimized according to the species, genotype, source of explant and the cell type. Helium pressure is the accelerating force for the macro- and microcarriers and is adjusted by using specific rupture discs. The acceleration pressure influences cell penetration and, most typically, rupture discs of 7.6 MPa (1100 psi) are used for plant transformation. The distance of target tissues from the microcarrier plate and the vacuum also influence cell penetration. Optimization of these parameters is critical for efficient transformation [30, 24].

Microprojectiles – material, size

Inert metal particles, such as tungsten and gold, are used as microcarriers. Tungsten has an irregular shape, which may enhance the formation of aggregates. Gold particles are spherical and hence reduce agglomeration [27]. The smaller the size of microparticles, the less tissue damage that results, but there is reduction in the amount of transferred DNA. Gold particles of 1.0 μm diameter are most widely used for gene transfer to the nucleus, and particles of 0.6 μm in diameter for gene transfer to plastids.

Coating microprojectiles with DNA – DNA precipitation, DNA concentration, amount of microcarriers

Precipitation conditions, including the concentration of calcium chloride and spermidine used to coat microprojectiles, have been optimized and it is important to vortex continuously during the addition of these components to support even coating and to prevent the formation of large aggregates. Recently, protamine has been suggested as an alternative to spermidine, since it supports better DNA protection and results in greater transient and stable transformation frequencies [31]. Excessive amounts of DNA enhance particle agglomeration [32, 29] and can increase the complexity of transgene loci [33, 34]. Excessive amounts of microcarriers can reduce even coating of particles with DNA, or increase tissue damage following bombardment [26]. Brief sonication may disperse particle aggregates, although excessive sonication may shear DNA [27].

12.2.3 Target tissues

The versatility of particle bombardment relies on its ability to transfer exogenous DNA into a range of plant organs, tissues and cells, including leaves, stems, immature embryos, immature inflorescence, microspores, meristems, callus and suspension cultures (for review see [3]). Plant regeneration through somatic embryogenesis supports the formation of non-chimeric stably transformed plants [35]. Direct regeneration through organogenesis may reduce somaclonal variation. However, organogenesis increases the regeneration of chimeric plants following gene transfer [36]. Tissue culture response, embryogenesis and regeneration efficiency are genotype dependent [37].

Embryogenic tissues

The totipotency of grass meristems supports the production of embryogenic tissues and subsequent plant regeneration by appropriate *in vitro* manipulation [38]. Actively dividing, undifferentiated embryogenic tissues are the preferred target for gene transfer because: (i) actively dividing cells enhance integration of exogenous DNA; (ii) undifferentiated cells allow effective selection in tissue culture; and (iii) following selection of transgenic events, embryogenic tissue is capable of plant regeneration in response to manipulation of growth regulators in the culture medium and growth conditions. Some limitations for the production of high quality embryogenic calli include: (a) genotype dependency; commercially important cultivars may not produce embryogenic calli and hence are not easily amenable to transformation; (b) generation and maintenance of embryogenic calli is time consuming; and (c) prolonged tissue culture may affect plant regeneration potential and also result in mutations [29]. Therefore, it is important to optimize protocols to minimize the culture time. The maturity of explants at the time of callus induction, typically affects the plant regeneration ability of callus. Good quality embryogenic tissue may be obtained from a range of tissues, including germinating seedlings (e.g. bahiagrass [39]), immature inflorescences (e.g. seashore paspalum [40]), immature embyros (e.g. rye [41], wheat [25]) and mature embryos (e.g. ryegrass [42]).

PROTOCOL 12.1 Preparation of Gold Stock (60 mg/ml)

Equipment and Reagents

- 1.0 μm gold particles (BioRad)
- 50% (v/v) glycerol (Sigma-Aldrich)
- 70% (v/v) ethanol[a]
- Autoclaved double distilled water (ddH_2O)
- 1.5 ml microfuge tubes
- Bench-top centrifuge (Model 5415 D; Eppendorf)

Method

1 Weigh 60 mg of 1.0 μm gold particles in a sterile 1.5 ml microfuge tube.

2 Vortex for 3–5 min after adding 1 ml of 70% (v/v) ethanol.

3 Centrifuge briefly (5 s) to pellet the microparticles.

4 Discard the supernatant; follow by three washes each with 1 ml autoclaved ddH_2O.

5 Vortex for 1 min.

6 Centrifuge briefly (3–5 s) and again remove the supernatant.

7 Add 1 ml sterile 50% (v/v) glycerol.

8 Store the gold stock at −20 °C.

Note

[a]70% (v/v) ethanol should be prepared with ddH_2O and ethanol.

PROTOCOL 12.2 Preparation of Minimal Linear Expression Cassettes

Equipment and Reagents

- 0.8% (w/v) agarose gel prepared using 0.5 × TBE (45 mM Tris, 45 mM boric acid, 1 mM EDTA, pH 8.0)
- Electrophoresis apparatus – gel tank, casting tray, combs and power pack (e.g. BioRad)
- Bench-top centrifuge (Model 5415 D; Eppendorf)
- Spectrophotometer (ND-1000; Nanodrop Technologies).
- Qiagen gel purification kit (Qiagen Inc.)
- UV transilluminator (FOTO/UV; Fotodyne Inc.)

Method

1 Digest 100 μg of plasmid DNA using restriction enzymes that excise the gene expression cassette (promoter, gene, 3′UTR) and that do not cleave within the gene expression cassette[a].

2 Check for complete digestion by running a 1 μl aliquot of the digested plasmid on a 0.8% (w/v) agarose gel. Complete digestion will be indicated by the expected size and number of bands.

3 Load 10–15 μg of DNA/well on a 0.8% (w/v) agarose gel and electrophorese at 70 V for 2 h to achieve good separation of bands.

4 Excise the band corresponding to the expression cassette with a scalpel during visualization with a UV-transilluminator[b].

5 Purify the excised band with the QIAquick gel purification kit following the instructions provided by the manufacturer.

6 Check DNA quality by electrophoresis of 2 μl of the eluted DNA.

7 Quantify the DNA concentration using the Nanodrop or alternative spectrophotometer.

Notes

[a]If the vector backbone and gene expression cassette have a similar size, choose a restriction enzyme which cleaves in the central region of the vector backbone, but which does not cleave within the gene expression cassette.

[b]Eye protection must be worn while working with the UV transilluminator.

PROTOCOL 12.3 Sterilization of the Biolistic Gene Delivery Device and Components

Equipment and Reagents

- 70% (v/v) ethanol
- Absolute ethanol
- Autoclave
- Biolistic gene delivery instrument (PDS-1000/He; BioRad) and components including rupture discs (7.6 MPa, 1100 psi), macrocarriers, macrocarrier holders, stopping screen
- Laminar air flow cabinet

Method

1 Clean the chamber of the biolistic gene delivery device (PDS-1000/He, BioRad) and laminar flow cabinet thoroughly by wiping with 70% (v/v) ethanol[a].

2 Autoclave macrocarrier holders, stopping screens and macrocarriers.

3 Sterilize rupture discs by briefly immersing (2–3 s) in absolute ethanol; allow the discs to air-dry in the laminar flow cabinet.

Note

[a]Allow sufficient time for ethanol to evaporate completely before using the device.

PROTOCOL 12.4 Preparation of DNA Coated Microparticles

Equipment and Reagents

- 0.1 M spermidine solution (Sigma-Aldrich)
- 2.5 M calcium chloride solution (Sigma-Aldrich)
- Sterile double distilled water (ddH_2O)
- Absolute ethanol

- Microfuge tubes (Fisher Scientific)
- Bench-top centrifuge (Model 5415 D; Eppendorf)
- Vortex (Vortex-Genie; Scientific Industries)
- Sonicator (Branson 2200; Branson Ultrasonics)

Method

1. Transfer 30 μl of the gold stock suspension after vortexing to a 1.5 ml sterile microfuge tube.
2. Place 20 μl 0.1 M freshly prepared spermidine and 50 μl 2.5 M $CaCl_2$ on the lid of the microfuge tube.
3. Add gene expression cassette (1–5 μg) and sterilized ddH_2O to a final volume of 60 μl to the lid.
4. Mix all components by closing the lid of the microfuge tube and vortexing for 1 min.
5. Centrifuge briefly (3–5 s) to pellet the gold.
6. Discard the supernatant without disturbing the pellet; add 250 μl absolute ethanol for washing.
7. Centrifuge briefly (3–5 s) and discard the supernatant.
8. Repeat the previous washing with ethanol.
9. Resuspend the gold pellet in 90 μl absolute ethanol by sonication for 1 s.
10. Keep the DNA coated microparticles on ice[a].

Note

[a]The DNA coated gold particles tend to settle quickly. Therefore, the microparticles should be resuspended by vortexing and then immediately pipetted onto the macrocarrier.

PROTOCOL 12.5 Particle Bombardment[a] using the PDS-1000/He Instrument

Equipment and Reagents

- Biolistic gene delivery PDS-1000/He instrument in a laminar flow cabinet; components including rupture disc retaining cap, microcarrier launch assembly, target plate shelf, macrocarrier rupture discs, stopping screens and macrocarriers (BioRad)[a]
- Forceps
- Seating tool for placing the macrocarriers on to the macrocarrier holders
- Vortex (Vortex-Genie; Scientific Industries)

- Vacuum pump
- Pressurized helium gas

Method

1 Turn on PDS-1000/He Particle Delivery System and vacuum pump; ensure the helium supply is at least 1.4 MPa (200 psi) above the desired pressure optimum and adjust the pressure regulator 345 kPa (50 psi) above the rupture point of the chosen rupture discs.

2 Place the rupture disc in the centre of the rupture disc retaining cap and secure correctly inside the chamber.

3 Place the macrocarriers into holders with forceps and push down with the seating tool to secure the macrocarriers in the holders.

4 Apply 5 µl of the suspension of DNA coated microparticles[b] into the centre (inner 5 mm diam.) of the macrocarrier[c].

5 Place the stopping screen in the microcarrier launch assembly; insert the inverted macrocarrier holder on top. Secure the macrocarrier cover lid on top of the launch assembly.

6 Place the macrocarrier plate containing the macrocarrier at the highest level of the inner chamber; place the Petri dish (lid removed) with the target tissue on the shelf at the second level from the bottom of the chamber.

7 Initiate a vacuum to 698 mmHg; press and hold the fire button until the disc ruptures at 7.6 MPa (1100 psi).

8 Vent[d] the chamber and remove the Petri dish. Place the lid on the dish; dismantle the assembly and prepare for the next shot.

Notes

[a]Particle bombardment of plant tissues is performed under aseptic conditions.
[b]Prior to use, resuspend the DNA-coated microparticles by vortexing briefly.
[c]Allow the ethanol to evaporate completely before use.
[d]Release the vacuum before attempting to open the chamber.

PROTOCOL 12.6 Preparation of Bahiagrass Seeds for Callus Induction [38]

Equipment and Reagents

- Commercially available seeds with at least 85% germination (The Scotts Co.)
- Concentrated sulfuric acid (Fisher Scientific)

- Cheese cloth
- Glacial acetic acid (Fisher Scientific)
- 6% (v/v) sodium hypochlorite solution (The Clorox Co.)
- Sterile ddH_2O
- Petri dishes, 9 cm diam. (Fisher Scientific)
- Glass beakers, 1000 ml and 50 ml (Fisher Scientific)
- Glass rod
- Glass desiccator
- Callus induction medium (CIM; Refer to Table 12.1)

Method

1 Weigh 2.0 g seed in a vial and treat with concentrated sulfuric acid[a] for 12–16 min (depending on seed size) in a fume hood.

2 Transfer the seeds to an empty 1000 ml glass beaker using 500 ml dH_2O. Stir the mixture with a glass rod to mix. Decant the liquid and floating debris.

3 Add 500 ml ddH_2O to the seeds and strain them through two layers of cheese cloth.

4 Gently rub the seeds in the cloth to remove the debris.

5 Wash the seeds with 500 ml dH_2O and repeat twice steps 3–5.

6 After the third wash, leave the seeds in the cheese cloth and dry them for 15 min.

7 Transfer the seeds to 9 cm diam. Petri dishes.

8 For sterilization, mix 20 ml of 6% (v/v) sodium hypochlorite solution with 10 ml glacial acetic acid[b] in a 50 ml glass beaker. Place the resulting solution at the bottom of a glass desiccator in the fume hood (avoid inhalation of fumes).

9 Place open Petri dishes with seeds and lid in the desiccator for 1 h. Chlorine fumes from the mixture will sterilize the seeds in the Petri dishes.

10 Add enough autoclaved ddH_2O to submerge the seeds and leave to soak for at least 1 h before placing the seeds on the surface of CIM[c].

Notes

[a]Sulfuric acid is a scarifying agent that breaks dormancy.

[b]Add glacial acetic acid slowly to the sodium hypochlorite solution.

[c]Place 20–50 seeds in each Petri dish depending upon the size of seeds (e.g. 20 seeds of bahiagrass cv. Argentine per dish).

Table 12.1 Culture media for tissue culture, selection and shoot regeneration in bahiagrass.

Callus induction medium (CIM)	Callus induction and osmoticum medium (CIOM)	Callus selection medium (CSM)	Callus regeneration and selection medium (CRSM)	Selection and rooting medium (SRM)
Sucrose[a] 30 g/l	Sucrose[a] 30 g/l	Sucrose 30 g/l	Sucrose 30 g/l	Sucrose 30 g/l
	Sorbitol[b] 72.9 g/l			
MS basal salts[c] 4.33 g/l (72)	MS basal salts 4.33 g/l	MS basal salts 4.33 g/l	MS basal salts 4.33 g/l	MS basal salts 4.33 g/l
$CuSO_4$[d] 12.45 mg/l	$CuSO_4$ 12.45 mg/l	$CuSO_4$ 12.45 mg/l	$CuSO_4$ 2.45 mg/l	$CuSO_4$ 12.45 mg/l
Phytagel[e] 3.0 g/l	Phytagel 3.0 g/l	Agarose[f] 6.0 g/l	Agarose 6.0 g/l	Agarose 6.0 g/l
MS vitamins[g] 103.12 g/l of 1000 × stock solution [72][h]	MS vitamins 103.12 g/l of 1000 × stock solution [72][h]	MS vitamins 103.12 g/l of 1000 × stock solution [72][h]	MS vitamins 103.12 g/l of 1000 × stock solution [72][h]	MS vitamins 103.12 g/l of 1000 × stock solution [72][h]
Dicamba[i] 3 mg/l	Dicamba 3 mg/l	Dicamba 3 mg/l	Dicamba 3 mg/l	
BAP[j] 1.1 mg/l	BAP 1.1 mg/l	BAP 1.1 mg/l	BAP 0.1 mg/l	BAP 0.1 mg/l
		Paromomycin[k] 50 mg/l	Paromomycin 50 mg/l	Paromomycin 50 mg/l

[a]Sucrose (Phytotechnologies).
[b]Sorbitol (Phytotechnologies).
[c]MS basal salts (Phytotechnologies).
[d]$CuSO_4$ (Sigma-Aldrich).
[e]Phytagel (Phytotechnologies).
[f]Agarose (Type 1; Sigma-Aldrich).
[g]MS vitamins (Phytotechnologies).
[h]MS vitamins, growth regulators and paromomycin are added after autoclaving as filter sterilized, concentrated stock solutions.
[i]Dicamba (Phytotechnologies).
[j]BAP (6-Benzylaminopurine; Phytotechnologies).
[k]Paromomycin sulfate (Phytotechnologies).

PROTOCOL 12.7 Production of Immature Embryo Explants from Rye Inbred Lines [41]

Equipment and Reagents

- Glassine bags (Seedburo Equipment Co.)
- Metal halide lamps (150 W; Philips Electronics)
- 2.4% (v/v) sodium hypochlorite solution (The Clorox Co.)
- 70% (v/v) ethanol
- 0.01% (v/v) Tween 20 (Sigma-Aldrich Inc.)
- Sterile dH_2O
- Petri dishes (9 cm diam.; Fisher Scientific)
- Callus induction and maintenance medium (CIMM; refer to Table 12.2)

Table 12.2 Media for rye tissue culture, selection and regeneration.

Callus induction and maintenance medium (CIMM)	Osmoticum medium (OM)	Regeneration medium (RM)	Regeneration and selection medium (RSM)
Sucrose 30 g/l	Sucrose 30 g/l	Sucrose 30 g/l	Sucrose 30 g/l
	Sorbitol[a] 72.9 g/l		
MS basal salts 4.33 g/l [72], supplemented with 100 mg/l casein hydrolysate[b], 500 mg/l glutamine[c]	MS basal salts 4.33 g/l [72], supplemented with 100 mg/l casein hydrolysate, 500 mg/l glutamine	MS basal salts 4.33 g/l [72], supplemented with 100 mg/l casein hydrolysate, 500 mg/l glutamine	MS basal salts 4.33 g/l [72], supplemented with 100 mg/l casein hydrolysate, 500 mg/l glutamine
2,4-D 2.5 mg/l			
Phytagel 3.0 g/l		Phytagel 3.0 g/l	Phytagel 3.0 g/l
Set pH at 5.8. Autoclave for 20 min			
MS vitamins 103.12 g/l of 1000 × stock solution [72]	MS vitamins 103.12 g/l of 1000 × stock solution [72]	MS vitamins 103.12 g/l of 1000 × stock solution [72]	MS vitamins 103.12 g/l of 1000 × stock solution [72]
		Paromomycin 100 mg/l	Paromomycin 100 mg/l

[a]Sorbitol (Phytotechnologies).
[b]Casein hydrolysate (Phytotechnologies).
[c]Glutamine (Phytotechnologies).

Method

1 At the time of flowering[a], bag the spikes in glassine bags to prevent cross-pollination.

2 Excise immature seeds at anthesis and surface sterilize with 70% (v/v) ethanol for 3 min, followed by 2.4% (v/v) sodium hypochlorite solution containing 0.01% Tween 20 for 20 min.

3 Rinse the immature seeds five times with ddH_2O.

4 Excise the immature embryos from immature seeds corresponding to development stage 3[b] [43].

Notes

[a]Two weeks after germination, vernalize rye seedlings for 50 days at 4 °C with a 8 h photoperiod (natural light intensity in a glasshouse, i.e. 260 μmol/m^2/s). Subsequently, move plants to controlled environment chambers with a 12 h photoperiod (equipped with 150 W metal halide lamps with 262 μmol/m^2/s light intensity) at 10 °C. Gradually increase the photoperiod and temperature to 16 h and 20 °C, respectively, at the time of flowering.

[b]Immature embryos at 10 days after pollination typically correspond to the most responsive development stage. It is important to follow embryo development, rather than time, as it may vary depending on genotype and growth conditions.

PROTOCOL 12.8 Tissue Culture of Seedling-Derived Calli of Bahiagrass [38]

Equipment and Reagents

- Callus induction medium (CIM; refer to Table 12.1)
- Callus induction and osmoticum medium (CIOM; refer to Table 12.1)
- Incubator (CU-36 L5; Percival Scientific Inc.) equipped with Phillips fluorescent light bulbs with dimmable ballast to provide 30–150 μmol/m^2/s illumination

Method

1 Initiate cultures on CIM[a] and subculture to a new medium of the same composition twice each week.

2 Maintain cultures under low light intensity (30 μmol/m^2/s) at 28 °C with a 16 h photoperiod in an incubator.

3 Continue the callus induction phase for 2–4 weeks. Bombard bahiagrass calli 6 weeks after culture initiation.

4 Place tissues on CIOM for 4–6 h prior to bombardment[b].

5 Immediately after bombardment, or up to 16 h after bombardment, transfer tissues to CIM and maintain at less than 30 μmol/m^2/s illumination at 28 °C with a 16 h photoperiod for 7 days before initiating selection.

Notes

[a]CIM should not be stored for more than 28 days before use.

[b]Callus should be placed in the centre of a 2 cm diam. circle for bombardment.

PROTOCOL 12.9 Tissue Culture of Immature Embryos of Rye [41]

Equipment and Reagents

- CIMM (refer to Table 12.2)
- Osmoticum medium (OM; refer to Table 12.2)
- Incubator (LTI 818; Fisher Scientific)

Method

1 Place the immature embryos with the scutellum side up on CIMM.

2 Culture the embryos for 5–7 days in the dark at 25 °C before bombardment.

3 Transfer the embryos to OM for 4–6 h prior to bombardment.

4 Immediately after bombardment, or up to 16 h after bombardment, transfer calli to CIMM and maintain in darkness for 7 days before initiating selection.

12.2.4 Reporter gene assays

Transient gene expression, detected 2–4 days after reporter gene transfer to target tissues, allows optimization of gene delivery parameters. Reporter genes (e.g. commonly used *GFP* or *GUS*) have also been used for analysing promoter efficacy [14, 44–47], protein localization [48, 49], viral infections [50] and the establishment of transformation protocols (e.g. bentgrass, [51]; perennial ryegrass, [52]). The *Escherichia coli uidA* gene, encoding β-glucuronidase (GUS; [53]) produces a blue colour after addition of the substrate X-gal. The reporter gene encoding the green fluorescent protein (GFP), in contrast to the GUS reporter system, supports a non-destructive assay [54, 55]. The majority of bombarded cells do not stably integrate transgenes into their genomes. Efficient selection protocols employing selectable marker/selective agents are therefore needed to identify transgenic events and to suppress proliferation and regeneration into shoots of non-transgenic tissues.

PROTOCOL 12.10 GUS Reporter Gene Assay [53]

Equipment and Reagents

- Solution 1: Add 70 mg X-gluc (Sigma-Aldrich) to 2 ml dimethyl sulfoxide (DMSO); cover the container with aluminium foil to prevent exposure to light
- Solution 2: Mix 150 ml of 100 mM Na_3PO_4 with 5 ml of 0.5 M EDTA; add 200 μl Triton X-100
- GUS assay solution: Mix solutions 1 and 2 and make the final volume up to 200 ml with ddH_2O. Aliquot into 15 ml tubes; store at −20 °C in the dark
- Stereo-microscope (Stemi SV6 stereomicroscope; Carl Zeiss)

Method

1. Maintain bombarded calli on CIM/CIMM media for 2–3 days.
2. Add enough GUS assay solution to completely submerse the calli and apply a vacuum of 27 mmhg for 10 min.
3. Incubate tissue for 16 h at 37 °C in the dark.
4. Observe the calli under a stereomicroscope and count the number of blue foci.

PROTOCOL 12.11 GFP Reporter Gene Assay [54]

Equipment and Reagents

- Stereomicroscope equipped with a fluorescent module (Stemi SV6 stereomicroscope; Carl Zeiss)

Method

1. Detect GFP expression as fluorescent signals visualized using a stereomicroscope[a].
2. Observe transient gene expression approx. 2 days after bombardment.

Note

[a]The microscope requires a mercury lamp, an excitation filter (e.g. BP470/20 nm) and a barrier filter (e.g. BP505–530).

12.2.5 Selection and plant regeneration

Selection systems using the antibiotics kanamycin and hygromycin, or the herbicide phosphinothricin, in combination with the respective selectable marker gene, are the most widely exploited systems for the identification of transgenic calli and plants.

The bar gene, conferring resistance to the herbicide phosphinothricin, has been introduced into both cereals (e.g. wheat [25]; rye [26]) and grasses (e.g. creeping bentgrass, [56]; tall fescue, [57]; bahiagrass, [58]). The neomycin phosphotransferase II (*npt*II)/paromomycin selectable marker/selection system reduces the number of non-transgenic plants escaping selection compared to the bar/phoshinothricin or bialaphos procedure. Paromomycin selection has been established in various turf and forage grasses such as perennial ryegrass [42], red fescue [59], bahiagrass [60] and cereals (rye; [61, 62]). The *hph* gene encoding hygromycin phosphotransferase, allows effective selection and is used at a range of concentrations from 20 mg/l in orchardgrass [63] to 250 mg/l in transformed tall fescue [57, 64–66]. Phosphomannose isomerase (PMI), an enzyme not present in plants, catalyses the reversible interconversion of mannose 6-phosphate and fructose 6-phosphate. Plant cells lacking this enzyme are unable to survive on culture medium containing mannose. Thus, PMI/mannose selection has supported the identification of transformed plant cells in monocotyledons such as maize [67], wheat [68], rice [69] and sugarcane [70]. A comprehensive review on various selection systems is provided by Miki and McHugh [71].

PROTOCOL 12.12 Selection and Regeneration of Bahiagrass using *npt*II/Paromomycin [38]

Equipment and Reagents

- Callus selection medium (CSM; refer to Table 12.1)
- Callus regeneration and selection medium (CRSM; refer to Table 12.1)
- Selection and rooting medium (SRM; refer to Table 12.1)
- Extra-deep Petri dishes (10 cm diam.; Fisher Scientific)
- Incubator (CU-36 L5; Percival Scientific Inc.) equipped with Phillips fluorescent bulbs and dimmable ballast to provide 30–150 μmol/m^2/s illumination
- Magenta boxes (Kraeckler Scientific)
- Fafard No. 2 mix (Fafard)

Method

1 Transfer calli onto CSM 7 days after gene transfer and maintain at 30 μmol/m^2/s illumination, with a 16 h photoperiod at 25 °C.

2 Subculture calli to new selection medium biweekly until 28 days after initiation of selection^b.

3 Transfer calli to CRSM; maintain on CRSM for 14 days and increase the light intensity from 30 μmol/m^2/s to 150 μmol/m^2/s.

4 Transfer calli with shoots to SRM in deep Petri dishes. Incubate under high light intensity (150 μmol/m²/s), with a 16 h photoperiod at 25 °C for 14 days.

5 Transplant elongated shoots with roots into Fafard No. 2 mix and maintain in a growth chamber with 12 h photoperiod at 25 °C temperature. Keep the regenerated plants covered during the first 4–6 days with a transparent container to maintain high humidity.

6 After 14 days of acclimatization in growth chambers, move the transgenic plants to glasshouses maintained at 28 °C day and 22 °C night, under natural illumination.

Notes

[a]The selection medium consists of CIM with 50 mg/l paromomycin. To prevent precipitation of paromomycin, replace Phytagel by agarose (Type I, Sigma).

[b]It is important to track the identity of independent callus lines through the selection and regeneration procedure to assure independent transformants.

PROTOCOL 12.13 Selection and Regeneration of Rye using *npt*II/Paromomycin [61]

Equipment and Reagents

- CIMM (refer to Table 12.2)
- Regeneration selection medium (RSM; refer to Table 12.2)
- Extra-deep dishes, 10 cm diam. (Fisher Scientific Inc.)
- Magenta boxes (Kraeckler Scientific)
- Incubator (CU-36L5; Percival Scientific) equipped with Phillips fluorescent light bulbs and dimmable ballast to provide 30–150 μmol/m²/s illumination
- Fafard No. 2 mix (Fafard)

Method

1 Transfer bombarded calli to CIMM and maintain in darkness at 25 °C for 14 days.

2 Transfer calli[a] 14 days after bombardment to RSM at 150 μmol/m²/s illumination, 16 h photoperiod at 25 °C.

3 Transplant the regenerated rooted plants into Fafard No. 2 mix, and maintain plants in a growth chamber under a 16 h photoperiod and 20 °C/15 °C (day/night) temperatures[b].

4 After 14 days, move the plants to the glasshouse with the same temperature regime and natural photoperiod.

Notes

[a]Calli cultured in darkness should be prevented from long exposure to light during the transfer process. Cultures should be taken out of the incubators in small batches.

[b]The newly transplanted shoots should be covered with large culture containers during the first for 2–4 days after transplantation. This will protect the shoots from excessive water loss and facilitate acclimatization.

PROTOCOL 12.14 Selection and Regeneration Protocol for Rye using *bar*/BASTA [26]

Equipment and Reagents

- CIMM (refer to Table 12.2)
- Regeneration medium (RM; refer to Table 12.2)
- Extra-deep Petri dishes (10 cm diam.; Fisher Scientific)
- Magenta boxes (Kraeckler Scientific)
- Incubator (CU-36L5 Percival Scientific Inc.) equipped with Philips fluorescent light bulbs and dimmable ballast to provide 30–150 $\mu mol/m^2/s$ illumination
- 0.05% (v/v) Basta solution (Bayer CropScience)
- Fafard No. 2 mix (Fafard)

Method

1. Transfer bombarded calli to CIMM 12–16 h after bombardment and culture in the dark at 25 °C.
2. Transfer calli 14 days after bombardment to Magenta boxes with RM and maintain under a 16 h photoperiod (150 $\mu mol/m^2/s$) at 25 °C.
3. After 28 days, spray elongated shoots (>2 cm) with filter sterilized (0.05% Basta) herbicide solution in Magenta boxes[a].
4. Transplant the regenerated, rooted, Basta resistant plants into Fafard No. 2 mix 21 d after herbicide application.

Note

[a]Herbicide application must be carried out in a vertical laminar flow fume hood to provide both product and personnel protection.

PROTOCOL 12.15 Confirmation of Putative Transgenic Plants using NPTII Enzyme-Linked Immunosorbent Assay (ELISA) Assay

Equipment and Reagents

- NPTII ELISA kit (Agdia Inc.)
- Polyvinyl polypyrrolidone (PVP; Sigma-Aldrich
- Bovine serum albumen (BSA; Sigma-Aldrich)
- Microfuge tubes (Fisher Scientific)
- Micropestles (Fisher Scientific)
- Centrifuge (Model 581; Eppendorf)
- Vortex (Vortex-Genie; Scientific Industries)
- Protein Determination Reagent (USB Corporation)
- Spectrophotometer (Smart Spec 3000; BioRad)

Method

Protein Extraction:

1 Harvest 600 mg of fresh leaf tissue; store samples on ice.

2 Add 10 mg PVP and 600 μl 10 × plant extraction buffer (PEB1 supplied with the NPTII Agdia ELISA kit) to each sample.

3 Grind the leaf samples using a micropestle. Keep the samples on ice.

4 Centrifuge the samples at 20,800 g at 4 °C for 15 min.

5 Transfer the supernatant to a new microfuge tube and store on ice.

Protein estimation:

6 Turn on the spectrophotometer 15 min before use.

7 Dilute the protein determination reagent 1:1 (v:v) using sterile ddH_2O. Prepare enough to use 1 ml per sample, including standards and blank.

8 Prepare a standard dilution series using BSA (0–20 μg).

9 Add 1 ml diluted protein determination reagent to each cuvette. Add 5 μl of sample to each cuvette and mix by vortexing.

10 Incubate at room-temperature while preparing the remaining samples.

11 Measure OD_{595} of each sample (ideally this should be 0.2–0.8).

12 Plot a standard curve using BSA and use it to estimate total protein concentration of the samples; calculate the volume of each sample required for 15 µg total protein per well.

Assay:

13 Prepare samples, including wild-type using 15 µg protein and the volume of PEB1 buffer required to make the total volume to 110 µl.

14 Prepare standards as follows: 110 µl buffer PEB1 (negative control) and 110 µl of the provided positive control.

15 Prepare a humid box by putting damp paper towel in a box with a lid.

16 Add 100 µl of each sample and standard in the ELISA microplate provided with the kit. The order of samples should be noted at this time.

17 Place the plate in the humid box and incubate for 2 h at room temperature.

18 Prepare the wash buffer (PBST supplied with the NPTII Agdia ELISA kit) by diluting 5 ml to 100 ml (20×) with ddH_2O.

19 Prepare the enzyme conjugate diluent by mixing 1 part (MRS-2 with 4 parts 1 × wash buffer PBST. Make enough to add 100 µl per well.

20 A few minutes before the incubation ends, add 10 µl from bottle A and 10 µl from bottle B per 1 ml of enzyme conjugate diluent to prepare the enzyme conjugate.

21 When incubation is complete, remove plate from humid box and empty wells.

22 Fill all wells with 1 × buffer PBST and then empty the wells again. Repeat five times.

23 Ensure complete removal of wash solution by tapping the frame firmly upside down on paper towels.

24 Add 100 µl of the prepared enzyme conjugate into each well and return the plate to the humid box for 2 h at room temperature.

25 Meanwhile, aliquot 100 µl substrate (TMB substrate supplied with the NPTII Agdia ELISA kit) and allow it to warm to room temperature.

26 When the incubation is complete, wash the plate with 1 × buffer PBST as before.

27 Add 100 µl of room temperature TMB substrate solution to each well and place the plate back in the humid box for 15 min. A blue colour will develop, the intensity of which will be directly proportional to the amount of NPTII protein in the sample; negative samples will remain white.

28 To terminate the reaction, add 50 µl 3 M sulfuric acid to each well. The substrate colour will change from blue to yellow.

29 The results must be recorded within 15 min after addition of the stop solution otherwise the reading will decline. Colour development can be scored visually or recorded with an ELISA plate reader.

12.3 Troubleshooting

- After pipetting DNA coated particles onto macrocarriers these should be used as soon as ethanol has evaporated. Excessive drying may result in DNA degradation.
- Rupture discs should not be stored after sterilization in ethanol since their rupture point may change over time. It is important to monitor that the rupture discs fracture at the desired pressure (7.6 MPa, 1100 psi) for efficient particle bombardment.
- For bombardments of more than 20 Petri dishes with callus, allow time to resterilize the biolistic gene delivery device (PDS-1000/He, BioRad) and its components with 70% (v/v) ethanol to reduce the risk of contamination.
- Donor plant quality is a critical factor if immature tissues are used for callus induction or gene transfer. Stress caused by heat or the use of systemic pesticides may have a negative influence on the results of callus induction and transformation.
- Growth regulators, vitamins and selective agents (e.g. paromomycin) are added to autoclaved media components as filter sterilized stock solutions to prevent their exposure to excessive heat.
- Paromomycin is used in combination with agarose (Type I, Sigma) to prevent precipitation of paromomycin.
- The absence of growth regulators in the selection rooting media will promote root growth and shoot elongation of transformed shoots.
- For semiquantitative ELISA, a plate reader can be used. It is important to sample the same developmental stage of leaves for all samples (e.g. the first fully expanded leaf).

References

1. Sanford JC (1988) *Trends Biotechnol*. **6**, 302–322.
2. Altpeter F, Baisakh N, Beachy R, *et al.* (2005) *Mol. Breed*. **15**, 305–327.
3. Taylor NJ, Fauquet CM (2002) *DNA Cell Biol*. **21**, 963–977.
4. Toriyama K, Arimoto Y, Uchimiya H, *et al.* (1988) Bio/Technology 6, 1072–1-74.
5. Datta SK, Peterhans A, Datta K, *et al.* (1990) Bio/Technology 8, 736–740.
6. Kaeppler H, Somers D, Rines H, *et al.* (1992) Theor. Appl. Genet. 84, 560–566.

7. Sautter C (1993) *Plant Cell Tissue Organ Cult*. **33, 251–257.

The first report describing microtargeting of intact tissues.

*8. Sanford JC, Klein T, Wolf E, *et al.* (1987) *J. Part. Sci. Tech.* **5**, 27–37.

The first report describing particle bombardment of living cells.

9. Finer J, Vain P, Jones M, *et al.* (1992) *Plant Cell Rep.* **11, 323–328.

The first report describing the particle inflow gun.

**10. Christou P, McCabe D (1992) Agracetus Inc., Middleton, WI.

The first report describing ACCELL technology.

11. Kikert J (1993) *Plant Cell Tiss. Org. Cult.* **33**, 221–226.

12. Koprek T, Hansch R, Nerlich A, *et al.* (1996) *Plant Sci.* **119**, 79–91.

13. Zhang P, Pounti-Kaerlas J (2000) *Plant Cell Rep.* **19**, 1041–1048.

14. Gondo T, Tsuruta S, Akashi R, *et al.* (2005) *J. Plant Physiol.* **162**, 1367–1375.

***15. McCabe D, Christou P (1993) *Plant Cell Tissue Organ Cult.* 33, 227–236.

16. McCabe DE, Swain WF, Martinell BJ, *et al.* (1988) *Bio/Technology* **6**, 923–926.

17. Christou P, Ford T, Kofron M (1991) *Bio/Technology* **9**, 957–962.

18. McCown BH, McCabe DE. Russell DR, *et al.* (1991) *Plant Cell Rep.* **9**, 590–594.

19. Keller G, Spatola K, McCabe D, *et al.* (1997) *Transgenic Res.* **6**, 385–392.

20. Sautter C, Leduc N, Bilang R, *et al.* (1995) *Euphytica* **85**, 45–51.

21. Zhang S, William CR, Lemaux PG (2002) *Plant Cell Rep.* **21**, 263–270.

22. Helenius E, Boije M, Niklander-Teeri V, *et al.* (2000) *Plant Mol. Biol. Rep.* **18**, 287–288.

23. Carsono N, Yoshida T (2008) *Plant Production Sci.* **11**, 88–95.

24. Rasco-Gaunt S, Riley A, Barcelo P, *et al.* (1999) *Plant Cell Rep.* **19**, 118–127.

25. Altpeter F, Vasil V, Srivastava V, *et al.* (1996) *Plant Cell Rep.* **16**, 12–17.

26. Popelka JC, Xu JP, Altpeter F (2003) *Transgenic Res.* **12**, 587–596.

27. Hunold R, Bronner R, Hahne G (1994) *Plant J.* **5**, 593–604.

28. Vain P, Keen N, Murillo J, *et al.* (1993) *Plant Cell Tiss. Org. Cult.* **33**, 237–246.

29. Southgate EM, Davey MR, Power JB, *et al.* (1995) *Biotechnol. Adv.* **13**, 631–651.

30. Christou P (1996) *Trends Plant Sci.* **1**, 423–430.

31. Sivamani E, DeLong RK, Qu R (2009) *Plant Cell Rep.* **28**, 213–221.

32. Klein T, Gradziel T, Fromm M, *et al.* (1988) *Bio/Technology* **6**, 559–563.

33. James VA, Avart C, Worland B, *et al.* (2002) *Theor. Appl. Genet.* **104**, 53–661

34. Sandhu S, Altpeter F (2008) *Plant Cell Rep.* **27**, 1755–1756.

35. Vasil V, Srivastava V, Castillo AM, *et al.* (1993) *Bio/Technology* **11**, 1553–1558.

36. Christou P (1990) *Ann. Bot.* **66**, 379–386.

37. Morrish F, Vasil V, Vasil IK (1987) *Adv. Genet.* **2**, 431–499.

38. Vasil V, Vasil, IK (1980). *Theor. Appl. Genet.* **56**, 97–99.

39. Altpeter F, James VA (2005) *Int. Turfgrass Soc. Res. J.* **10**, 485–489.

40. Neibaur I, Altpeter F, Gallo M (2008) *In Vitro* Cell. Dev. Biol.-Plant **44**, 480–486.

41. Popelka JC, Altpeter F (2001) *Plant Cell Rep.* **20**, 575–582.

42. Altpeter F, Xu J, Ahmed S (2000) *Mol. Breed.* **6**, 519–528.

43. Zimny J, Lörz H (1989) *Plant Breed.* **102**, 89–100.

44. Toyama K, Bae CH, Kang JG, *et al.* (2003) *Mol. Breed.* **11**, 203–211.

45. Newell CA, Birch-Machin I, Hibberd JM, *et al.* (2003) *Transgenic Res.* **12**, 631–634.

46. Basu C, Kausch AP, Chandlee JM (2004) *Biochem. Biophys. Res. Comm.* **32**, 7–10.

47. Lee SH, Lee DG, Woo HS, *et al.* (2006) *Plant Sci.* **141**, 408–414.

48. Hibberd JM, Linley PJ, Khan MS, *et al.* (1998) *Plant J.* **16**, 627–632.

49. Jang IC, Nahm BH, Kim JK (1999) *Mol. Breed.* **5**, 453–461.

50. Oparka KJ, Roberts AG, Santacriz S, *et al.* (1997) *Nature* **388**, 401–402.

51. Yu TT, Skinner DZ, Liang GH *et al.* (2000) *Hereditas* **133**, 229–233.

52. Altpeter (2006) In: *Methods in Molecular Biology*, Vol. 343. *Agrobacterium Protocols*, 2nd edn. Edited by K. Wang. Humana Press, Totowa, NJ, USA, pp. 223–231.

53. Jefferson R, Kavanagh TA, Bevan MW (1987) *EMBO J.* **6**, 3901–3907.

54. Chalfie M, Tu Y, Euskirchen G, *et al.* (1994) *Science* **263**, 802–805.

55. Niwa Y, Hirano T, Yoshimota K, *et al.* (1999) *Plant J.* **18**, 455–463.

56. Hartman CL, Lee L, Day PR, *et al.* (1994) *Bio/Technology* **12**, 919–923.

57. Dalton SJ, Bettany AJE, Timms E, *et al.* (1995) *Plant Sci.* **108**, 63–70.

58. Smith RL, Grando MF, Li YY, *et al.* (2002) *Plant Cell Rep.* **20**, 1017–1021.

59. Altpeter F, Xu J (2000) *J. Plant Physiol.* **157**, 441–448.

60. James VA, Altpeter F, Positano MV (2004) Poster Abstract 149. Plant Biology Conf. Lake Buena Vista, FL, 24–28 July 2004. Am. Soc. Plant Biologists, Rockville, MD.

61. Popelka JC, Altpeter F (2003) *Theor. Appl. Genet.* **106**, 53–590.

62. Altpeter F, Popelka JC, Weiser H (2004) *Plant Mol. Biol.* **54**, 783–792.

63. Horn ME, Shillito RD, Conger BV, *et al.* (1988) *Plant Cell Rep.* **7**, 469–472.

64. Wang ZY, Takamizo T, Iglesias VA, *et al.* (1992) *Bio/Technology* **10**, 691–699.

65. Ha SB, Wu FS, Thorne TK (1992) *Plant Cell Rep.* **11**, 601–604.

66. Spangenberg G, Wang ZY, Wu XL, *et al.* (1995) *J. Plant Physiol.* **145**, 693–701.

67. Negrotto D, Jolley M, Beer S, *et al.* (2000) *Plant Cell Rep.* **19**, 798–803.

68. Wright M, Dawson J, Dunder E, *et al.* (2001) *Plant Cell Rep.* **20**, 429–436.

69. Lucca P, Ye X, Potrykus I (2001) *Mol. Breed.* **7**, 43–49.

70. Jain M, Chengalrayan K, Abouzid A, *et al.* (2007) *Plant Cell Rep.* 26, 581–590.

71. Miki B, McHugh S (2004) *J. Biotech.* **107**, 193–232.

72. Murashige T, Skoog F (1962) *Physiol. Plant.* **15**, 473–479.

13

Plastid Transformation

Bridget V. Hogg[1], Cilia L.C. Lelivelt[2], Aisling Dunne[1], Kim-Hong Nguyen[1] and Jacqueline M. Nugent[1]

[1] *Institute of Bioengineering and Agroecology, National University of Ireland, Maynooth, Ireland*
[2] *Rijk Zwaan Breeding B.V., Fijnaart, The Netherlands*

13.1 Introduction

Plastid transformation is becoming an increasingly important tool for plant biologists. This technology has been used in basic and applied science [1] including assessing plastid gene function by targeted gene knockout (reviewed in [2]), engineering improved photosynthetic efficiency [2] and metabolic pathways [3–5], studying RNA editing [6, 7], expressing agriculturally important traits (e.g. insect resistance, herbicide resistance [2, 8], and producing biopharmaceuticals, vaccine antigens, biopolymers and enzymes in plants [8–10]). Some of the advantages afforded by plastid transformation, compared to nuclear transformation, include the very high levels of protein accumulation that can be achieved in plastids, the absence of any position effects, the ability to express multiple genes together in operons, and the enhanced level of transgene containment afforded by the mostly maternal mode of inheritance of plastids [2, 10]. However, the technology is not without its limitations. For example, plastid transformation is still possible only in a restricted number of flowering plant species (Table 13.1), and as protein glycosylation does not take place in plastids, the system is not suitable for expressing proteins that require glycosylation to be functional. Although strict maternal inheritance of plastids has been emphasized as an advantage of the system, it is now emerging that a low level of paternal

Plant Cell Culture Edited by Michael R. Davey and Paul Anthony

Table 13.1 Selectable marker genes and methods used for plastid transformation in flowering plants.

Species	Selectable marker gene	Method	Reference
Tobacco	*rrn*16 mutation[a]	Biolistics	[11]
	aadA[b]	Biolistics	[12]
	aadA	PEG	[13–15]
	aphA-6[c]	Biolistics	[16]
Nicotine-free tobacco	*aadA*	Biolistics	[17]
Potato	*aadA*	Biolistics	[18, 19]
Tomato	*aadA*	Biolistics	[20]
	*rrn*16 mutation	PEG	[21]
Petunia	*aadA*	Biolistics	[22]
Arabidopsis	*aadA*	Biolistics	[23]
Lesquerella	*aadA*	Biolistics	[24]
Oilseed rape	*aadA*	Biolistics	[25]
Cauliflower	*aadA*	PEG	[26]
Cabbage	*aadA*	Biolistics	[27]
Lettuce	*aadA*	PEG	[28]
	aadA	Biolistics	[29, 30]
Cotton	*aphA*-6 and *npt*II[d]	Biolistics	[31]
Carrot	*aadA*	Biolistics	[32]
Soybean	*aadA*	Biolistics	[33]
Rice	*aadA*	Biolistics	[34]
Poplar	*aadA*	Biolistics	[35]

[a] *rrn*16 mutation confers resistance to spectinomycin.
[b] *aad*A confers resistance to spectinomycin and streptomycin.
[c] *aphA*-6 confers resistance to kanamycin.
[d] *npt*II confers resistance to kanamycin.

plastid inheritance occurs in a background of mostly maternal inheritance in several flowering plants [36–38]. This chapter presents sample protocols for two methods that have been used to achieve plastid transformation in angiosperms, namely biolistic-mediated transformation of leaf explants (focusing on tobacco plastid transformation) and polyethylene glycol (PEG)-mediated transformation of protoplasts (focusing on lettuce plastid transformation). Information on plastid transformation in algae and non-flowering plants is given in references [39–41].

13.2 Methods and approaches

13.2.1 Principles of plastid transformation

Plastid transformation proceeds by homologous recombination and results in targeted integration of transgenes into the plastid genome [2]. A plastid transformation vector typically consists of a selectable marker gene (usually *aad*A; Table 13.1) and a transgene (gene of interest) flanked on both sides by 1–2 kb of plastid-targeting DNA, cloned into an *Escherichia coli* plasmid backbone. After DNA delivery, the transgene and selectable marker gene become integrated into the plastid genome by two homologous recombination events that occur between the two plastid-targeting sequences on the vector and the corresponding target region on the plastid genome (Figure 13.1a). Homoplastomic transformed cells (where all plastid genome copies

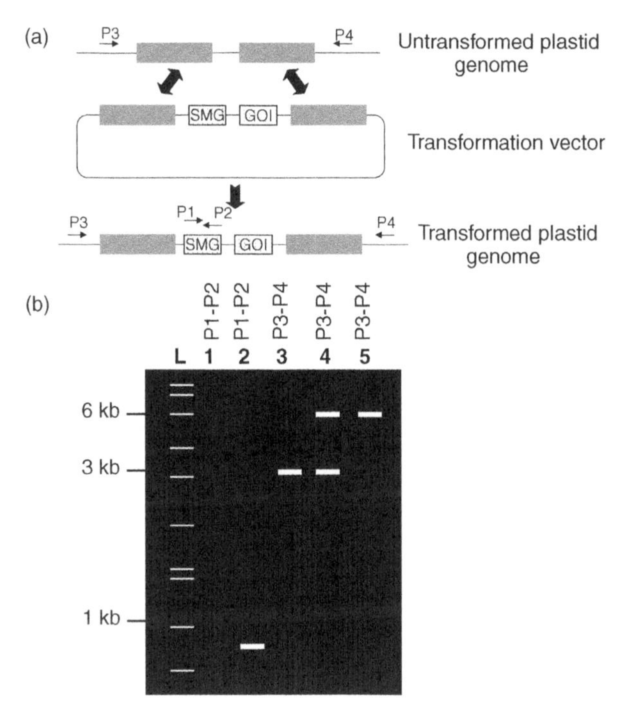

Figure 13.1 Plastid transformation is achieved by targeted genome integration. (a) A plastid transformation vector typically consists of a selectable marker gene (SMG) and a gene of interest (GOI) flanked by plastid-targeting DNA (solid grey boxes), cloned into an *Escherichia coli* plasmid backbone. Homologous recombination across vector and target genome sequences (thick arrows) gives rise to targeted transgene integration. Selected shoots, or cell lines, can be assessed for plastid transformation by PCR using a primer pair specific for the SMG (P1–P2) and a primer pair specific for the region of the plastid genome flanking the vector integration site (P3–P4). (b) PCR results show the presence of the SMG in a transformed plant (lane 2), a transformed plant that is heteroplastomic (lane 4) and a transformed plant that is homoplastomic (lane 5) for the transgene insertion. Results of control PCR reactions carried out with wild-type DNA template are indicated in lanes 1 (P1–P2) and 3 (P3–P4). A DNA ladder (L) allows for accurate sizing of PCR products.

are transformed) are achieved through a process of plastid DNA replication, sorting and selection of transformed genomes, by maintaining the transformed tissue/cells on selective culture medium [2]. Recent reviews have discussed plastid transformation [2], the vector systems that are available for this process [42–44] and the potential applications of plastid transformation [8, 10] in much greater detail than is possible in this chapter. The reader is encouraged to consult these publications for broader background information.

13.2.2 Biolistic-mediated plastid transformation

The biolistic delivery method is the most commonly used procedure for achieving plastid transformation (Table 13.1). This method involves coating microcarriers (gold or tungsten particles) with transforming DNA and firing these coated particles at the tissue to be transformed (usually leaf tissue) using a particle gun. This approach was adopted for tobacco plastid transformation nearly 20 years ago and has since been used to transform the plastid genome of several flowering plants (Table 13.1). However, to date, tobacco still remains the species most amenable to plastid transformation. The biolistic method of tobacco plastid transformation is robust and reproducible; many researchers have adopted the method in their laboratories and transplastomic plants can be obtained relatively rapidly, usually in about 5 months. The main disadvantage of this method is that it is relatively expensive and requires access to a particle gun and associated materials. The procedure described here is a modified version of procedures published previously [7, 45]. The method specifically applies to plastid transformation of *Nicotiana tabacum* cv. Petite Havana. The expected tobacco plastid transformation frequency, using the Bio-Rad PDS1000/He biolistic gun, is one transplastomic shoot per bombardment. Typically, 20–30 leaf samples are bombarded per construct [7, 45].

PROTOCOL 13.1 Preparation of Tobacco Leaf Material for Bombardment

Equipment and Reagents

- Laminar air flow cabinet for aseptic work
- Tobacco seed (cv. Petite Havana)
- 70% (v/v) ethanol
- 1.5 ml microfuge tubes
- Sterile distilled water
- 10% (v/v) bleach (sodium hypochlorite active ingredient, e.g. 'Domestos' bleach)
- Tube rotator (MSE Centaur)
- Sterile Petri dishes (shallow–100 mm/15 mm, Greiner)

- 1/2 strength Murashige and Skoog-based MS10 medium [46]: 2.3 g/l MS salts and vitamins (Duchefa), 10 g/l sucrose, 8 g/l microagar (Duchefa), pH 5.8
- Illuminated growth chamber (fluorescent illumination; 60 μmol/m^2/s, 25 °C)
- Sterile clear plastic Pet Cups and lids (10 oz; Dart Container Corp.)
- MS30 medium: 4.6 g/l MS salts and vitamins, 30 g/l sucrose, 8 g/l microagar, pH 5.8
- RMOP medium (MS medium supplemented with 1 mg/l BAP and 0.1 mg/l NAA): 4.6 g/l MS salts and vitamins, 30 g/l sucrose, 0.1 g/l myoinositol, 1 mg/l thiamine HCl, 1 mg/l 6-benzylaminopurine (BAP), 0.1 mg/l naphthaleneacetic acid (NAA), 8 g/l microagar, pH 5.8

Method

This work should be carried out in a laminar airflow cabinet.

1 Wash tobacco seeds in 70% (v/v) ethanol for 1 min in a 1.5 ml microfuge tube.

2 Rinse seeds three times with sterile distilled water.

3 Wash seeds in 1 ml of 10% (v/v) bleach for 10 min; keep the seeds in continuous suspension using a tube rotator.

4 Rinse seeds four times in sterile distilled water and sow onto semisolid 1/2 strength MS10 medium in Petri dishes (30 ml medium/dish).

5 Incubate seeds for 1 week at 25 °C (16 h photoperiod).

6 Transfer seedlings to Pet Cups each containing 60 ml semisolid MS30 medium; culture for 4–6 weeks at 25 °C (16 h photoperiod).

7 The day before bombardment, excise leaves from 4–6-week-old axenic, *in vitro* propagated, seedlings and place the leaves abaxial side up on semisolid RMOP medium in Petri dishes (30 ml medium/dish). Place one leaf in the centre of each dish[a].

8 Leave the dishes at 25 °C overnight (16 h photoperiod).

Note

[a]Surface sterilized leaves from glasshouse-grown plants may also be used [45].

PROTOCOL 13.2 Preparation of Gold Particles and Coating with DNA

Equipment and Reagents

- 0.6 μm gold microcarriers (Bio-Rad)
- 1.5 ml microfuge tubes
- Ice-cold 70% (v/v) ethanol

- Vortex mixer (Vortex Genie-2)
- Microcentrifuge (MSE, Micro Centaur)
- Pasteur or Gilson-type pipettes
- Ice-cold sterile distilled water
- Sterile 50% (v/v) glycerol (Sigma)

Method

1. Weigh 30 mg of 0.6 μm gold microcarriers into a 1.5 ml microfuge tube[a].
2. Add 1 ml of 70% (v/v) ethanol and vortex vigorously for 2 min to suspend the particles.
3. Centrifuge for 20–30 s (600 g).
4. Remove the ethanol with a Pasteur or Gilson-type pipette and discard.
5. Add 1 ml of sterile distilled water.
6. Vortex vigorously for 2 min.
7. Centrifuge for 20–30 s (600 g).
8. Remove the water with a pipette and discard.
9. Repeat steps 5–8 three times.
10. Add 500 μl of 50% (v/v) glycerol and vortex for 1 min to resuspend the particles (final concentration 60 mg/ml).
11. Aliquot 50 μl of the gold particles into each of 10 tubes, keeping the gold in suspension[b].

Notes

[a]Tungsten microcarriers may also be used [2].

[b]The sterile gold particles can be used directly for coating with DNA or stored at −20 °C until use.

PROTOCOL 13.3 Coating Gold Particles with DNA

Equipment and Reagents

- Sterile gold particles (60 mg/ml, see Protocol 13.2)
- Vortex mixer
- Plasmid DNA (transformation vector) 1 μg/μl
- Sterile 2.5 M $CaCl_2$ solution

- Sterile 0.1 M spermidine-free base (Sigma)
- Microcentrifuge
- Ice-cold absolute ethanol

Method

1. Take 50 μl of the gold particles (60 mg/ml) and vortex for 1 min[a].
2. Add 5–10 μl of plasmid DNA (1 μg/μl); vortex gently to mix.
3. Add 50 μl of 2.5 M $CaCl_2$; vortex gently.
4. Add 20 μl of 0.1 M spermidine; vortex gently.
5. Allow contents to settle at room temperature for 10 min.
6. Centrifuge for 10 s (600 g).
7. Remove the supernatant and discard.
8. Wash particles with 150 μl of ice-cold absolute ethanol.
9. Centrifuge for 10 s (600 g).
10. Remove the supernatant and discard.
11. Gently resuspend the DNA-coated particles in 50 μl of ice-cold absolute ethanol.
12. Use the DNA-coated particles immediately[b].

Notes

[a]This amount of coated gold particles is sufficient for five bombardments.

[b]If there is any delay before bombardment, replace the ethanol in the tube with new absolute ethanol.

PROTOCOL 13.4 Particle Bombardment of Leaves

Equipment and Reagents

- Laminar air flow cabinet
- 70% (v/v) ethanol
- Absolute ethanol
- Bio-Rad PDS1000/He biolistic gun[a,b,c]
- Particle gun macrocarrier holders (Bio-Rad)
- Particle gun stopping screens (Bio-Rad)
- Particle gun macrocarriers (Bio-Rad)

- Sterile Whatman filter paper, No. 4
- Vacuum pump (Vacuubrand)
- Helium supply and gas regulator
- DNA-coated gold particles (see Protocol 13.3)
- Particle gun rupture discs (7.6 MPa, 1100 psi; Bio-Rad)
- 70% (v/v) isopropanol
- Pre-prepared tobacco leaves (see Protocol 13.1)
- Parafilm (Pechiney Plastic Packaging)
- Illuminated growth chamber (60 μmol/m^2/s)
- Sterile Petri dishes (deep – 100 mm/20 mm, Greiner)
- Spectinomycin (Duchefa)
- RMOP medium with 500 mg/l spectinomycin
- Sterile clear plastic Pet Cups and lids
- MS30 medium with 500 mg/l spectinomycin

Method

1 Sterilize the gun chamber, rupture disc retaining cap, microcarrier launch assembly and target shelf with 70% (v/v) ethanol; allow to dry.

2 Autoclave macrocarrier holders and stopping screens[d].

3 Soak macrocarriers in absolute ethanol (5 min); air dry on sterile filter paper.

4 Switch on the gene gun and the vacuum pump.

5 Open the regulator valve on the helium tank and set to 8.9 MPa (1300 psi; 1.4 MPa or 200 psi above the rupture disc value).

6 Insert macrocarriers into macrocarrier holders.

7 Spread 8–10 μl of freshly prepared DNA-coated gold particles onto each macrocarrier. Air dry for 5–10 min (up to five samples can be made at a time).

8 Dip the rupture discs in 70% (v/v) isopropanol (remove excess isopropanol).

9 Immediately place the sterilized rupture discs into the retaining cap and screw in place.

10 Place a stopping screen and a macrocarrier holder (face down) into the microcarrier launch assembly and screw in place.

11 Place the microcarrier launch assembly into the gun chamber (just below the rupture disc).

12 Remove the lid from the dish containing the preprepared target leaf on semisolid RMOP medium (remove any excess moisture from the leaf surface with sterile filter paper).

13 Place the dish on the target shelf and close the chamber door[e].

14 Open the vacuum valve and evacuate the chamber. When the vacuum reaches 711 mmHg (28 inches of Hg), press the fire button and continue to hold until the rupture disc breaks (a 'pop' is heard); release the fire button.

15 Vent the chamber immediately and remove the leaf sample.

16 Replace the lid on the dish and seal with Parafilm.

17 Repeat steps 8–16 for subsequent leaf bombardments.

18 After bombardments, turn off the helium tank regulator and release the pressure by pressing the fire button while the chamber is under vacuum. Vent the chamber and turn off the vacuum pump and gene gun.

19 Place the sealed Petri dishes at 25 °C for 2 days (16 h photoperiod).

20 Cut the bombarded leaves into small pieces (0.5 cm^2 each) and place the leaf pieces abaxial side down on semisolid RMOP medium with 500 mg/l spectinomycin in deep Petri dishes (50 ml medium/dish)[f,g].

21 Seal dishes with Parafilm and incubate the cultures for 4–12 weeks at 25 °C (16 h photoperiod).

22 Transfer the bombarded leaf pieces to new RMOP medium with 500 mg/l spectinomycin every 4 weeks.

23 Green, spectinomycin-resistant, shoots should appear 4–12 weeks after bombardment[h,i].

24 Transfer spectinomycin-resistant shoots to semisolid MS30 medium with 500 mg/l spectinomycin in Pet Cups and incubate at 25 °C (16 h photoperiod).

Notes

[a]The particle gun must be stored and used in a laminar air flow cabinet.

[b]Consult the Bio-Rad Biolistic PDS-1000/He Particle Delivery System Manual for user instructions.

[c]Using a Hepta-adaptor version of the Bio-Rad PDS1000/He biolistic gun can significantly increase the transformation efficiency [7].

[d]These can also be soaked in absolute ethanol (5 min) and air dried.

[e]The target shelf can be 6 or 9 cm below the stopping screen (level 2 or 3).

[f]No more than five to seven explants on each Petri dish.

[g]Using streptomycin for selection delays shoot formation.

[h]Spectinomycin inhibits protein synthesis on plastid prokaryotic-type 70S ribosomes, but has no affect on cytoplasmic 80S ribosome function. In the presence of the antibiotic, untransformed cells survive but become bleached (due to inhibition of chlorophyll biosynthesis) when maintained on sucrose supplemented culture medium; plastid transformed cells and shoots appear green on the same medium.

[i]Only shoots that originate from separate regions of a leaf piece, or from different leaf pieces, are recorded as independently-derived transformants.

13.2.3 PEG-mediated plastid transformation

The PEG-mediated method of plastid transformation is less widely used than the biolistic method (Table 13.1), but has been exploited to transform tobacco [13–15], tomato [21], cauliflower [26], and lettuce plastid genomes [28]. This method involves the enzymatic removal of plant cell walls to obtain protoplasts that are exposed to transforming DNA in the presence of PEG. PEG treatment causes reversible disturbance of the protoplast membrane allowing the transforming DNA to enter the protoplasts [2]. A review of this method and a detailed protocol for PEG-mediated tobacco plastid transformation has been published [47, 48]. A detailed protocol for PEG-mediated lettuce plastid transformation is presented here, and the expected transformation frequency is one to two spectinomycin-resistant cell lines per 10^6 viable protoplasts [28]. The main advantage of this system is that it is relatively inexpensive. The disadvantages are that it requires protoplast culture experience, it can take 12 months to obtain transplastomic lettuce plants, and many of the transformed plants that regenerate are polyploid.

PROTOCOL 13.5 Lettuce Protoplast Preparation

Equipment and Reagents

- Laminar air flow cabinet
- Lettuce seeds (cv. Flora; Leen de Mos)
- 10% (v/v) bleach solution (see Protocol 13.1)
- Illuminated growth chamber (60 µmol/m^2/s)
- Sterile glass culture jars and lids (175 ml; Sigma)
- Semisolid MS30 medium (see Protocol 13.1)
- Sterile Petri dishes (100 mm/15 mm)
- PG solution: 5.47 g sorbitol, 735 mg $CaCl_2.2H_2O$ in 100 ml water, autoclave; store at 4 °C
- Parafilm
- Enzyme solution: 2.5 ml 40 × B5 macro elements, 100 µl 1000 × B5 micro elements, 150 mg $CaCl_2.2H_2O$, 100 µl 1000 × NaFeEDTA, 100 µl 1000 × B5 vitamins, 13.7 g sucrose, 1 g Cellulase R10 (Duchefa), 250 mg Macerozyme R10 (Duchefa) in 100 ml water, pH 5.6 with 0.5 M KOH, filter sterilize; store at −20 °C
- Orbital shaker (Thermo Scientific)
- CPW16S solution: 1 ml 100 × CPW salts, 16 g sucrose, 148 mg $CaCl_2.2H_2O$ in 100 ml water, pH 5.8 with 0.5 M KOH, autoclave; store at 4 °C
- Sterile nylon mesh (41 µm pore size; PlastOk Ltd.)
- Sterile 15 ml tubes (Sarstedt)

- W5 solution: 4.5 g NaCl, 9.2 g $CaCl_2.2H_2O$, 185 mg KCl, 495 mg glucose, 50 mg MES in 500 ml water, pH 5.8 with 0.5 M KOH, autoclave, store at 4 °C
- Bench top centrifuge (MSE Centaur)
- Transformation buffer: 7.29 g mannitol, 304 mg $MgCl_2.6H_2O$, 1 g MES in 100 ml water, pH 5.8 with 10 M KOH, autoclave; store at 4 °C
- Haemocytometer (Sigma)
- 40 × B5 macro elements: 25 g KNO_3, 1.215 g $MgSO_4$, 1.7 g $NaH_2PO_4.2H_2O$, 1.34 g $(NH_4)_2SO_4$ in 250 ml water; store at −20 °C
- 1000 × B5 micro elements: 25 mg $Na_2MoO_4.2H_2O$, 1 g $MnSO_4.H_2O$, 200 mg $ZnSO_4.7H_2O$, 300 mg H_3BO_3, 75 mg KI, 100 µl $CuSO_4$ (16 mg/ml), 100 µl $CoCl_2.6H_2O$ (25 mg/ml) in 100 ml water; store at −20 °C
- 1000 × NaFeEDTA: 367 mg NaFeEDTA in 10 ml water; store at −20 °C
- 1000 × B5 vitamins: 1.12 g B5 vitamin powder (Duchefa) in 10 ml water; store at −20 °C
- 100 × CPW salts: 1.01 g KNO_3, 2.46 g $MgSO_4.7H_2O$, 272 mg KH_2PO_4, 100 µl KI (16 mg/ml), 10 µl $CuSO_4$ (16 mg/ml) in 100 ml; store at −20 °C

Method

1. Surface-sterilize lettuce seed in 10% (v/v) bleach for 20 min; sow the seeds as described in Protocol 13.1.
2. Incubate seeds for 10 days at 25 °C (16 h photoperiod).
3. Transfer seedlings to semisolid MS30 medium in glass culture jars (60 ml of medium/jar) and culture for 3–4 weeks at 25 °C (16 h photoperiod).
4. Remove four to five leaves and cut into small pieces (each approx. 25–50 mm^2)[a,b].
5. Float the leaf pieces on 20 ml of PG solution in sterile Petri dishes; seal the dishes with Parafilm and leave at 4 °C, in the dark, for 1 h.
6. Remove the PG solution and discard.
7. Add 20 ml of enzyme solution, reseal the dishes, incubate at 25 °C in the dark for 16 h.
8. Shake the dishes (60 rpm) for 2 h at 25 °C.
9. Add 10 ml of CPW16S solution and swirl gently to release the protoplasts.
10. Filter the protoplast suspension through a sterile nylon mesh (41 µm pore size); collect the filtered suspension.
11. Divide the suspension between three 15 ml tubes; overlay each sample with 1 ml of W5 solution.
12. Centrifuge at 70 g for 8 min.
13. Collect the protoplasts from the CPW16S/W5 interface[c].

14 Divide the protoplasts between two 15 ml tubes.

15 Gently add 9 ml of W5 solution to each tube.

16 Centrifuge at 70 g for 5 min.

17 Resuspend each protoplast pellet in 5 ml of W5 solution.

18 Centrifuge at 70 g for 5 min.

19 Resuspend the protoplast pellet in transformation buffer and adjust the protoplast density to 2×10^6 protoplasts/ml. Calculate the protoplast density using a haemocytometer[d].

20 Divide into 600 µl aliquots in 15 ml tubes.

Notes

[a]4–5 leaves should yield sufficient protoplasts for one transformation experiment.

[b]Cut the leaves under PG solution to prevent drying out.

[c]A large pellet at the bottom of the tube, rather than a thick band of protoplasts at the CPW16S/W5 interface, indicates the protoplasts have burst.

[d]Protoplast viability may be checked by fluorescein diacetate staining. Adjust 50 µl of protoplasts to a concentration of 0.002% (w/v) FDA and view with a microscope under UV and white light. Only viable protoplasts fluoresce under UV light.

PROTOCOL 13.6 PEG-Mediated Protoplast Transformation

Equipment and Reagents

- Laminar air flow cabinet
- Plasmid DNA (transformation vector) 1 µg/µl
- PEG solution: Dissolve 80 g of PEG 6000 (Sigma) in 100 ml of buffer (4.72 g $Ca(NO_3)_2.4H_2O$, 14.57 g mannitol in 100 ml water) by heating gently; adjust volume to 200 ml with distilled water and divide into two 100 ml aliquots. Adjust the pH of one aliquot to pH 10 with 1 M KOH and store both aliquots overnight at 4 °C. The following day, bring both aliquots to room temperature. Readjust the pH of the aliquot previously set to pH 10 to pH 8.2 using the non-adjusted aliquot; filter sterilize; store at −20 °C
- Prepared protoplasts (see Protocol 13.5)
- Wash solution: 36.45 g mannitol, 1.176 g $CaCl_2.2H_2O$ in 500 ml; autoclave and store at 4 °C
- Bench top centrifuge
- B5 solution: 12.5 ml/l 40 × B5 macroelements, 500 µl/l 1000 × B5 microelements, 375 mg/l $CaCl_2.2H_2O$, 500 µl/l 1000 × NaFeEDTA, 500 µl/l 1000 × B5 vitamins, 270 mg/l Na succinate, 103 g/l sucrose, 300 µl/l BAP (1 mg/ml), 100 µl/l 2,4 D

(1 mg/ml), 100 mg/l MES, pH 5.8 with 0.5 M KOH, filter sterilize; store at 4 °C (keep no longer than 1 week)

- B5 agarose medium: 2% (w/v) Sea Plaque agarose (Duchefa) in B5 solution
- Sterile Petri dishes (3.5 cm, 6 cm, 9 cm diam., shallow and deep, Greiner)
- Parafilm
- Spectinomycin
- Illuminated growth chamber (60 μmol/m^2/s)
- SH2 medium: 3.184 g/l SH salts (Duchefa), 10 ml/l 100 × SH vitamins, 30 g/l sucrose, 5 g/l agarose (Sigma), 100 μl/l NAA (1 mg/ml), 100 μl/l BAP (1 mg/ml), pH 5.8 with 0.5 M KOH
- SHREG medium: 3.184 g/l SH salts (Duchefa), 10 ml/l 100 × SH vitamins, 15 g/l sucrose, 15 g/l maltose, 5 g/l agarose, 100 μl/l NAA (1 mg/ml), 100 μl/l BAP (1 mg/ml), pH 5.8 with 0.5 M KOH
- Sterile glass culture jars and lids (175 ml; Sigma)
- SH30 medium: 3.184 g/l SH salts (Duchefa), 10 ml/l 100 × SH vitamins, 30 g/l sucrose, 8 g/l microagar (Duchefa), pH 5.8 with 0.5 M KOH
- SH30 + IBA medium: SH30 medium, indole-3-butyric acid 1 mg/l (Sigma)
- 100 × SH vitamins: 1.1 g of SH vitamins (Duchefa) in 10 ml of water; store at −20 °C

Method

1. Add 10 μl of plasmid DNA (1 μg/μl) and 400 μl of PEG solution (pH 8.2) to each protoplast aliquot; mix gently.
2. Incubate at room temperature for 10 min.
3. Add 9 ml of wash solution and mix gently[a].
4. Centrifuge at 70 g for 5 min.
5. Remove the supernatant and discard[b].
6. Gently resuspend the protoplast pellet to a density of 12×10^4 protoplasts/ml with B5 solution.
7. Add an equal volume of B5 agarose medium[c].
8. Pipette 1.5 ml aliquots of the agarose mixture into 3.5 cm Petri dishes and allow to set. Cut the agarose into quarters and transfer the quarters to 6 cm Petri dishes (two per dish); overlay with 4 ml of B5 solution (no antibiotic).
9. Seal the dishes with Parafilm and culture at 25 °C, in the dark, for 6 days.
10. After 6 days, add spectinomycin to the B5 solution in the dishes to a final concentration of 500 mg/l; culture at 25 °C, in the dark, for 1 day.
11. Remove 2 ml of the B5 medium from the dishes and replace with 2 ml of new B5 liquid medium containing 500 mg/l spectinomycin.

12 Seal the dishes and culture for 1 week at 25 °C (16 h photoperiod).

13 Remove the agarose quarters, cut into small pieces, place on semisolid SH2 medium with 500 mg/l spectinomycin and culture at 25 °C (16 h photoperiod).

14 Transfer the agarose pieces onto new SH2 medium with 500 mg/l spectinomycin every 2 weeks until green calli appear (6–8 weeks).

15 When the microcalli reach about 0.5 mm in diam., transfer onto semisolid SHREG medium with 500 mg/l spectinomycin in deep Petri dishes (9 cm diam.); culture at 25 °C (16 h photoperiod).

16 Spectinomycin-resistant shoots should regenerate after approx. 6 weeks[d].

17 Transfer shoots onto semisolid SH30 medium with 500 mg/l spectinomycin, but lacking growth regulators, in deep Petri dishes[e].

18 Transfer established shoots to semisolid SH30 medium with IBA (1 mg/l), but without spectinomycin, in glass culture jars for rooting.

Notes

[a]The wash solution terminates PEG treatment.

[b]Do not disturb the pellet as significant protoplast loss can occur at this stage.

[c]Melt and maintain at 50 °C until required.

[d]All shoots that regenerate from the same callus material are subclones.

[e]Reduce the selection to 250 mg/l spectinomycin, if there is difficulty in establishing shoots.

13.2.4 Identification and characterization of transplastomic plants

Spectinomycin-resistant cell lines, or shoots, may be true transformants or spontaneous spectinomycin-resistant mutants (a mutation in the plastid small ribosomal RNA gene, *rrn*16, can also confer spectinomycin resistance [11, 12]). It is possible to distinguish between these two possibilities using an antibiotic sensitivity assay [45]. However, this assay may be omitted and transformation can be assessed directly by polymerase chain reaction (PCR). Two separate reactions are used to characterize each of the primary transformed shoots/cell lines. One uses a primer pair specific for the selectable marker gene (SMG), while the other reaction uses a primer pair specific for the region of the plastid genome flanking the vector integration site (Figure 13.1a). The latter reaction can confirm targeted integration of the vector into the plastid genome, eliminating the small chance that spectinomycin resistant shoots are the result of *aad*A integration into the nuclear genome and expression of *aad*A in the nucleus. It can also indicate whether the transformed shoots are homoplastomic or heteroplastomic for the transgene insertion (Figure 13.1b). The primary regenerated shoots obtained using the biolistic method are usually heteroplastomic, containing cells with transformed and non-transformed plastid genomes [2]. Homoplastomic shoots are obtained by regenerating new shoots from heterplastomic leaf explants, under antibiotic selection. Homoplasmy can be confirmed by PCR (using the primer pair flanking the vector integration site), or by Southern Blot hybridization [45]. Primary regenerated lettuce shoots obtained using the PEG method are generally homoplastomic and subsequent rounds of regeneration under selection are not required with this method [28].

PROTOCOL 13.7 PCR Characterization of Transplastomic Shoots or Calli

Equipment and Reagents

- Putatively transformed shoot or callus material
- Mini-prep DNA isolation kit (Qiagen)
- SMG-specific PCR primers (e.g. *aad*A P1 – 5′-CGCCGAAGTATCAACTCAAC-3′; P2 – 5′-CTACATTTCGCTCATCGCC-3′)
- PCR primers specific for the region of the plastid genome flanking the vector integration site (choice of primers will depend on the vector integration site)
- PCR tubes (0.2 ml volume)
- 10 mM dinucleotide triphosphates (dNTPs): 2.5 mM each dATP, dCTP, dGTP and dTTP (Promega)
- AccuTaq la 10 × PCR buffer (Sigma)
- AccuTaq la DNA polymerase (Sigma)
- Sterile distilled water
- Thermocycler
- Equipment and reagents for agarose gel electrophoresis and EtBr staining
- DNA ladder (DirectLoad Wide Range DNA Marker; Sigma)
- UV gel imaging system

Method

1 Remove 50–100 mg leaf tissue from each shoot and isolate total DNA using a mini-prep DNA isolation kit according to the manufacturer's instructions[a,b].

2 Set up two PCR reactions per DNA sample

   (i) Using a primer pair specific for the SMG (P1, P2; Figure 13.1a).

   (ii) Using a primer pair specific for the region of the plastid genome flanking the vector integration site (P3, P4; Figure 13.1a).

3 Combine the following in a 0.2 ml PCR tube (total volume 50 μl) for each set of primers, include appropriate negative controls:

0.5 µl DNA (50 ng)

4 µl 10 mM dNTPs (2.5 mM each dNTP)

5 µl AccuTaq la 10 × PCR buffer

1 µl each primer (10 pmol/µl)

0.5 µl AccuTaq la DNA Polymerase

38 µl dH_2O

4 Place tubes in a thermocycler and use the following cycling conditions:

Number of cycles	Programme
1	3 min at 94 °C
34	60 s at 94 °C; 60 s at 55 °C[c]; 60 s at 68 °C[d]
1	10 min at 68 °C
1	Hold at 4 °C

5 Run 5–10 µl of the PCR products on a 0.8% (w/v) agarose gel alongside a DNA ladder.

6 Stain the gel with ethidium bromide and observe and photograph the gel under UV illumination.

Notes

[a]Alternatively, use the CTAB method for DNA isolation [45, 49].

[b]Callus, generated using the PEG method, can be assessed for plastid transformation before shoots regenerate from the tissue.

[c]The annealing temperature may need to be adjusted depending on the T_m of the primers.

[d]The extension time may need to be adjusted depending on the length of the expected PCR product (allow at least 60 s/kb).

The PCR results may show that some of the plants are homoplastomic (all plastid DNA copies are transformed), some are heteroplastomic (contain transformed and wild-type copies), and some contain only wild-type plastid DNA copies (Figure 13.1b). If the primary regenerants are heteroplastomic, then homoplastomic lines are obtained by regenerating new shoots, under antibiotic selection, from small leaf explants taken from primary transformants.

PROTOCOL 13.8 Generating Homoplastomic Plant Lines

Equipment and Reagents

- Laminar air flow cabinet

- Heteroplastomic, spectinomycin-resistant tobacco plants
- Sterile Petri dishes (9 cm diam.)
- Spectinomycin
- Semisolid RMOP medium containing 500 mg/l spectinomycin
- Illuminated growth chamber (60 µmol/m^2/sec)
- Sterile clear plastic Pet Cups and lids
- Semisolid MS30 medium containing 500 mg/l spectinomycin

Method

1 Remove two to four leaves from each of the heteroplastomic plant lines[a,b].

2 Cut each leaf into 5 mm^2 pieces and place on semisolid RMOP medium containing 500 mg/l spectinomycin (30 ml of medium/Petri dish) for shoot formation.[c]

3 Culture the leaf pieces for 3–4 weeks at 25 °C (16 h photoperiod).

4 Transfer individual, regenerated shoots to semisolid MS30 medium containing 500 mg/l spectinomycin in Pet Cups.

5 When the regenerated shoots are established with leaves and roots, remove a leaf from each shoot and repeat the regeneration process.

6 Check if third round regenerated plants are homoplastomic by PCR as described in Protocol 13.7[d].

7 Transfer rooted plants to compost, grow to flowering, allow flowers to self-pollinate and harvest seed.

8 Seed derived from homoplastomic lines should give 100% spectinomycin-resistant (green) seedlings.

Notes

[a]Take three to four transformed lines through to homoplastomy.

[b]All shoots derived from a primary transformant are considered subclones.

[c]Culture 10 explants per leaf on selection medium because not all leaf pieces will form shoots (five to seven leaf pieces/Petri dish).

[d]Two to three rounds of regeneration are usually sufficient to achieve homoplastomy.

13.3 Troubleshooting

13.3.1 Biolistic-mediated transformation

- Too many tobacco leaf pieces/dish on selection medium can inhibit shoot formation.

- If all spectinomycin-resistant shoots obtained are spontaneous mutants, the transforming DNA is not being delivered to plastids. Repeat the DNA coating of gold particles. Repeat transformation using a vacuum of at least 711 mmHg (28 inches Hg); check DNA delivery into plastids using a vector containing a reporter gene construct (*uid*A or *gfp*).
- If wild-type and transformed plastid DNA is detected in spectinomycin-resistant plants, the plants are heteroplastomic. Take the plants through another round of regeneration on selection medium.
- Persistent, faint, wild-type plastid DNA bands detected in plants after two to three rounds of regeneration on selection medium may be due to plastid DNA in nuclear and/or mitochondrial genomes and can be ignored [50–52]. Assess seed progeny for antibiotic resistance. If all progeny are spectinomycin-resistant (green), assume the parent lines are homoplastomic.
- If transformed plants do not set seed, due to culture-induced male sterility, attempt hand pollination with wild-type pollen to obtain seed.

13.3.2 PEG-mediated transformation

- Leaf age can affect protoplast yield. If the number of isolated protoplasts is low, repeat the isolation using younger leaves from *in vitro* propagated shoots.
- Once the enzymatic digestion of the cell walls is complete, handle the protoplasts gently to prevent bursting.
- If spectinomycin-resistant calli are not obtained, the transforming DNA is not being delivered to plastids. Check the number and viability of the protoplasts, the plasmid DNA, and the concentration of spectinomycin used for selection.
- Accurate determination of protoplast numbers is essential. Spectinomycin-resistant calli will not grow if the protoplast plating density is too low.
- If the transformed plants do not set seed they may be polyploid (PEG treatment can cause protoplast fusion events). Check the ploidy level of plants by flow cytometry; discard tetraploid or polyploid plants.

Acknowledgements

The authors thank John Gray and Phil Dix for their help in establishing tobacco plastid transformation. Plastid transformation in J.N.'s laboratory was funded by the EU 6th Framework Programme, Pharmaplanta (B.H. and A.D.), Science Foundation Ireland (A.D.) and the Health Research Board, Ireland (K.-H.N.).

References

1. Bock R (2001) *J. Mol. Biol*. **312**, 425–438.

2. Maliga P (2004) *Annu. Rev. Plant Biol*. **55, 289–313.

Useful review of plastid transformation in flowering plants.

3. Viitanen PV, Devine AL, Khan MS, Deuel DL, Van Dyk DE, Daniell H (2004) *Plant Physiol*. **136**, 4048–4060.
4. Wurbs D, Ruf S, Bock R (2007) *Plant J*. **49**, 276–288.
5. Ruhlman T, Daniell H (2007) In: *Applications of Plant Metabolic Engineering*. Edited by R Verpoorte, AW Alfermann and TS Johnson. Springer-Verlag, Berlin, Heidelberg, pp. 79–108.
6. Bock R (1998) *Methods* **15**, 75–83.
7. Lutz K, Maliga P (2007) *Methods Enzymol*. **424**, 501–518.
8. Bock R (2007) *Curr. Opin. Biotech*. **18**, 100–106.
9. Nugent JM, Joyce SM (2005) *Curr. Pharm. Des*. **11**, 2459–2470.
10. Daniell H (2006) *Biotechnology J*. **1**, 1071–1079.

*11. Svab Z, Hajdukiewicz P, Maliga P (1990) *Proc. Natl Acad. Sci. USA* **87**, 8526–8530.

The original publication describing biolistic-mediated tobacco plastid transformation.

12. Svab Z, Maliga P (1993) *Proc. Natl Acad. Sci. USA* **90**, 913–917.

*13. Golds T, Maliga P, Koop H-U (1993) *Bio/Technology* **11**, 95–97.

The original publication describing PEG-mediated tobacco plastid transformation.

14. O'Neill C, Horváth GV, Horváth E, Dix PJ, Medgyesy P (1993) *Plant J*. **3**, 729–738.
15. Koop H-U, Steinmuller K, Wagner H, Rossler C, Eibl C, Sacher L (1996) *Planta* **199**, 193–201.
16. Huang FC, Klaus SM, Herz S, Zou Z, Koop HU, Golds TJ (2002) *Mol. Genet. Genomics* **268**, 19–27.
17. Yu L-X, Gray BN, Rutzke CJ, Walker LP, Wilson DB, Hanson MR (2007) *J. Biotechnol*. **131**, 362–369.
18. Sidorov VA, Kasten D, Pang SZ, Hajdukiewicz PT, Staub JM, Nehra NS (1999) *Plant J*. **19**, 209–216.
19. Nguyen TT, Nugent G, Cardi T, Dix PJ (2005) *Plant Sci*. **168**, 1495–1500.
20. Ruf S, Hermann M, Berger IJ, Carrer H, Bock R (2001) *Nat. Biotechnol*. **19**, 870–875.
21. Nugent GD, ten Have M, van der Gulik A, Dix PJ, Uijtewaal BA, Mordhorst AP (2005) *Plant Cell Rep*. **24**, 341–349.

The original publication describing PEG-mediated lettuce plastid transformation.

22. Sikdar SR, Serino G, Chaudhuri S, Maliga P (1998) *Plant Cell Rep*. **18**, 20–24.
23. Skarjinskaia M, Svab Z, Maliga P (2003) *Transgenic Res*. **12**, 115–122.
24. Hou B-K, Zhou Y-H, Wan L-H *et al*. (2003) *Transgenic Res*. **12**, 111–114.
25. Nugent GD, Coyne S, Nguyen TT, Kavanagh TA, Dix PJ (2006) *Plant Sci*. **170**, 135–142.
26. Liu C-W, Lin C-C, Chen JJW, Tseng M-J (2007) *Plant Cell Rep*. **26**, 1733–1744.

*27. Lelivelt CLC, McCabe MS, Newell CA, *et al.* (2005) *Plant Mol. Biol.* **58**, 763–774.

28. Zubko MK, Zubko EI, van Zuilen K, Meyer P, Day A (2004) *Transgenic Res.* **13**, 523–30.

The original publication describing PEG-mediated lettuce plastid transformation.

29. Kanamoto H, Yamashita A, Asao H *et al.* (2006) *Transgenic Res.* **15, 205–217.

Biolistic-mediated lettuce plastid transformation.

30. Ruhlman T, Ahangari R, Devine A, Samsam M, Daniell H (2007) *Plant Biotech. J.* **5, 495–510.

Biolistic-mediated lettuce plastid transformation.

31. Kumar S, Dhingra A, Daniell H (2004) *Plant Mol. Biol.* **56**, 203–216.

32. Kumar S, Dhingra A, Daniell H (2004) *Plant Physiol.* **136**, 2843–2854.

33. Dufourmantel N, Pelissier B, Garçon F, Peltier G, Ferullo J-M, Tissot G (2004) *Plant Mol. Biol.* **55**, 479–489.

34. Lee SM, Kang K, Chung H *et al.* (2006) *Mol. Cells* **21**, 40–410.

35. Okumura S, Sawada M, Park YW, *et al.* (2006) *Transgenic Res.* **15**, 637–646.

36. Azhagiri AK, Maliga P (2007) *Plant J.* **52**, 817–823.

37. Ruf S, Karcher D, Bock R (2007) *Proc. Natl Acad. Sci. USA* **104**, 6998–7002.

38. Svab Z, Maliga P (2007) *Proc. Natl Acad. Sci. USA* **104**, 7003–7008.

39. Purton S (2007) *Adv. Exp. Med. Biol.* **616**, 34–45.

40. Sugiura C, Sugita M (2004) *Plant J.* **40**, 314–321.

41. Chiyoda S, Linley PJ, Yamato KT, Fukuzawa H, Yokota A, Kohchi T (2007) *Transgenic Res.* **16**, 41–49.

42. Maliga P (2005) *Photochem. Photobiol. Sci.* **4**, 971–976.

43. Lutz KA, Azhagiri AK, Tungsuchat-Huang T, Maliga P (2007) *Plant Physiol.* **145**, 1201–1210.

44. Verma D, Daniell H (2007) *Plant Physiol.* **145**, 1129–1143.

45. Lutz KA, Svab Z, Maliga P (2006) *Nature Protocols* **1, 900–910.

Detailed protocol for biolistic-mediated tobacco plastid transformation.

46. Murashige T, Skoog F (1962) *Physiol. Plant.* **15**, 473–497.

47. Koop HU, Kofer W (1995) In: *Gene Transfer to Plants*. Edited by I Potrykus and G Spangenberg. Springer-Verlag, Berlin, Heidelberg, New York, pp. 75–82.

48. Kofer W, Eibl C, Steinmuller K, Koop H-U (1998) *In Vitro Cell Dev. Biol.-Plant* **34**, 303–309.

49. Murray MG, Thompson WF (1980) *Nucl. Acids Res.* **8**, 4321–4325.

50. Ayliffe MA, Scott NS, Timmis JN (1998) *Mol. Biol. Evol.* **15**, 738–745.

51. Matsuo M, Ito Y, Yamauchi R, Obokata J (2005) *Plant Cell* **17**, 665–675.

52. Cummings MP, Nugent JM, Olmstead R, Palmer JD (2003) *Curr. Genet.* **43**, 131–138.

14

Molecular Characterization of Genetically Manipulated Plants

Cristiano Lacorte, Giovanni Vianna, Francisco J.L. Aragão and Elíbio L. Rech
Embrapa Recursos Geneticos e Biotecnologia, Parque Estação Biologica – PqEB, Brazil

14.1 Introduction

Plant cell transformation can be achieved through the soil bacterium *Agrobacterium* or by 'direct' methods, which include techniques such as biolistics, electroporation and microinjection [1, 2]. The choice of transformation method depends largely on the species, due mainly to technical reasons. *Agrobacterium*-mediated transformation may be the most convenient choice for model plants such as tobacco and *Arabidopsis*, as well as several other species for which the protocols are well established. However, for many important crops, particularly cereals and legumes, protocols based on the biolistic method have been exploited extensively [1–3].

Regardless of the transformation method adopted, the initial transformation event is only the beginning of a long and cumbersome process that involves tissue culture and cell selection, followed by plant regeneration and acclimatization. If the goal is the development of a commercial variety, the transgenic plants and their progeny are subjected to a series of tests and molecular analyses that are essential to check the genetic stability and assure field performance, together with product quality and safety, in order to comply with both market demands and the relevant regulatory processes [4]. From the technical viewpoint, molecular analysis is an important issue because the integration pattern of the transgene(s) can have direct implications on genetic stability and expression level in subsequent generations, either through segregation of the transgenes, if integrated in different loci (i.e. different chromosomes or unlinked loci), or as a result of deleterious rearrangements that

Plant Cell Culture Edited by Michael R. Davey and Paul Anthony

could generate aberrant RNA molecules, which in turn can trigger gene silencing [1].

There is no well established method currently available to achieve transformation in a way that copy number, integration site or gene expression are accurately predetermined. Nevertheless, several studies on the molecular analysis of transgenic plants allow some general assumptions on the expected integration patterns according to the transformation method used. *Agrobacterium*-mediated transformation tends to generate less complex integration patterns, but the presence of sequences that are outside the T-DNA is frequently observed [5]. Biolistics (particle bombardment) can generate complex patterns and concatemers, particularly if intact plasmids are used for transformation. However, if a fragment is used for transformation, single copy integration events are most commonly observed [6–8].

This chapter outlines some essential protocols for basic molecular analysis of transgenic plants, including two commonly used protocols for the extraction of plant DNA. A standard protocol for polymerase chain reaction (PCR) is provided to allow the screening of putative transgenic plants and segregation analysis in the progeny. A method for Southern blot is presented and can be used as proof of integration and to determine the copy number of a transgene. For more refined analysis, the protocols for inverse PCR and Tail-PCR are described for the amplification of sequences flanking the integration site.

14.2 Methods and approaches

The usual approach for the molecular analysis of genetically modified plants commences with the *in vitro* regenerated plants themselves, from which tissue samples are collected and tested by PCR, indicating the presence of the foreign gene sequence within the genome of the transgenic plants. These PCR positive plants are transferred to soil and acclimated under glasshouse conditions. In some cases, it is wise to micropropagate these lines and keep some plants *in vitro* as a 'backup'. Acclimated plants (T_0 generation) may be tested directly for the desired trait (e.g. pathogen resistance, herbicide tolerance), although T_0 plants are often weak in growth and may show epigenetic variation due to their time in culture. Chimeric plants are frequently obtained, depending on the species and the selection method. These plants may harbour the foreign DNA in some tissues, but not in the cell lines that give rise to the germinative cells and, consequently, the transgene will not be transmitted to the offspring. The offspring of plants that do transmit the transgene may contain an array of copy numbers and integration patterns for the transgene. The presence of the transgene in these T_1 plants may also be tested by PCR and this segregation analysis provides important information that, together with Southern blot analysis, can indicate the expected integration pattern. Mendelian segregation does not imply a single copy integration event, as tandem arrays and interspersed copies, even by megabase sized pieces of DNA, will be linked at the genomic level and will not segregate independently. Further tests for the validation of an elite transformation event may still be undertaken throughout several generations to assure transgene homozygosity and stability of expression [3, 4].

The sequence flanking the integration site may be required by regulatory agencies and also allows traceability of a single transformation event. These sequences may be amplified from the plant genome by PCR-based techniques, such as inverse PCR and Tail-PCR, as outlined in Protocols 14.5 and 14.6.

14.2.1 Plant DNA extraction

Plant tissues are amongst the most difficult material from which to extract high quality DNA. Difficulties in disrupting cell walls and potential contamination with polyphenols and polysaccharides, can affect yield and quality [9–11]. A reliable and rapid method for the isolation of small amounts of DNA to be used in PCR analysis is based on the CTAB method [9], described in Protocol 14.1. The other widely used method, described in Protocol 14.2, is based on the procedure of Dellaporta *et al.* [10], and yields DNA suitable for Southern blot analysis.

PROTOCOL 14.1 Small-Scale Plant DNA Extraction: The CTAB Method

Equipment and Reagents

- Extraction buffer: 0.8 g of hexadecyltrimethylammonium bromide (CTAB), 16 ml of 5 M NaCl, 4.4 ml of 500 mM ethylenediaminetetraacetic acid (EDTA), 22 ml of 1 M Tris-HCl pH 8.0, 1 g of sarcosyl, 2.55 g of sorbitol, 20 μl of β-mercaptoethanol; autoclaved distilled water to 100 ml
- 1.5 ml microfuge tubes (Axygen)
- Plastic micropestles (Eppendorf)
- Water bath (55 °C)
- Chloroform : isoamyl alcohol (24 : 1; v : v)
- Vortex mixer (Vortex Gene 2, Scientific Industries)
- Microcentrifuge (Model 5415C; Eppendorf)
- Isopropanol
- 70% (v/v) ethanol
- Autoclaved distilled water

Method

1 Macerate 20–40 mg fresh weight of tissue in a microcentrifuge tube using a plastic pestle.

2 Add 200 μl of extraction buffer.

3 Incubate for 20 min at 55 °C.

4 Add 200 μl of chloroform : isoamyl alcohol; mix vigorously by vortexing for approx. 3 min.

5 Centrifuge at 12 000 g (~13 000 rpm) for 5 min. Transfer the supernatant to a new tube.

6 Add 1 vol. (~150 μl) of isopropanol. Mix by inverting the tube.

7 Centrifuge at 12 000 g (~13 000 rpm) for 5 min.

8 Carefully discard the supernatant and add 200 μl of 70% (v/v) ethanol.

9 Centrifuge 12 000 g (~13 000 rpm) for 5 min. Carefully discard the supernatant; leave the pellet to dry.

10 Resuspend the pellet in 30–50 μl of autoclaved, distilled water.

PROTOCOL 14.2 Plant DNA Extraction: Modified Dellaporta Method

Equipment and Reagents

- Extraction buffer: 50 ml of 5 M NaCl, 50 ml of 50 mM EDTA, pH 8.0, 50 ml of 1 M Tris-HCl, pH 8.0, 1 ml of β-mercaptoethanol; autoclaved distilled water to 500 ml
- Liquid nitrogen
- Pestle and mortar (Branson)
- 50 ml polypropylene tubes (Nunc)
- Sodium dodecyl sulfate (SDS; 20% w/v)
- Water bath (65 °C)
- 5M potassium acetate: 60 ml 5 M of potassium acetate, 11.5 ml acetic acid; autoclaved distilled water to 100 ml
- Ice bath
- Centrifuge (5810R; Eppendorf)
- Miracloth (Millipore)
- Isopropanol
- TE buffer: 10 mM Tris-HCl, 20 mM EDTA
- 1.5 ml microfuge tubes (Axygen)
- RNAse: 10 mg/ml; dissolved in 50 mM sodium acetate and boiled for 20 min to eliminate contaminating DNase
- Phenol : chloroform : isoamyl alcohol (25 : 24 : 1, v : v : v)

- Chloroform : isoamyl alcohol (24 : 1, v : v)
- 3 M sodium acetate, pH 5.2
- 70% (v/v) ethanol
- Spectrophotometer (ND-1000; NanoDrop)

Method

1. Collect 3–5 g of leaf material (preferably young leaves); freeze the material in liquid nitrogen and store at −80 °C.
2. Transfer the frozen leaves to a mortar and add liquid nitrogen. Using a pestle, macerate the leaf material to a fine powder.
3. Transfer the powder to a polypropylene tube containing 15 ml of extraction buffer.
4. Add 1 ml of 20% (v/v) SDS and mix gently by inverting the tube. Incubate for 15–20 min at 65 °C; mix by inverting the tube after 10 min of incubation.
5. Add 5 ml of 5 M potassium acetate; mix by inverting the tube, and incubate on ice for 20 min. Mix by inverting the tube every 5 min.
6. Centrifuge for 20 min (20 000 g, at 4 °C); transfer the supernatant to a new tube, filtering through Miracloth.
7. Add 15 ml of cold isopropanol to the filtered supernatant; mix gently by inversion and incubate for at least 2 h at −20 °C.
8. Centrifuge at 20 000 g for 20 min; discard the supernatant and dry the pellet at room temperature.
9. Dissolve the dried pellet in 500 μl of TE buffer; transfer to a microcentrifuge tube. Add 10 μl of RNaseA and incubate at 37 °C for 30 min.
10. Extract the RNaseA by adding 500 μl (1 vol.) of phenol : chloroform : isoamyl alcohol. Mix by vortexing (5–10 s) and centrifuge at 12 000 g for 5 min. Carefully transfer the upper phase to a new tube and 500 μl of chloroform–isoamyl alcohol. Carefully transfer the upper phase to a new tube. Mix by vortexing (5–10 s) and centrifuge at 12 000 g for 5 min.
11. Precipitate the DNA by adding 50 μl (1/10 vol.) of 3 M sodium acetate (pH 5.2) and 385 μl (0.7 vol.) of cold isopropanol; mix and incubate at −20 °C for at least 2 h.
12. Centrifuge at 12 000 g for 15 min. Decant the supernatant; add 1 ml of 70% (v/v) ethanol. Centrifuge at 12 000 g for 2 min.
13. Carefully discard the supernatant; dry the pellet at room temperature.
14. Dissolve the pellet in 200–500 μl of TE buffer or sterile autoclaved distilled water. Take 1 μl of this DNA preparation and estimate the concentration spectrophotometrically[a].

Note

[a]The Nanodrop spectrophotometer requires only 1 μl of undiluted DNA for accurate reading and quantification. If a conventional spectrophotometer is used, the DNA sample should

be diluted 500–1000× in water. Measure the sample at A_{260} and A_{280} and calculate the DNA concentration considering the dilution factor and that a 1.0 reading at A_{260} nm corresponds to 50 μg/ml of DNA. The A_{260}/A_{280} ratio can provide an estimate of the sample purity. As a general rule, a $A_{260/260}$ ratio of 1.8–2.0 should be obtained for a 'pure' sample.

14.2.2 Polymerase chain reaction

PCR is a simple, but powerful technique. Besides its application in cloning and fingerprinting analysis, PCR is used to detect specific sequences in DNA samples. The essential requirement of the amplification process involves the denaturing of DNA to generate single strands, the annealing of primers to the DNA strands and extension of the DNA, generating a double stranded DNA molecule that can be denatured again, starting a new cycle of amplification [11]. The use of heat resistant polymerases, the ability to synthesize sequence-specific primers and the increasing sequence information available on public data bases, has made this a standard technique. Improvements and new developments are broadening its application [11, 12].

Protocol 14.3 describes a standard PCR procedure to screen putative transgenic plants (T_0 plants) and/or the study of segregation of the transgene throughout generations. A list of primers specific to commonly used regulatory sequences and selection and marker genes is presented.

PROTOCOL 14.3 PCR Amplification

Equipment and Reagents

- Taq polymerase: 5 U/μl (Invitrogen)
- 10 × reaction buffer (Invitrogen)
- 50 mM magnesium chloride solution
- 10 mM dinucleotide triphosphate (dNTP) mix: 2.5 mM each of dATP, dCTP, dGTP, dTTP (Invitrogen)
- Primer – sense[a]: 10 μM (Invitrogen)
- Primer – antisense[a]: 10 μM (Invitrogen)
- Autoclaved distilled water
- 0.2 ml PCR microtube (Axygen)
- Thermocycler (iCycler; Bio-Rad)

- 1% (w/v) agarose gels (Ultrapure; Invitrogen) containing 10 ng/ml ethidium bromide
- 6 × loading buffer (Invitrogen)
- DNA kb ladder (1 kb plus DNA ladder; Invitrogen)
- 1 × TBE running buffer: 89 mM Tris-HCl, pH 8.5, 89 mM boric acid, 2 mM EDTA
- Horizontal electrophoresis system (Fisher Brand)
- Power supply (Basic; Bio-Rad)
- Gel photo documentation apparatus (Gel Doc XR, Bio-Rad)

Method

1 Prepare a PCR 'Master Mix' with the following reagents for each reaction, including the controls and one extra reaction to compensate for pipetting errors:

2.5 µl of 10× reaction buffer

0.5 µl of Primer-sense[a] (10 µM)

0.5 µl of Primer-antisense[a] (10 µM)

0.3 µl of dNTP

0.6 µl of magnesium chloride (50 mM)

0.3 µl of Taq polymerase (5 U/µl)

18.5 µl of autoclaved distilled water

2 Homogenize by inverting the tube and transfer 25 µl of the reaction mix to each tube; add 1.5 µl of genomic DNA.

3 Place the tubes in a thermocycler with the following programme:

Number of cycles	Programme
1	3 min at 95 °C
35	1 min at 95 °C → 1 min at 55 °C → 1 min at 72 °C[b]
1	7 min at 72 °C

4 When the programme terminates, analyse the amplified product by agarose gel electrophoresis. Add 5 µl of loading buffer to each reaction tube and load the samples on an agarose gel. Leave one well to add 2–5 µl of the kb ladder. Place the gel in the electrophoresis tank and set the voltage for 6–7 V/cm. Analyse and photograph the gel under UV light using a gel documentation system.

Notes

[a]The following primers, specific to marker and selection genes and regulatory sequences, are commonly utilized to analyse transgenic plants by PCR:

Gene	Position/sequence (5′–3′)	Size of the amplified fragment
gusA	401 - CGTCTGGTATCAGCGCGAAG 858c - TCACGCAGTTCAACGCTGAC	456
*npt*II	60 - GAGGCTATTCGGCTATGACT 470c - TCGACAAGACCGGCTTCCAT	411
bar	58 - GCGGTCTGCACCATCGTCAA 516c - TACCGGCAGGCTGAAGTCCA	459
egfp	228-CGACCACATGAAGCAGCACG 667c - CCAGCAGGACCATGTGATCG	440
*hpt*II	222 - TCCGGAAGTGCTTGACATTG 695 - ATGTTGGCGACCTCGTATTG	474

Promoters/terminators	
CaMV35S	64 - ATCCTTCGCAAGACCCTTCC
t-nos	223c - AGTAACATAGATGACACCGC
t-CaMV35S	148c - AGGGTTTCGCTCATGTGTTG

gusA: β-glucuronidase; *npt*II: neomycin phosphotransferase II; *bar*: phosphinothrycin acetyl transferase; *hpt*: hygromycin phosphotransferase; *egfp*: enhanced green fluorescent protein; CaMV 35S and t-CaMV35S: promoter and polyadenylation sequence, respectively, from the Cauliflower mosaic virus; t-nos: polyadenylation sequence from the nopaline synthase gene.

[b]The elongation time depends on the length of the fragment. An elongation time of 1 min for each kb of PCR fragment is usually recommended.

14.2.3 Southern blot technique

The Southern blot technique is used extensively to detect specific sequences in DNA samples. In a transgenic plant, besides detecting a transgene, Southern blot analysis can also indicate the number of copies of the foreign sequence integrated into the genome [13]. The principle of this simple procedure is according to the following steps. DNA from the plant to be tested is digested with a suitable restriction enzyme that cuts once the inserted transgene. The fragments are separated by agarose gel electrophoresis and this size-separated DNA is transferred from the gel to a membrane by capillarity, or by an electric field (see Figure 14.1). The membrane is incubated with a labelled single stranded DNA fragment that may hybridise with complementary DNA immobilized on the membrane and is detected by the signal emitted by the probe (see Figure 14.2). Protocol 14.4 describes the three basic steps of this technique, namely, DNA transfer from agarose gel to a membrane, hybridization with the probe and exposure and development. An example of an assembled Southern blot system and a schematic diagram is shown in Figure 14.1.

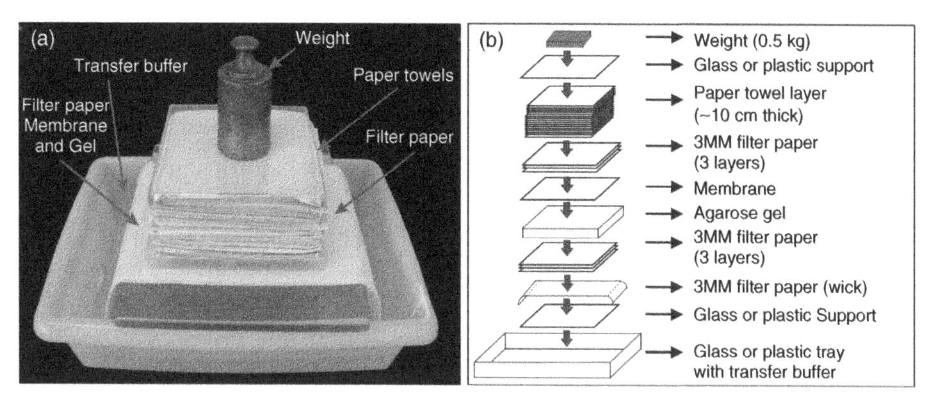

Figure 14.1 Southern blot. (a) Assembled Southern blot system for capillary transfer of DNA from an agarose gel to a membrane. (b) Schematic diagram of a Southern blot assembly.

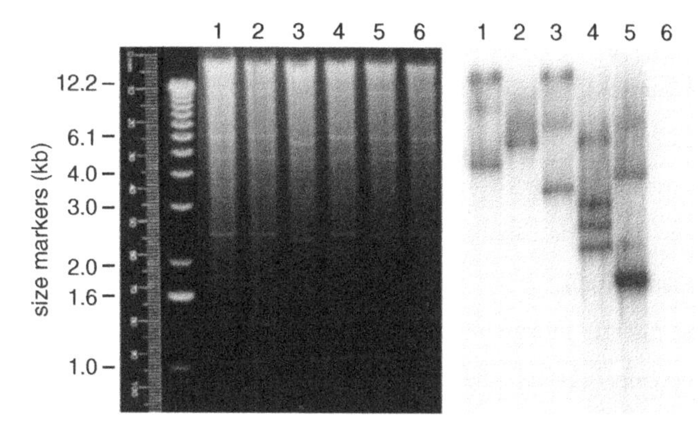

Figure 14.2 Agarose gel electrophoresis of digested DNA. Samples from transgenic lines (1–5) and a non-transformed plant (6) stained with ethidium bromide (left) and the DNA transferred to a membrane after Southern hybridization with a radioactive probe (right).

PROTOCOL 14.4 Southern Blot

Equipment and Reagents

- 1.7 ml Microfuge tubes (Axygen)
- Denaturing solution: 0.5 M NaOH, 1.5 M NaCl
- Neutralization solution: 0.5 M Tris-HCl, pH 7.4, 1.5 M NaCl
- Southern blot apparatus (plastic or glass tray, two glass plates, paper towels and a weight (0.5 kg) (see Figure 14.1)
- Filter paper (3 MM; Whatman)
- Nylon membrane (Hybond-N; GE Healthcare)
- Transfer buffer (20 × SSC): 0.3 M sodium citrate, 3 M NaCl

- UV cross-linking equipment (Stratalinker 2400; Stratagene)
- 100 × Denhardt's solution (10 g polyvinylpyrrolidone, 10 g Ficoll 400, 10 g bovine serum albumin; autoclaved distilled water to 500 ml)
- Salmon sperm DNA: 10 mg/ml
- Pre-hybridization solution: 2.5 ml of 20 × SSC; 500 μl of 100 × Denhardt's solution, 500 μl of 10% SDS, 20 μl of 10 mg/ml salmon sperm DNA; autoclaved distilled water to 10 ml
- Primer labelling kit ('Ready-To-Go DNA Labeling Beads', GE Healthcare)
- [α^{32}P]-labelled dCTP, 3000 Ci/mmol (Perkin Elmer)
- Ice bath
- Hybridization bottle (35 × 300 mm; Shel Lab)
- Rotary hybridization oven (1012, SL; Shel Lab)
- Washing solution I: 2 × SSC with 0.1% (w/v) SDS
- Washing solution II: 1 × SSC and 0.1% (w/v) SDS
- Washing solution III: 0.1 × SSC and 0.1% (w/v) SDS
- Geiger counter (900 series; Mini-instruments)
- PVdC wrap (Saran Wrap)
- Photographic cassette with an intensifier screen (GE Healthcare)
- X-ray film, developer (Dektol) and fixative (Kodak)

Method

DNA transfer from an agarose gel to a membrane:

1. Digest 10–30 μg of plant genomic DNA with 5–10 U of restriction enzyme[a] per μg of DNA. Use suitable reaction conditions (buffer and temperature) according to the manufacturer's instructions in a final volume of 50 μl; incubate for 4–16 h.
2. Load the digested DNA onto an agarose gel (13 × 16 cm). After electrophoresis (6–20 h at 1–3 V/cm), photograph the gel under UV light (see Protocol 14.3) with a ruler along the side of the gel, in order to determine the position of the bands after blotting the membrane. The 'zero' of the ruler should be aligned with the wells (see Figure 14.2).
3. (Optional) Incubate the gel in depurination solution (0.2 M HCl), agitating gently for 10 min in a tray. Rinse briefly twice in distilled water. This treatment partially hydrolyses the DNA by depurination, facilitating transfer of fragments larger than 10 kb to the nylon membrane. This step is recommended only if the target fragment is expected to be longer than 15 kb for the detection of high molecular weight fragments.
4. Incubate the gel in denaturing solution; agitate gently for 30 min.
5. Rinse the gel briefly with distilled water and incubate in neutralization solution; agitate for 30 min.

6 To assemble the transfer apparatus, place a plastic or glass support plate on the edges of a plastic or glass tray (see Figure 14.1). Cut a piece of filter paper to make a bridge over the glass plate and the bottom of each side of the tray. Fill the tray with transfer buffer (20 × SSC), making sure the paper bridge (wick) is thoroughly wet.

7 Cut three pieces of filter paper having the exact size of the gel and place them over the wick. Carefully place the gel upside down, on top of the filter paper layer. Remove any air bubbles by rolling a glass rod over the papers.

8 With a ruler, measure the gel and cut a piece of nylon membrane to the same size; soak the membrane in 20 × SSC buffer and place over the gel. Remove any air bubbles.

9 Cut 3 sheets of filter paper to the exact size of the membrane; soak the papers in 20 × SSC and place them over the membrane. Remove any air bubbles.

10 Place a layer of approx. 10 cm of paper towels on top of the filter paper layer. The size of these paper towels should be about the size of the membrane and gel. Place a glass or plastic plate on top of this layer and a weigh of approximately 0.5 kg on the glass (see Figure 14.1).

11 Leave the assembly undisturbed for 12–16 h.

12 After transfer, carefully dismantle the apparatus. With a pencil, mark on the membrane the position corresponding to the wells.

13 Cross-link the DNA to the membrane using a UV cross-linking instrument[a,b,c].

Prehybridization and hybridization:

14 Carefully roll up the membrane and place it in a hybridization tube.

15 Add 10–20 ml of prehybridization solution pre-heated to 65 °C. Place the bottle in the rotary support in a hybridization oven. Incubate for 2 h to overnight at 65 °C, rotating at 6–7 rpm.

16 Meanwhile, proceed with the probe labelling using the 'Ready-To-Go DNA Labeling Beads', or any available commercial kit for random primer labelling[d].

Denature the DNA at 95–100 °C for 2–3 min and immediately place on ice for 2 min. Centrifuge briefly and transfer 25–50 ng ($\leq$45μ l) to the tube containing the Reaction Mix bead. Add 5 μl of [α^{32}P]dCTP (50 μCi) and distilled water to final volume of 50 μl. Mix gently, centrifuge briefly and incubate at 37 °C for 5–15 min.

(Optional) Removal of unincorporated nucleotides can lead to reduced background. Several commercial columns are available (e.g. 'Qiaquick Nucleotide Removal Kit', Qiagen).

17 Denature the labelled probe at 95–100 °C, for 5 min, and place it immediately on ice, for 3–5 min. Spin briefly and transfer the denatured probe to the prehybridization solution. The total volume of the hybridization solution should be just enough to cover the membrane.

18 Incubate in hybridization solution at 65 °C in a rotary hybridization oven for 12–16 h, rotating at 6–7 rpm.

Washing, exposure and detection:

19 Wash the membrane twice with washing solution I for 15 min at room temperature.

20 Wash the membrane once with washing solution II for 15 min at 65 °C.

21 Place the membrane on a plastic film and monitor the radioactivity on the membrane using a Geiger counter. If necessary (based on the signal from areas corresponding to the negative controls or areas that should not contain any DNA), wash the membrane once or twice with washing solution III, for 15 min at 65 °C.

22 Remove excess solution leaving the membrane slightly wet. Wrap the membrane in PVdC film.

23 Place the wrapped membrane in a photographic cassette on an intensifier screen. In a dark room, place an X-ray film in contact with the membrane and leave it at −80 °C.

24 After 1–7 days, develop the autoradiograph, using film developer and fixative.

Notes

[a]Nitrocellulose membranes can also be used as an alternative to nylon membrane. DNA fixation to the nitrocellulose membrane is carried out by incubating the membrane at 80°C in a vacuum oven for 2 h.

[b]DNA cross-linking can also be effected on a transilluminator. The UV exposure time must be adjusted empirically (e.g. three pulses of 1 min at 2 min intervals).

[c]The blot can be used immediately or stored after being dried thoroughly.

[d]Several commercial kits that use non-radioactive detection methods are also available. These methods, generally based on digoxigenin- or biotin-modified nucleotides, have become an attractive alternative to radioactive labelled probes. The main advantages in relation to the traditionally used ^{32}P-labelled probes are the lower costs, the long shelf life and avoidance of the hazards associated with handling of isotopes [14].

14.2.4 Analysis of the integration site: inverse PCR (iPCR) and thermal asymmetric interlaced PCR (Tail-PCR)

Determining the sequences of the integrated insert as well as the flanking regions at the integration site in the plant genome are often required by regulatory agencies as part of the application process for the commercial release of transgenic plants, as this information allows traceability of a specific transformation event [1]. Several approaches, such as plasmid rescue, genomic walking, iPCR and Tail-PCR [5,15–18] have been used to determine the plant genome flanking the site of integration of the insert. The general principles of inverse PCR and for Tail-PCR are detailed below.

Inverse PCR and Tail-PCR methodology

iPCR [15, 16] and Tail-PCR [18] methodologies are both based on PCR amplification using a series of insert-specific nested primers. The final PCR product can

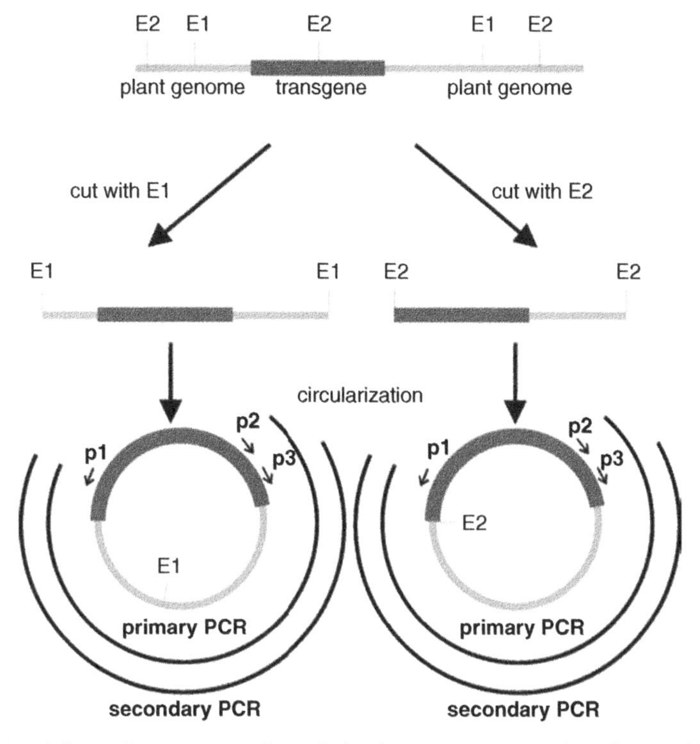

Figure 14.3 Schematic representation of the inverse-PCR procedure for amplification of plant genomic sequences flanking an inserted transgene. Genomic DNA is digested with restriction enzymes that do not cut (E1) or cut once (E2) the inserted DNA. After digestion, fragments of the genomic DNA are circularized and two rounds of PCR are carried out with the nested primers (small arrows) at higher annealing temperatures. When an E2 enzyme is used, only one border can be obtained for each PCR cycle.

be cloned and used for sequencing. Analysis of these sequences should reveal part of the inserted transgene and a segment of the plant genome sequence that should contain the integration site. The principle of iPCR is shown in Figure 14.3 and the method is described in Protocol 14.5. The Tail-PCR method is explained in Figure 14.4 and described in Protocol 14.6.

PROTOCOL 14.5 iPCR

Equipment and Reagents

- 1.7 ml microtubes (Axygen)
- T4 DNA ligase (Invitrogen)
- 0.2 ml PCR microtube (Axygen)
- Platinum Taq DNA Polymerase High Fidelity: 5 U/µl (Invitrogen)
- 10 × reaction buffer (Invitrogen)

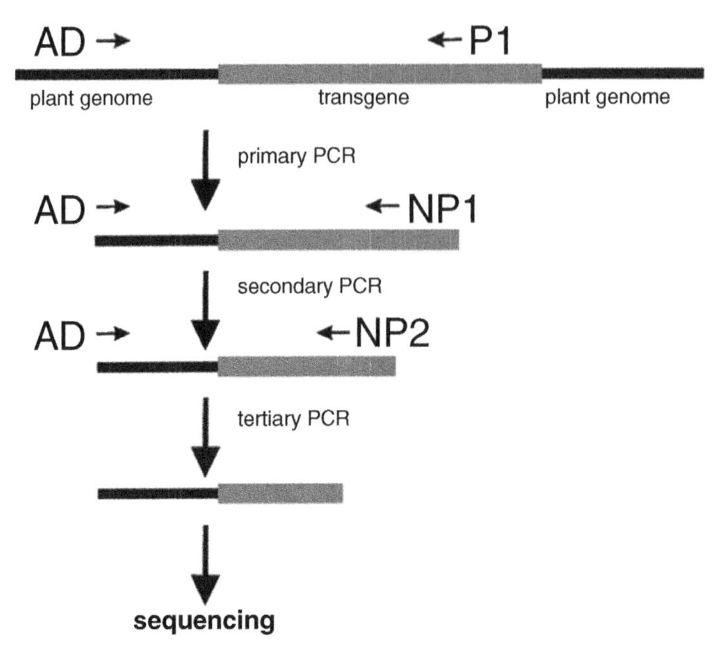

Figure 14.4 Tail-PCR procedure for specific amplification of genomic sequence flanking the transgene. Three reactions are carried out using transgene-specific primers (P1, and nested primers NP1 and NP2) annealing on one flanking side and an 'arbitrary degenerated' (AD) primer on the other side.

- 4 mM dNTP mix: 1.0 mM each of dATP, dCTP, dGTP, dTTP (Invitrogen)
- Primers P1, P2 and nested primer P3[a]: 10 μM (Invitrogen)
- Autoclaved distilled water
- Thermocycler (iCyclerl; Bio-Rad)
- UV transilluminator (2000; Bio-Rad)
- Scalpel or razor blade
- Commercial gel extraction kit (QIAquick-Gel Extraction kit; Qiagen)
- Vector for cloning PCR products (pGEM-T Easy; Promega)

Method

1 Isolate genomic DNA from leaf material of transgenic plants using the CTAB method (see Protocol 14.1).

2 Digest 2–5 μg (~20 μl) of the extracted genomic DNA with an appropriate restriction enzyme (Figure 14.3) for 16 h. Use 100 U of restriction enzyme in a final volume of 100–200 μl. Use the correct reaction conditions (i.e. buffer and temperature) according to the manufacturer's instructions.

3 Inactivate the restriction enzyme by heating (65 °C, 20 min). Alternatively, inactivate the enzyme with 1 vol. of phenol : chloroform : isoamyl alcohol and extract the DNA

by isopropanol precipitation. Resuspend the dried pellet in 100–200 μl TE buffer (see Protocol 14.2, steps 10–13).

4 Ligate the fragments of the genomic DNA (100 μl) with T4 DNA ligase in a volume of 200 μl, at 12 °C, overnight, in order to obtain circularized molecules.

5 For Inverse PCR, 20–50 ng (~2–4 μl) of the ligation reaction product is used as template.
Prepare a reaction mix with the following reagents:

5 μl of 10 × reaction buffer

2 μl of 50 mM $MgSO_4$

3.1 μl of 4 mM dNTP mix

0.5 μl of primer P1

0.5 μl of primer P2

0.3 μl of Platinum Taq DNA Polymerase High Fidelity (5 U/μl)

2–4 μl (20 ng) of genomic DNA

Autoclaved distilled water to 50 μl

6 Place the tube in a thermocycler and set the following programme for the first round of PCR, with a relatively low annealing temperature (≤55 °C):

Number of cycles	Programme
1	3 min at 95 °C
35–40	30 s at 95 °C → 30 s at 55 °C → 3–6 min at 72 °C
1	10 min at 72 °C

7 For the second round of nested PCR, transfer 1 ml of the first round PCR product in a final volume of 100 ml with the nested primer P3 at higher annealing temperatures (≥60 °C).

8 Analyse the PCR products by electrophoresis on 1% (w/v) agarose gels (see Protocol 14.3). Using a scalpel or a razor blade cut the bands from the gel and purify the DNA using a gel extraction kit, following the manufacturer's instructions (e.g. QIAquick-Gel Extraction kit; Qiagen).

9 Clone the purified fragments into a vector for PCR products (e.g. pGEM-T Easy) following the manufacturer's instructions. Select 5–10 clones and send them to a sequencing facility to be sequenced using either the nested or vector universal primers.

Note

[a]Primer P1 and nested primer P3 (Figure 14.3) should be complementary to the transgene, at a position 50–100 bp from the *Agrobacterium* T-DNA Left or Right borders, or from the site where the insert has been cut, if the plant has been transformed by a direct transformation method. Primer P2 should be 100–200 bp upstream of nested primer P3.

PROTOCOL 14.6 Tail-PCR

Equipment and Reagents

- 0.2 ml PCR microtubes (Axygen)
- Platinum Taq DNA Polymerase High Fidelity: 5 U/μl (Invitrogen)
- 10 × reaction buffer
- 4 mM dNTP mix (1.0 mM each of dATP, dCTP, dGTP, dTTP)
- 'Arbitrary degenerated' (AD) primers[a] (10 μM), as designed by Liu *et al.* (18): 5′-NTCGASTWTSGWGTT-36′ (AD1, $T_m = 45.3\,^{\circ}C$); 5′-NGTCGASWGANAWGAA-3′ (AD2, $T_m = 45.3\,^{\circ}C$); 5′-WGTGNAGWANCANAGA-3′ (AD3, $T_m = 45.3\,^{\circ}C$). Where S = G or C and W = A or T (see Figure 14.4)[b]
- Primer P1: 10 μM (see Figure 14.4)
- Nested specific primers NP1 and NP2: 10 μM (see Figure 14.4)
- Autoclaved distilled water
- Thermocycler (iCycler; Bio-Rad)

Method

1 Isolate genomic DNA from leaf material of transgenic plants using the CTAB method (see Protocol 14.1).

2 For the primary PCR, add ~20 ng (0.5–1.0 μl) of genomic DNA to the following reaction in a 0.2 ml PCR tube:

5 μl of 10 × reaction buffer

2 μl of 50 mM $MgSO_4$

3.1 μl of 4 mM dNTP mix

0.5 μl of P1 primer

0.5 μl of 'arbitrary degenerated' primers (10 μM) (AD1, AD2 or AD3)

0.3 μl of Platinum Taq DNA Polymerase High Fidelity (5 U/μl) (Invitrogen)

0.5–1.0 μl of genomic DNA

Autoclaved distilled water to 50 μl

Prepare one reaction of the primary PCR for each AD primer in combination with the P1 primer.

3 Place the tubes in a thermocycler with the following programme:

Number of cycles	Programme
1	95 °C, 1 min
5	95 °C, 60 s → 62 °C, 1 min → 68 °C 2.5 min
1	95 °C, 60 s → 25 °C, 3 min, ramping to 68 °C, over 3 min → 68 °C 2.5 min
15	94 °C, 30 s → 66 °C, 1 min → 68 °C, 2.5 min; → 94 °C 30 s; → 66 °C, 1 min → 68 °C, 2.5 min → 94 °C, 30 s → 44 °C, 1 min → 68 °C, 5 min
1	68 °C, 5 min

4 Dilute the primary PCR product 50-fold with water and transfer 1 μl to a tube containing the secondary PCR reaction mix. Carry out this reaction with a nested specific primer (NP1) and each of the 'arbitrary degenerated' primers (AD1, AD2 or AD3). Place the tubes in a thermocycler with the following programme:

Number of cycles	Programme
1	95 °C, 1 min
5	95 °C, 60 s → 62 °C, 1 min → 68 °C 2.5 min
1	95 °C, 60 s → 25 °C, 3 min, ramping to 68 °C, over 3 min → 68 °C 2.5 min
15	94 °C, 30 s → 66 °C, 1 min → 68 °C, 2.5 min; → 94 °C 30 s; → 66 °C, 1 min → 68 °C, 2.5 min → 94 °C, 30 s → 44 °C, 1 min → 68 °C, 5 min
1	68 °C, 5 min

5 Dilute the secondary PCR product 10-fold with water; transfer 1 μl to a reaction tube containing the secondary PCR reaction mix. Carry out this reaction with a nested specific primer (NP2) and each of the 'arbitrary degenerated' primers (AD1, AD2 or AD3). Place the tubes in a thermocycler with the following programme:

Number of cycles	Programme
12	94 °C, 30 s → 64 °C, 1 min → 68 °C, 2.5 min → 94 °C, 30 s → 64 °C, 1 min → 68 °C, 2.5 min → 94 °C, 30 s → 44, 1 min → 68 °C, 2.5 min
1	68 °C, 5 min

6 Analyze PCR fragments by electrophoresis on 1% (w/v) agarose gels (see Protocol 14.3). Cut the bands from the gel and purify the DNA using a commercial gel extraction kit (see Protocol 14.5).

7 Clone the purified fragments into a vector for PCR products (see Protocol 14.5). Send 5–10 clones to a sequencing facility (see Protocol 14.5).

Note

[a]The 10 nucleotide arbitrary primer kits (OP primers) from Operon (www.operon.com) may also be utilized.

[b]Tm is the melting temperature of the primer. It strongly influences the specificity of the PCR.

14.3 Troubleshooting

- PCR is a very sensitive technique. Therefore, contaminating DNA can interfere with the results [11]. To minimize the risk of contamination, work in a designated, separate area of the laboratory and retain materials, pipettes and reagents exclusively for PCR manipulation [11]. Always add the positive control as the last reaction tube. Autoclave all reagents and wear disposable gloves [11] (Protocols 14.3, 14.5 and 14.6).

- The optimal temperatures for pre-hybridization and hybridization should be the same and must be determined empirically. However, a temperature of 65 °C for probes that share 100% identity is generally effective. If the sequence identity is less than 100%, the hybridization temperature should be reduced (Protocol 14.4).

- The washing step with solution III is highly stringent and should be avoided if the radioactive signal is weak, as monitored by a Geiger counter, or if the probe identity to the target sequence is less than 100% (Protocol 14.4).

- Southern blot membranes can be stripped using a hot SDS procedure. Place the moist membrane in a glass or plastic tray and pour a boiling solution of 0.1% (w/v) SDS onto the blot. Allow to cool; briefly rinse the blot in 2 × SSC and check the removal of the probe with a Geiger counter. Proceed with hybridization according to Protocol 14.4.

- Analysis of the integration site from plants with larger genomes (greater than 10^9 bp) can be difficult, particularly if iPCR is used. In that case, Tail-PCR has been shown to be more efficient [18] (Protocol 14.6).

- Amplification by Tail or iPCR may result in an array of sequences, both specific and non-specific (Protocols 14.5 and 14.6). To reduce this effect, the annealing temperature for the primer should be increased. Alternatively, Southern blot analysis (see Protocol 14.4) of the PCR products can be performed to identify the fragments that contain transgene-specific sequences.

References

1. Altpeter F, Baisakh N, Beachy R, *et al.* (2005) *Mol. Breed.* **15, 305–327.

A critical and broad review gathering data of integration analysis and transformation methods.

2. Sharma KK, Bhatnagar-mathur P, Thorpe TA (2005) *In Vitro Cell Dev. Biol.-Plant* **41**, 102–112.

*3. Rech EL, Vianna GR, Aragão FJL (2008) *Nature Protocols* **3**, 410–418.

A detailed protocol for biolistic transformation of embryo apical meristems.

4. Ramessar K, Peremarti A, Gómez-Galera S, *et al.* (2007) *Transgenic Res.* **16**, 261–280.

5. Zhang J, Cai L, Cheng J, *et al.* (2008) *Transgenic Res.* **17**, 293–306.

6. Bonfim K, Faria JC, Nogueira EOPL, Mendes EA, Aragão FJL (2007) *Mol. Plant Microbe Interact.* **20**, 717–726.

7. Vianna GR, Albino MMC, Dias BBA, Silva LM, Rech EL, Aragão FJL (2004) *Sci. Hort.* **99**, 371–378.

8. Xiangdong F, Duc LT, Fontana S, *et al.* (2000) *Transgenic Res.* **9**, 11–19.

9. Doyle JJ, Doyle JL (1987) *Phytochem. Bul.* **19**, 11–15.

10. Dellaporta SL, Wood J, Hicks JB (1983) *Plant Mol. Biol. Rep.* **1**, 19–21.

*11. Dieffenbach CW, Dveksler GS (2003) *PCR Primer: A Laboratory Manual*, 2nd edn. Cold Spring Harbor Laboratory Press, Cold Spring Harbor, NY, USA.

A comprehensive book covering several PCR-related techniques.

12. Hamill JD, Rounsley S, Spencer A, Todd G, Rhodes JC (1991) *Plant Cell Rep.* **10**, 221–224.

13. Mansfield ES, Worley JM, McKenzie SE, *et al.* (1995) *Mol. Cell Probes* **9**, 145–56.

14. Sambrook J, Russell DW (2001) *Molecular Cloning: A Laboratory Manual*. Cold Spring Harbor Laboratory Press, Cold Spring Harbor, NY, USA.

15. Ochman H, Gerber AS, Hartl DL (1988) *Genetics* **120**, 621–623.

16. Cottage A, Yang A, Maunders H, Lacy RC, Ramsay NA (2001) *Plant Mol. Biol. Rep.* **19**, 321–327.

17. Lisauskas SFC, Rech EL, Aragão FJL (2007) *Cloning and Stem Cells* **9**, 456–460.

18. Liu Y-G, Mitsukawa N, Oosumi T, Whittier RF (1995) *The Plant J.* **8, 457–463.

The original publication describing the use of Tail-PCR and the AD primers for the amplification of inserted sequences and their flanking genome regions.

15
Bioreactors

Spiridon Kintzios
Agricultural University of Athens, Athens, Greece

15.1 Introduction

The bioreactor is an automated culture system whose main function is to provide a controlled environment in order to achieve optimal conditions for cell growth and/or product formation. Plant cells were first grown in bioreactors, in the 1960s, using various commercial or non-commercial designs adapted from the culture of animal cells [1–3]. The bioreactor is the main part of any biological process in which microbial, mammalian or plant cell systems are employed for the commercial manufacture of a wide range of useful biological products [4]. Bioreactors have two advantages over Erlenmeyer flasks for culturing plant cells, namely improved control of the culture environment and scalability.

Bioreactors are used mainly for:

- Large-scale culture of plant cells for biomass or metabolite production.
- Continuous or semicontinuous control of the internal parameters of the culture.
- Automatization of the cultures.

The performance of any bioreactor depends on several functions, including:

- Biomass concentration, which must remain high.
- Maintenance of axenic (sterile) conditions.
- Uniform distribution of nutrients and living materials in the reactor through effective agitation.

Plant Cell Culture Edited by Michael R. Davey and Paul Anthony

- Addition or removal of heat depending on the requirement of the cultures.
- Creation of the correct shear conditions. High shear rates may be harmful to cultured cells, but low shear rates may also be undesirable, because of unwanted flocculation or attachment of cell aggregates to the stirrer and the walls of the bioreactor [5].

There are several basic bioreactor configurations that can be used for culture, but to choose an optimal bioreactor configuration for any specific process depends on a number of parameters [6]. Some of the most important parameters are oxygen transfer, mixing, and the magnitude of the acceptable shear stress.

Bioreactors can be classified into six categories, depending mainly on the method of agitation:

1 *Mechanical agitation.* This is the most common type of bioreactor used in cell culture, including microbial and animal cells. It uses mechanical energy for gas–liquid mass transfer and mixing by means of impellers of various types. Plant cells are sensitive to the high shear associated with the bursting of air bubbles at high agitation speeds [7]. Consequently, mixing may become a serious problem [8]. The rotation of the impeller can lead to the formation of a vortex which can be eliminated by the use of baffles [1].

   Temperature, pH, the amount of dissolved oxygen and nutrient concentration can be more easily controlled within this reactor than with other types [9–11]. In microbial cultures, agitation serves the dual purpose of mixing and oxygen transfer. Because of the sensitivity of plant cells to shear forces, agitation speeds appropriate for plant cell cultivation are generally insufficient to break the incoming gas stream into small bubbles. The gas stream may be dispersed as fine bubbles by using an appropriate gas distributor [1].

2 *Pneumatic agitation.* Contrary to mechanical agitation, pneumatically agitated reactors utilize air for gas-liquid mass transfer and mixing. They are also taller and thinner and have no moving parts. This type of bioreactor is of the bubble column or the airlift design [1, 2, 12, 13].

   The advantages of the airlift bioreactor are as follows:

   - Lower shear, which means that such a bioreactor can be used for growing shear-sensitive plant cells.
   - In the core of large vessels, which can be several meters in height, the pressure at the bottom of the vessel increases oxygen solubility.
   - It is easier to maintain sterility due to the absence of the agitator shaft.

   The disadvantages of the airlift bioreactor are:

   - High energy costs, since the air has to be forced into the medium under a greater pressure.
   - The reactor is not suitable for high-density culture because of insufficient mixing inside the reactor.

- The separation of gas from the liquid is not very efficient when foam is present.

The bubble column design is characterized, as the name suggests, by the formation of bubbles. It provides low shear and a satisfactory biomass production and oxygen transfer [6]. However, at high inocula densities, the bubble column has been observed to reduce growth performance [14].

3 *Vibratory agitation.* These are characterized by the reduction of foam and clusters of living cells [15].

4 *Rotating drum agitation.* Such bioreactors are characterized by the rotation of the whole culture vessel and are used mainly for metabolite production. A greater biomass yield and productivity can be obtained in rotating drum fermenters than in other types of instruments. Tanaka *et al.* [16] and Fujita and Tabata [17] reported the use of a rotating drum bioreactor to cultivate *Catharanthus roseus* and *Lithospermum erythrorhizon* for shikonin production.

5 *Spin filter agitation.* The spent medium is removed, without an outlet, from the cells. The spin filter is permeable to the medium but not to the cultured cells, allowing removal of the spent medium. However, it is commonly observed that the cells aggregate and block the filter. This type of reactor is used to maximize the production of somatic embryos, biomass and metabolites [18].

6 *Gaseous phase or liquid-dispersed bioreactor.* The culture medium is sprayed or misted onto filters carrying the living cells, avoiding the requirement for agitation. This type of reactor has the optimum low shear environment with maximum oxygen transfer [19]. The sprayed liquid and mist are drained from the bottom of the reactor to a reservoir and are recirculated [20–22].

In the following sections, the application of pneumatically agitated bioreactors for the scale-up culture of plant cells is presented, particular for cells of sweet basil (*Ocimum basilicum* L.). Different protocols are described in order to demonstrate the versatility of the system for different purposes. The desired application will affect the selected mode of operation, basically depending on whether growth and production occur simultaneously or sequentially. For example, maximum cell growth is required for micropropagation, while an extended, often non-growth phase, is desired for the accumulation of secondary metabolites, such as the phenolic antioxidant rosmarinic acid (α-*O*-caffeoyl-3,4-dihydroxyphenyllactic acid) (RA).

15.2 Methods and approaches

15.2.1 Medium scale disposable or semidisposable airlift reactors

The construction and operation of two different airlift bioreactors is presented in Protocols 15.1 and 15.2, which have been developed specifically for plant cell culture. The bioreactor systems discussed are of medium scale, having a vessel volume of 1–5 l and are commercially available at competitive cost. Due to their intrinsic configuration, they are more suitable for research-oriented applications, although mass micropropagation of many species is also feasible, provided an appropriate protocol exists. Larger systems (up to 20 000 l or more) [13, 23] are specifically designed, home-built models, which are operated to fulfill particular requirements and, therefore, are not widely available. It should be remembered that as scale-up of the culture increases, considerable compromises to shear stress, mixing and oxygen transfer are required, the calculation of which lies beyond the scope of this chapter.

15.2.2 The RITA temporary immersion reactor

The RITA system is a low cost, semi-dispensable airlift plant bioreactor. Its simple operating principle is based on the combination of culture nutrition and aeration through the temporary flooding of a chamber, located in the lower part of the reactor, with liquid nutrient medium [24–28]. By pumping air into the lower chamber, the liquid medium is forced into the upper chamber of the reactor, containing the cultured plant tissue. The chambers are separated from each other by means of a sieve. In this way, temporary immersion is achieved of the culture in the nutrient medium. The time required for filling the lower chamber with air is 30–60 s and depends on the density of the liquid medium and the volume of air remaining in the chamber. After this time elapses, air bubbles pass through the sieve from the lower to the upper chamber, thus aerating the culture.

In a more detailed description, a standard RITA reactor is composed of the following parts (Figure 15.1):

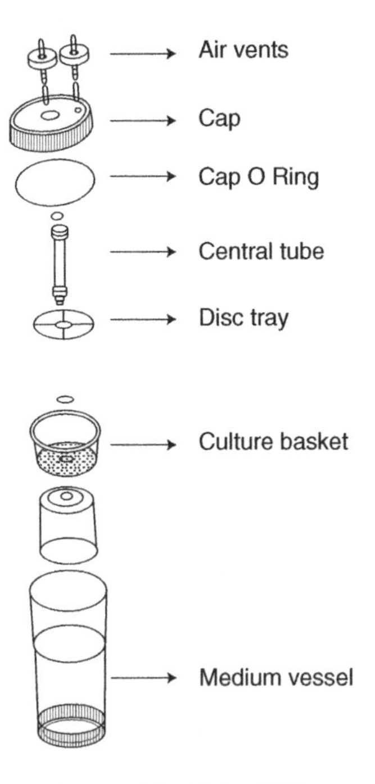

Figure 15.1 Components and assembly of the RITA temporary immersion reactor.

- a medium vessel (the lower chamber, into which air is pumped)
- a culture basket (the upper chamber, containing the cultured plant cells)
- a disc-shaped, plastic sieve (holding the culture inoculum)
- tubing allowing pumped air to reach the lower vessel
- air vents (bearing 0.2 μm sterilizing filters) at air input/output sites
- air pump (not shown in Figure 15.1).

All bioreactor parts are autoclavable, either individually or as parts of an assembled kit.

PROTOCOL 15.1 Micropropagation of Sweet Basil (*Ocimum basilicum* L.) in a RITA Bioreactor

Equipment and Reagents

- Laminar air flow cabinet
- Disposable RITA bioreactor (Vitropic; www.vitropic.fr)
- Sterile filters (0.22 μm pore size)
- Air pump or compressor capable of supplying 1 l/min at 20 kPa (0.2 bar)
- Two autoclavable airvents
- Thermoresistant (autoclavable) silicone tubing (internal diam. 5 mm)
- Timer with minimum of 1 min/day, a manifold equipped with a three-way solenoid valve and nozzles
- Murashige and Skoog (MS) liquid culture medium
- Young sweet basil plants

Method

1 Prepare MS basal liquid medium supplemented with 3 g/l sucrose, 0.1 g/l *meso*-inositol, 1 mg/l (4.4 μM) 6-benzyladenine (BA) and 1 mg/l (0.4 μM) 3-indolebutyric acid (IBA) as plant growth regulators.

2 Remove the shoots from young sweet basil plants.

3 Surface-sterilize for 12 min in 0.1% (w/v) $HgCl_2$, containing 1% (v/v) Tween 80; rinse three times in sterile distilled water, working in the laminar air flow cabinet.

4 Excise 1 cm long nodal segments under sterile conditions.

5 Transfer the autoclaved RITA bioreactor to the laminar flow cabinet.

6 If individual components have been autoclaved, but the bioreactor is not yet assembled, then assemble the reactor in the laminar flow cabinet.

7 Place the nodal segments (explants) on the disc sieve.

8 Pour a maximum of 300 ml of culture medium through the sieve into the lower vessel.

9 Screw the bioreactor cap in place and seal the rim of the cap with Parafilm or similar expandable sealing tape (e.g. Nescofilm).

10 Connect one end of an autoclaved silicone tube with the central air vent (air input).

11 Transfer the bioreactor (containing the explants and the culture medium) to the growth room.

12 Connect the other (free) end of the silicone tube to the air pump.

13 Switch on the pump.

14 Incubate the cultures at 23 ± 2 °C, under a photosynthetic photon flux density (PPDF) of 150–200 μmol/m^2/s (16 h photoperiod, from Cool White fluorescent lamps).

15 After 21 days, transfer the regenerated plants to the glasshouse for acclimatization.

15.2.3 The LifeReactor

The plastic film bioreactor (LifeReactor, Osmotek; www.osmotek.com) is a 5 l working volume vessel fabricated from clear plastic film with an inoculation port (diam. 6 cm). The vessel includes an autoclaveable port cap with two port channels for air inlet and air outlet, two additional channels for the control of culture conditions (e.g. pH, nutrient supplement, sampling) and medium recirculation through a filter as described above. A multiple use glass sparger is connected by silicon tubing and connectors to the air inlet port [29]. The vessels are 10 × 10 cm polypropylene containers with a 16 mm or 40 mm microporous membrane vent with a nominal pore size of 0.3 mm. Since each of the 5 l Lifereactor vessels can hold as many as 6000 plants, this means that 10 000 to 12 000 plants can be moved from multiplication to growth phase by a single operator in less than 1 day of labour. Moreover, the continuing growth phase is totally automated, unless the operator wishes to intervene in order to replenish the culture medium, or to make specific additions at different points in the growth phase. This type of highly efficient plant growth automation system allows both large and small laboratories in high-wage countries to compete effectively with lower cost production facilities in countries with minimal labour costs [30–37].

A standard LifeReactor is composed of the following parts (Figure 15.2):

- A presterilized, disposable flexible plastic culture vessel containing a disposable sparger to produce gas bubbles.
- A large inoculation port located near the top of the vessel for filling the reactor with medium and plant material.
- Air is pumped into the LifeReactor at 0.8–1.5 vvm (air volume/medium volume/min) through autoclaved silicone tubes connected to a carbon filter (to

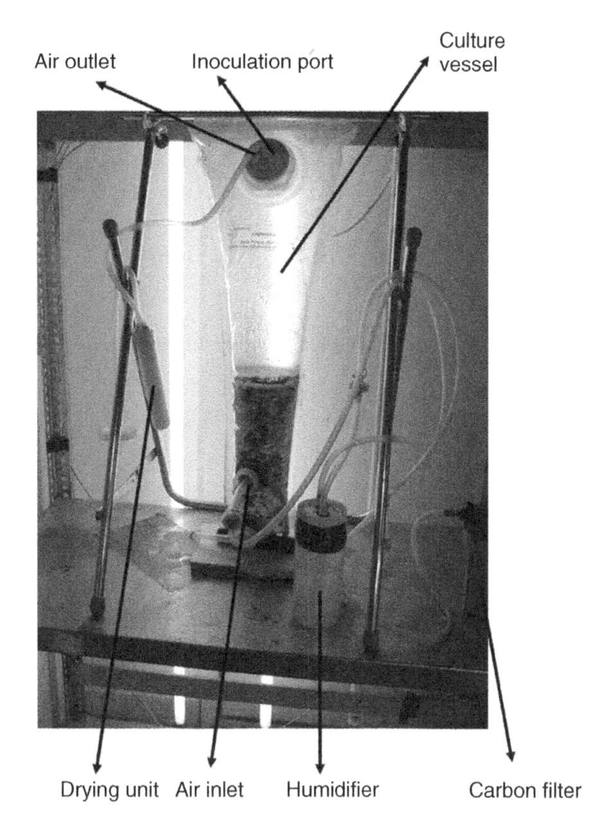

Figure 15.2 Basic set-up of the LifeReactor.

remove any airborne phytotoxic compounds) and two 0.2 μm sterilizing filters. Overpressure in the LifeReactor is vented through the air outlet, which is one of the ports located in the threaded cap. The air outlet is connected to a drying unit (a plastic tube filled with cotton), which removes excess moisture from the air exhausted from the reactor vessel to prevent water droplets from reaching the filters. This is, in turn, connected to two 0.2 μm sterilizing filters.

- The outlet of the second sterilizing filter is connected through a one-way check valve to a humidifier vessel, which reduces evaporation of medium from the reactor.

PROTOCOL 15.2 Culture of Sweet Basil Cell Suspensions and Plant Micropropagation in a 5 l Airlift Bioreactor

Equipment and Reagents

- Laminar flow cabinet

- Dispensable LifeReactor kit
- Sterile filters (0.22 μm pore size)
- Air pump or compressor capable of supplying 1 l/min at 20 kPa (0.2 bar)
- Two autoclavable airvents
- Thermoresistant (autoclavable) silicone tubing (diam. 5 mm)
- Rotary shaker
- Timer with minimum of 1 day, a manifold equipped with a three-way solenoid valve and nozzles
- Murashige and Skoog (MS) liquid medium
- Callus cultures of sweet basil

Method

1 For callus induction, surface-sterilize leaves from young sweet basil plants for 12 min in 0.1% (w/v) $HgCl_2$[a], containing 1% (v/v) Tween 80; rinse three times in sterile distilled water. Excise 1 cm^2 leaf pieces and place the explants on MS basal medium [38] supplemented with 5 g/l sucrose, 10 mg/l ascorbic acid, 0.1 g/l *meso*-inositol, 0.5 g/l L-phenylalanine, 1 mg/l (4.5 μM) kinetin (kin) and 2 mg/l (9 μM) α-naphthaleneacetic acid (NAA) as plant growth regulators and semisolidified with 0.8% (w/v) agar. For suspension culture, aseptically transfer callus, grown for 8 weeks on semisolid medium, into 250 ml Erlenmeyer flasks each containing 50 ml of liquid medium of the same composition and shaken at 100 rpm, under a PPDF of 150–200 μmol/m^2/s (16 h photoperiod, from Cool White fluorescent lamps).

2 For plant micropropagation, apply the previously described Protocol 15.1 (steps 1–4).

3 To inoculate the bioreactor, prepare 2 l of culture medium containing either 150 callus tissues (approx. weight 100 g, for suspension culture) or 150 nodal explants (for micropropagation).

4 After autoclaving, place all of the multiple use autoclavable LifeReactor components and accessories in a laminar flow cabinet and set up the apparatus for holding the LifeReactor vessel (commercially available with the reactor kit).

5 Pour the 2 l of medium containing plant material into the LifeReactor vessel.

6 After the LifeReactor is inoculated with culture medium and plant material, tightly secure the two-port cap in place.

7 Complete the assembly of the LifeReactor system by connecting: (a) *under sterile conditions*, the air drying unit to the air outlet port of the cap and the second check valve for the air inlet components to the air inlet port; (b) *under non-sterile conditions (outside the laminar flow cabinet, in the growth room)*, the inlet of the first sterilizing filter of the air inlet components to the carbon filter.

8 Switch on the air pump.

9 Incubate the cultures at 23 ± 2 °C, under a photosynthetic photon flux density of 150–200 μmol/m^2/s (16 h photoperiod, from Cool White fluorescent lamps)[b].

10 After 3 weeks, transfer the regenerated plants to the glasshouse for acclimatization.

11 Rosmarinic acid can be extracted in 80% (v/v) methanol from the cultured cell suspensions and regenerated plants and purified by means of high pressure liquid chromatography (HPLC), as described previously [37].

Notes

[a]Mercuric chloride is extremely toxic. Ensure that it is handled according to the local safety regulations.

[b]If medium is to be exchanged during operation, care must be taken with regards to pressure buildup in the vessel. In addition, introducing air into the filter capsule can cause a problem with exit of the medium. Therefore, it is recommended that the air supply be shut temporarily during exchange. Vacuum or suction from a pump should be applied to the medium exit port filter, while the entry of new medium can be facilitated by a gravity feed attached to the medium addition port filter (while the new medium should be sterile in order to reduce the microbiological load on the 0.2 μm pore size filter, this operation does not require a sterile environment, because the filter is aseptic from its exit and into the vessel).

15.2.4 Immobilized cell bioreactors

Bioreactors can also be used for the culture of cells immobilized in various substrates. Immobilisation helps in stabilising the cultured biomaterial for reuse. Whole-cell immobilisation can be defined as the physical confinement or localisation of intact cells to a certain defined region of space, with the preservation of some desired activity [39, 40]. The successful application of an immobilised cell system as a biocatalyst relies on the correct choice of the main components of the system, namely the matrix material, the cell type and the configuration of the immobilization system. Among the desirable characteristics for immobilized cell systems are a high surface area-to-volume ratio, chemical and mechanical stability and optimum diffusion of oxygen and nutrients [41]. However, the use of bioreactors specifically for immobilized cells is still in its infancy. Reports in the literature commonly refer to the application of basically pneumatically agitated reactors for the scale-up culture of immobilized cells from various plant species, such as *Daucus carota* [42], *Capsicum frutescens* [43] and, more recently, *Taxus cuspidata* [44]. In the case of sweet basil, there is only one published report on RA production in immobilized cell cultures, with rather disappointing results. Sweet basil cells immobilized in 1.5, 2 or 3% (w/v) calcium alginate beads accumulated RA at a much-reduced rate (<15 μg/g) compared to cell suspension cultures [45].

15.2.5 Mini-bioreactors

Mini-bioreactors have a volume of less than 100 ml and are available for various purposes, since they perform in a similar way to large-scale bioreactors as far as most of the process parameters are concerned. Thus, they offer the advantage for fast and direct scale-up, which reduces development time and costs [46].

An example of mini-bioreactors is the test tube reactor, which is a useful system for developing inocula for small scale fermentation. Usually, 20% of the total tube volume (2–25 ml) is filled with culture media. Test-tube reactors are very simple, low-cost systems, but have the disadvantage of low oxygen transfer [47].

Recently, a remarkably high RA production (21 mg/g dry weight) was achieved by immobilizing sweet basil cells at a high density (approx. 25×10^4 cells/ml) in specially designed, solid-state bioreactors (Georgia Moschopoulou, personal communication, 2006), a production performance that was 1400 times greater than in basil cells immobilized in beads. More significantly, RA was excreted into the culture medium, where it was collected without terminating the culture of immobilized cells. In both cases, RA accumulation in sweet basil did not require cessation of cell growth, as reported for other species. Protocol 15.3 presents the construction and operation of such a solid-state mini-bioreactor.

PROTOCOL 15.3 Culture of Immobilized Sweet Basil Cells in a Solid-State Bioreactor

Equipment and Reagents

- Laminar flow cabinet
- 15 ml Cellstar tubes (Greiner bio-one)
- Sterile filters (0.22 μm pore size)
- Air pump or compressor capable of supplying 1 l/m at 20 kPa (0.2 bar)
- Two autoclavable airvents
- Thermoresistant (autoclavable) silicone tubing (diameter 0.5 cm)
- Filter paper
- Murashige and Skoog (MS) liquid medium
- Sodium alginate (2% w/v)
- Calcium chloride ($CaCl_2$) solution (0.8 M)
- Callus cultures of sweet basil

Method

1 Prepare MS basal liquid medium supplemented with 3 g/l sucrose, 0.1 g/l *meso*-inositol, 10 mg/l of ascorbate, 1 mg/l (4.4 μM) kinetin and 2 mg/l (10.8 μM) NAA as plant growth regulators.

2 Induce callus cultures from leaves of sweet basil as described in Protocol 15.1 and culture cell suspensions in liquid medium for 7 days.

3 Configure a 15 ml sterile Cellstar tube as a bioreactor, by opening: (i) a hole at its bottom, so that an autoclaved silicone tube can be attached to it, then seal the opening around the tube with heat-resistant silicone paste; (ii) a hole in its screw cap, so that an air vent can be attached with a 0.2 μm pore size sterilizing filter (Figure 15.3).

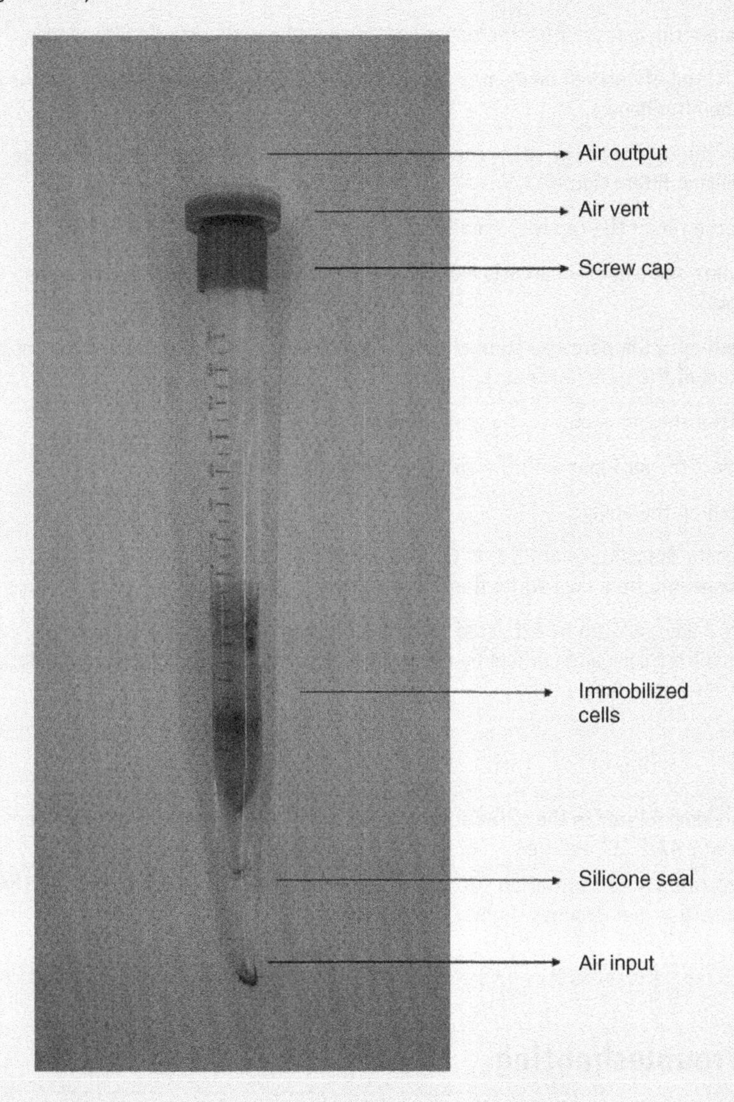

Figure 15.3 Basic set-up of the solid-state mini-reactor.

4 Fill the tube with 1.5 ml of liquid nutrient medium lacking phosphate and Fe-EDTA, but supplemented with 0.8 M $CaCl_2$. Clamp the silicone tube at the bottom of the reactor tube during the filling process, in order to avoid leakage of the medium from the reactor tube[a].

5 Separate the $CaCl_2$ solution from the upper and middle part of the tube with a sterile filter paper.

6 Mix 1 ml of cell suspension with 4 ml of 2% (w/v) sodium alginate.

7 Transfer the mixture to the tube.

8 Separate the mixture from the upper part of the tube with a sterile filter paper.

9 Add 1.5 ml of nutrient medium on top of the cell suspension (separated from the latter by the filter paper).

10 Close the tube with the screw cap bearing the air vent with the 0.2 μm pore size sterilizing filter (Figure 15.3).

11 Seal the rim of the cap with Parafilm.

12 Connect one end of an autoclaved silicone tube to the air vent on the screw cap (air output).

13 Attach a 0.2 μm pore size sterilizing filter to the free end of the silicone tube at the bottom of the tube (air input).

14 Transfer the bioreactor to the growth room.

15 Connect the air input with the air pump, through a silicone tube.

16 Switch on the pump.

17 Incubate the cultures at $23 \pm 2\,^{\circ}C$, under a PPDF of 150–200 μmol/m^2/s (16 h photoperiod, from Cool White fluorescent lamps).

18 After 7 days, RA can be extracted in 80% (v/v) methanol from the cultured immobilized cells and purified by means of high pressure liquid chromatography, as described previously [37][b].

Notes

[a]Cells are immobilized in the cylindrically shaped calcium alginate gel matrix at an optimum density of 2.47×10^5 cells/ml.

[b]The greatest RA concentration (20 mg/g dry weight) may be produced during the first week of culture/but declines slightly thereafter.

15.3 Troubleshooting

- A major concern in bioreactor-assisted micropropagation is contamination. From a practical point of view, contamination of a bioreactor vessel is much more detrimental than in smaller-scale vessels, such as Erlenmeyer flasks. This is due to two reasons: (i) a considerably larger volume of culture is wasted and (ii) liquid medium represents a favourable environment for microorganisms with a potential for fast growth. Consequently, sterile conditions must be applied rigorously and culture inoculation and transfer phases must be conducted in a laminar air flow cabinet, or an equivalent sterile room. Potential microorganism entry points on the bioreactor must be sealed with Parafilm or a similar compound. These are, for example, the assembly joints of the individual bioreactor parts as well as the points where microbial filters or silicone tubing are attached to each air vent. Since Parafilm, or similar sealing agents, tends to expand with changes in temperature, correct regulation of temperature in the growth room is strongly recommended.

- Plants regenerated in a liquid medium in flasks or bioreactors frequently appear malformed due to hyperhydricity, with a low survival rate after transplanting to compost. Hyperhydricity can result from several stress reactions caused by abnormal environmental conditions imposed simultaneously *in vitro* [48]. This problem can be bypassed, in part, by the use of acclimatization bioregulators, such as growth retardants (e.g. paclobutrazol, ancymidol, flurprimidol). Roots may thicken in the presence of growth retardants [49], although *ex vitro* produced root systems are better adapted to survive acclimatization. Growth retardants were reported to reduce shoot elongation and leaf area and to improve stress tolerance in a number of micropropagated plants, such as philodendron, chrysanthemum, rose, grapevine and gladiolus [see [30] for a review].

- Although many researchers consider bioreactors (and, in general, liquid culture systems) more appropriate for plant regeneration via somatic embryogenesis, this involves the risk of obtaining somaclonally variant offspring. Therefore, regeneration via direct organogenesis (e.g. from nodal explants) is generally preferred for commercial, mass propagation of true-to-type offspring.

- In pneumatically agitated culture systems, such as the ones described here, gas flow rate may be related to shear stress at the surface of plant cells or tissues. Since plant cells have low oxygen transfer requirements compared to microbial cultures, it is recommended that the gas flow rate is maintained at a reasonably low level (not greater than 1 l/min).

- Plant cells tend to aggregate when cultured in bioreactors. Therefore, the operating conditions of the reactor may have to be changed in order to reduce aggregate size. Larger aggregates may be more sensitive to shear stress or may negatively affect productivity (e.g. the synthesis of RA). In addition, cells may attach themselves onto inner surfaces of the reactor vessel, above the level of the liquid medium. Hence, they become deprived of nutrients and undergo growth retardation or even death. Consequently, aeration of the medium must be adjusted so that cultured cells remain within the liquid medium throughout the incubation period.

- The viscosity of the culture may increase dramatically with the concentration of cultured cells, leading to considerable reduction of oxygen and mass transfer. Therefore, the total cell concentration in the reactor should not exceed a certain percentage of the packed cell volume. This limit depends on the culture, but generally lies in the range of 45–60%.

- Depending on the cell line and culture conditions, especially the composition of the liquid medium, RA accumulation may be or may not be related to the growth rate of the culture, i.e. it can either parallel or be inversely related to tissue growth.

- In an ebb and flow or temporary immersion bioreactor, plant tissues are not constantly immersed and grown in the liquid medium. Following a preprogrammed operational cycle, the surface of the tissues is wetted with a thin film of liquid medium. This film is drained away and the tissues are exposed to air, followed again by a wetting cycle. This strategy is useful in promoting *in vitro* responses, such as the tuberization of potato or bulb formation in narcissus, and also the growth of crop plants such as coffee, pineapple, banana and sugarcane.

- It is very important to ensure that tubing connections between the bioreactor vessel and air pumps, as well as between the air outlet and the air drying unit (LifeReactor) are not bent, so that air flow to and from the reactor is unrestricted.

References

***1. Payne G, Bringi V, Prince C, Shuler M (1991) *Plant Cell and Tissue Culture in Liquid Systems*. Hanser Publishers, Munich.

Excellent review of the principles for the design and operation of bioreactors specifically for plant cell culture.

**2. Shuler ML, Kargi F (2002) *Bioprocess Engineering, Basic Concepts*, 2nd edn. Prentice Hall, NJ, USA.

Useful review of principles for the design and operation of bioreactors for all types of cultures; suggested for advanced reading.

3. Wang SJ, Zhong JJ (2007) *Bioproc. Value-Added Prod. Ren. Res.* **1**, 131–161.

4. Fowler MW, Warren GS, Moo-Young M (1988) *Plant Biotechnology*. Pergamon Press, New York.

5. Lipksy AKh (1992) *J. Biotechnol.* **26**, 83–97.

6. Panda AK, Mishra S, Bisaria VS, Bhojwani SS (1989) *Enz. Microb. Technol.* **11**, 386–397.

7. Wagner F, Vogelmann H (1977) In: *Plant Tissue and its Biotechnological Application*. Edited by W Barz, E Reinhard and MH Zenk. Springer-Verlag, Berlin, Germany, pp. 245–252.

8. Tanaka K (1987) *Process. Biochem.* **9**, 106–113.

9. Hilton MG, Rhodes MJ (1990) *Appl. Microbiol. Biotechnol.* **33**, 132–138.

10. Kim YH, Yoo YJ (1993) *Biotechnol. Lett.* **7**, 859–862.

11. Nuutila AM, Lindqvist AS, Kauppinen V (1994) *Biotechnol. Techn.* **11**, 363–366.

12. Smart NJ, Fowler MW (1984) *J. Exp. Bot.* **55**, 531–537.

13. Choi YE, Kim YS, Paek KY (2006) In: *Plant tissue culture engineering*. pp. 161–172. Edited by S D Gupta and Y Ibaraki. Springer-Verlag, Berlin.

14. Kwok KH, Doran PM (1995) *Biotechnol. Prog.* **11**, 429–435.

15. Titze IR, Hitchcock RW, Broadhead K, *et al.* (2004) *J. Biomech.* **37**, 1521–1529.

16. Tanaka HF, Suwa MN, Iwanoto T (1983) *Biotechnol. Bioeng.* **25**, 2359–2370.

17. Fujita Y, Tabata M (1987) In: *Plant Tissue and Cell Culture*. Edited by CE Green, DA Somers, WP Hackett and DD Biesboer. Alan R. Liss, New York, NY, USA, pp. 169–185.

18. Janes DA, Thomas NH, Callow JA (1987) *Biotechnol. Technol.* **1**, 257–262.

19. Weathers PJ, Zobel RW (1992) *Biotechnol. Adv.* **10**, 93–115.

20. McKelvey SA, Gehrig JA, Hollar KA, Curtis WR (1993) *Biotechnol. Prog.* **9**, 317–322.

21. Kim YJ, Weathers PJ, Wyslouzil BE (2002) *Biotechnol. Bioeng.* **80**, 454–464.

22. Souret FF, Kim Y, Wyslouzil BE, Wobbe KK, Weathers PJ (2003) *Biotechnol. Bioeng.* **83**, 653–667.

23. Anonymous (2006) VitroSys Inc, Republic of Korea (www.vitrosys.com).

24. Anonymous (2007) VitroPic Inc., France (www.vitropic.fr).

25. Hajari E, Watt MP, Mycock DJ, McAlister B (2006) *S. Afr. J. Bot.* **72**, 195–201.

26. Hanhineva K, Kokko H, Kärenlampi S (2007) *In Vitro Cell Dev. Biol.–Plant* **41**, 826–831.

27. Zhu LH, Li XY, Welander M (2005) *Plant Cell Tissue Organ Cult.* **81**, 313–318.

28. McAlister B, Finnie J, Watt MP, Blakeway F (2005) *Plant Cell Tissue Organ Cult.* **81**, 347–358.

29. Anonymous (2007) Osmotek Inc., Israel.

30. Ziv M (1992) In: *High-tech and Micropropagation. Biotechnology in Agriculture and Forestry*. Edited by YPS Bajaj. Springer-Verlag, Berlin, Germany, Vol. 19, pp. 72–90.

31. Ziv M, Gadasi G (1986) *Plant Sci.* **47**, 115–122.

32. Ziv M, Ronen G, Raviv M (1998) *In Vitro Cell Dev. Biol.–Plant* **34**, 152–158.

33. Katapodis P, Kavarnou A, Kintzios S, *et al.* (2002) *Biotechnol. Lett.* **24**, 1413–1416.

*34. Konstas J, Kintzios S (2003) *Plant Cell Rep.* **21**, 538–548.

A representative example of the application of bioreactors for scale-up micropropagation of a commercial species (cucumber).

35. Kintzios S, Konstas J (2004) *Acta Hort.* **616**, 95–104.

*36. Kintzios S, Kollias Ch, Straitouris Ev, Makri O (2004) *Biotechnol. Lett.* **26**, 521–523.

Original protocol for the micropropagation of sweet basil in a LifeReactor.

37. Kintzios S, Makri O, Pistola E, Matakiadis T, Shi HP, Economou A (2004) *Biotechnol. Lett.* **26**, 1057–1059.

38. Murashige T, Skoog F (1962) *Physiol. Plant.* **15**, 472–497.

39. Karel SF, Briasco CA, Robertson CR (1987) *Ann. NY Acad. Sci.* **506**, 84–105.

40. Willaert R, Baron G (1996) *Rev. Chem. Engin.* **12**, 5–205.

41. Looby D, Griffiths B (1990) *Trends Biotechnol.* **8**, 204–209.

42. Majerus F, Pareilleux A (1986) *Biotechnol. Lett.* **8**, 863–866.

43. Lindsey K, Yeoman MM (1984) *Planta* **162**, 495–501.

44. Han RB, Yuan YJ (2004) *Biotechnol Prog.* **20**, 507–509.

45. Kintzios S, Makri O, Panagiotopoulos EM, Scapeti M (2003) *Biotechnol. Lett.* **25**, 405–408.

***46. Kumar S, Wittmann Ch, Heinzle E (2004) *Biotechnol. Lett.* **26**, 1–10.

Excellent review on minibioreactors.

47. Tanaka H, Nakanishi M, Ogbonna JC, Ashiara Y (1993) *Biotechnol. Tech.* **7**, 189–192.

*48. Chen J, Ziv M (2001) *Plant Cell Rep* **20**, 22–27.

The original protocol relating to the application of LifeReactor™.

49. Roberts AV, Walker S, Horan I, Smith EF, Mottley J (1992) *Acta Hort.* **319**, 153–158.

16

Secondary Products

Kexuan Tang[1], Lei Zhang[2], Junfeng Chen[3], Ying Xiao[3], Wansheng Chen[3] and Xiaofen Sun[4]

[1]*Plant Biotechnology Research Center, Fudan-SJTU-Nottingham Plant Biotechnology R&D Center, School of Agriculture and Biology, Shanghai Jiao Tong University, Shanghai, China*

[2]*Department of Pharmacognosy, School of Pharmacy, Second Military Medical University, Shanghai, China*

[3]*Department of Pharmacy, Changzheng Hospital, Second Military Medical University, Shanghai, China*

[4]*State Key Laboratory of Genetic Engineering, School of Life Sciences, Fudan University, Shanghai, China*

16.1 Introduction

Plant secondary metabolism has multiple functions throughout the plant's life cycle. These functions can be classified as mediators in the interaction of the plant with its environment, such as plant–insect, plant–microorganism and plant–plant interactions [1, 2]. Plant secondary metabolites are frequently regarded as extravagances that serve no obvious biological purpose for the plant that produces them. However, it is becoming increasingly clear that these molecules may play important roles in plant signalling and defence mechanisms [3]. Besides their importance for the plant itself, secondary metabolites also determine important aspects of human food quality such as colour, taste and aroma, flower colour and scent of ornamental plants [4]. Moreover, many plant secondary metabolites are used for the production of medicines, dyes, insecticides, flavours and fragrances [5]. However, some of these phytochemicals are expensive because of their low abundance in plants. It is thus of interest to engineer secondary metabolite production and to exploit plants as cell factories. In the past, such redirection has been obtained through random mutagenesis and subsequent selection of improved strains [6], but with recent developments in plant biology it has become possible to apply more effective approaches for secondary metabolite accumulation, such as plant cell, tissue and organ (transformed roots) culture. Furthermore, cultures may be scaled-up in a

Plant Cell Culture Edited by Michael R. Davey and Paul Anthony

bioreactor to make secondary accumulation an acceptable biotechnological process for further application, such as producing valuable metabolites, increasing the production of secondary metabolites by chemical means, introducing foreign genes into plant genomes to produce recombinant proteins, or over-expressing proteins that otherwise have limiting metabolic pathways. The rapidly expanding field of secondary product exploitation has also been driven in recent years by advances in analytical methods such as high performance liquid chromatography (HPLC), gas chromatography–mass spectrometry (GC-MS) and HPLC-MS.

This chapter outlines different methods that can be applied to generate or to increase plant secondary products of interest, including cell, tissue and organ (transformed roots) culture, and procedures to maximize secondary product biosynthesis and scale up. The methods for secondary product extraction and further quantitative analysis are also discussed.

16.2 Methods and approaches

16.2.1 Plant cell cultures

Plants represent the most important source of natural products. All kinds of secondary metabolites, such as pigments, perfumes and pharmaceutical compounds, are extracted from plants. For decades, many effective anti-tumour compounds have been discovered in plants; the demands for such secondary plant compounds are also increasing. Secondary metabolites are in very low concentrations in plants, and some exist only in specific organs such as roots. Thus, dependence on natural sources cannot meet the increasing demands for secondary metabolites. This has led researchers to consider plant cells and tissues in culture combined with bioreactor technology, as alternative ways to produce secondary metabolites. The chemical totipotency of plant cells gives undifferentiated plant cells the potential to synthesize all metabolites. In exploiting chemical totipotency, many rare secondary metabolites have been obtained by cell culture, such as taxol from *Taxus media* [7]. For more utility and genetic engineering, hairy roots represent a superior source to produce metabolites. In addition to producing metabolites, hairy roots of *Catharanthus roseus* and tropane alkaloid producing species [8–10] have been considered as effective models with which to investigate biosynthetic pathways. In this text, the two most widely studied culture systems for producing useful compounds are discussed, namely suspension cells and hairy roots.

Induction of callus

Initiation of plant cell cultures begins with callus induction. Chemical screening is beneficial in selecting those genotypes producing the greatest concentration of target secondary metabolites, as a source of explants. Optimization of the callus induction medium necessitates consideration of the mineral composition and organic constituents, and the concentration of growth regulators. The latter determine dedifferentiation and differentiation of cells in culture. Optimization of the medium is often complex and empirical; the work is facilitated by factorial experiments.

PROTOCOL 16.1 Preparation of MS-Based Culture Medium [11]

Equipment and Reagents

- Clean bench (laminar air flow cabinet) and inoculation room
- Sterilizer (Labtech)
- Erlenmeyer flasks (150 ml, 2 l; Brand)
- Murashige and Skoog (MS) basal salts mixture (Sigma)
- Phytagel (Sigma)
- Plant growth regulators (Amresco)
- NaOH (Amresco)
- HCl (Amresco)

Method

1 Add 4.3 g[a] MS basal salts mixture to 800 ml deionized water in a 2 l flask.

2 Add 30 g (3%) sucrose.

3 Add growth regulators as required.

4 Adjust the mixture to pH 5.6 using 1 M NaOH or 1 M HCl.

5 Dilute the final volume to 1 l with deionized water.

6 Add 2.3 g Phytagel[b].

7 Seal the flask and sterilize in an autoclave at 121 °C, under steam pressure of 1.4 kg/cm^2 for 20 min.

8 Divide the sterilized medium into 150 ml capacity sterile flasks in a laminar air flow cabinet or clean bench (50 ml medium/flask).

9 Allow the medium to gel at room temperature.

Notes

[a]The amount of MS basal salts mixture is proportional to the volume of medium.

[b]Phytagel or agar is not required for liquid medium.

PROTOCOL 16.2 Induction of Callus from *Catharanthus Roseus* [12, 13]

Equipment and Reagents

- Clean bench (laminar air flow cabinet) and inoculation room

- Erlenmeyer flasks (500 ml; Brand)
- Ethanol (Amresco)
- $HgCl_2$ (Sigma)
- Filter paper (Whatman)
- Murashige and Skoog (MS) basal salts mixture (Sigma)
- Phytagel (Sigma)
- Plant growth regulators (Amresco)

Method

1 Cut explants[a] and wash in running water for at least 4 h. Choose mature and robust leaves as explants.

2 Immerse the explants in 75% (v/v) ethanol in a 500 ml Erlenmeyer flask for 15 s to remove surface contaminants and to eliminate surface tension.

3 Immerse the explants in 0.1% (w/v) $HgCl_2$ in a 500 ml Erlenmeyer flask for 3–10 min[b]; shake the flask constantly to ensure completely sterilization.

4 Wash the explants with sterile water (three to five changes); blot the explants dry with sterile filter-paper[c].

5 Cut the leaf explants into 1 cm lengths to wound the tissues.

6 Place the explants on callus induction medium[d] [MS + naphthleneacetic acid (NAA; 0.5 mg/l) + 2,4-dichlorophenoxyacetic acid (2,4-D; 0.5 mg/l) + 6-benzyladenine (BA; 2.0 mg/l)] and culture at 25 °C for at least 2 weeks.

Notes

[a]All plant organs including roots, stems, leaves, flowers and embryos can be used as explants. The ability to form callus depends upon the organs and their developmental stages. Callus from different organs also has a different potential to produce secondary metabolites.

[b]The time of surface sterilization is determined by the status of the explants; a longer time is generally required for mature than for young, tender tissues.

[c]Mercuric chloride is extremely toxic. All washing solutions must be collected and discarded according to the local laboratory regulations.

[d]The type and concentration of plant growth regulators, and the ratio of auxin to cytokinin, are best optimized by factorial experiments.

Optimization of cell lines

Once callus is obtained, the cells may undergo somaclonal variation during subsequent subcultures. Genetically stable cell lines should be obtained to avoid the erratic production of secondary metabolites. Usually, after several subcultures, callus can be considered a homogeneous cell line, with growth parameters of the cell line being repeated during subsequent transfers to new culture medium. Some cell lines can remain stable for 2 years or more in terms of the biosynthesis of secondary products [10].

When genetic stability is attained, chemical screening should be carried out to select the best cell line with the optimum metabolite production for the initiation of suspension cultures. The rate of growth and status of the callus should also be considered.

Cell suspension cultures

In 1953 Muri first reported cells of tobacco and *Tagetes erecta* as cell suspensions [14]. Suspensions are usually initiated by transferring established callus to liquid medium, which is agitated by shaking, rotating or spinning. Callus tissues should be friable to give rise to suspension cultures with the greatest degree of cell dispersion. The culture of cell suspensions in liquid media offers a unique system for detailed studies of growth and metabolite production. Compared to callus cultures, suspensions produce large quantities of cells from which metabolites can be more easily extracted. Ideally, suspensions should consist of single cells, but these are difficult to achieve. So far, most suspension cultures that have been established consist of both single cells and cell aggregates. The cells in the aggregates are in a different microenvironment, which contributes to the non-uniformity of metabolism.

In general, the media suitable for callus cultures are also suitable for suspension cultures, with the gelling agent being omitted. However, in some cases, the concentrations of auxins and cytokinins are more exacting. Usually, the growth kinetics of suspension cells are an exponential curve; secondary metabolites are produced mainly during the plateau phase, corresponding to a decline in primary metabolism and cell division.

Hairy root cultures

Bioreactor studies represent the final step that leads to possible commercial production of secondary metabolites. Differentiated cells generally produce the most secondary products. Consequently, most workers that use differentiated cultures in preference to cell suspensions have focused on transformed ('hairy') roots. Hairy roots grow as fast as, or faster, than cell suspensions, and can be subcultured and propagated without growth regulators in the medium. The greatest advantage of hairy roots is that they often exhibit about the same or greater biosynthetic capacity for secondary metabolite production as their mother plants. Hairy roots also have long-term genetic stability, a period of 5 years of genetic stability being reported for a hairy root line of *Catharanthus roseus* [14]. Unlike cell cultures, hairy roots are able to produce secondary metabolites concomitantly with growth, and represent a superior culture system compared to cell suspensions. Hairy root is a plant disease caused by *Agrobacterium rhizogenes*, a Gram-negative soil bacterium. When the bacterium infects the plant, the T-DNA of the Ri-plasmid in the bacterium is transferred and integrated into the nuclear genome of the host plant. The transformation process, involving the expression of bacterial genes in plant tissues, induces the development of hairy roots.

PROTOCOL 16.3 Induction of Hairy Root Cultures from *Catharanthus Roseus* [16]

Equipment and Reagents

- Clean bench and inoculation room
- Centrifuge (Eppendorf)
- Rotary shaker (Abbota)
- UV visible spectrophotometer (Thermo)
- 1.5 ml centrifuge tubes (Axygen)
- Erlenmeyer flasks (150 ml, 500 ml; Brand)
- Filter paper (Whatman)
- MS basal salts mixture (Sigma)
- Sterile distilled water
- YEB medium: beef extract (5 g), yeast extract (1 g), $MgSO_4.7H_2O$ (0.4 g), peptone (5 g), sucrose (5 g), water to 1 l
- Antibiotics (e g. Cefotaxime, rifampicin; Amresco)
- Acetosyringone (Sigma)
- *Agrobacterium rhizogenes* strain A_4

Method

1 Inoculate a single colony of *A. rhizogenes* strain A_4 into 1 ml of YEB liquid medium with antibiotic (rifampicin, 100 mg/l)[a] in a sterile 1.5 ml tube; incubate on a rotary shaker at 200 rpm, 28 °C in the dark, overnight.

2 Inoculate the bacteria (100 µl) into 50 ml of YEB liquid medium in a 150 ml sterile flask; incubate by shaking at 200 rpm, 28 °C, in the dark, overnight.

3 Centrifuge at 4000 rpm for 10 min to collect the bacteria. Decant the supernatant and resuspend the bacteria in 250 ml ½ strength MS liquid medium in a 500 ml flask. Inocubate bacterial cultures for 4 h at 28 °C on a rotary shaker at 120 rpm until $OD_{600} = 0.5$.

4 Cut the explants (*C. roseus* leaves)[b] into small pieces (each approx. 1 cm^2 in size); inoculate the explants by immersion in the bacterial suspension for 5 min. Decant the suspension and remove excess suspension by blotting the explants with sterile filter paper. Place the explants onto ½ strength hormone-free MS semisolid medium without growth regulators; co-culture the explants with bacteria under light (Warm White fluorescent illumination, 200 µmol/m^2/s, 16 h photoperiod) at 25 °C for 3 days. Rinse

the explants with sterile water, blot dry and transfer onto 1/2 strength hormone-free MS medium containing 500 mg/l cefotaxime[c]; culture the explants in the dark.

5 After 20 days, hairy roots should appear on the cut edges of the explants. Detach and transfer these roots when 2–3 cm in length onto new 1/2 strength hormone-free MS medium containing 500 mg/l cefotaxime at 25 °C in the dark.

6 Eliminate *Agrobacterium* from the cultures by subculturing the roots onto 1/2 strength semisolid MS medium containing cefotaxime, gradually decreasing the concentration from 500 to 100 mg/l every 7 days for at least five transfers.

7 Culture the bacteria-free roots on medium lacking antibiotics. Use these hairy roots to initiate cultures in liquid medium.

Notes

[a]The antibiotic(s) that are added to the bacterial culture medium are determined by the resistance gene(s) on the vector carried by the bacteria. Antibiotics maintain the bacteria under stress and ensure that the bacteria retain the plasmid(s) of interest.

[b]Leaves and stems are normally selected as explants, young leaves being the optimum material in most species. The lower regions of leaves and veins readily produce hairy roots in response to inculation with *A. rhizogenes*. Preculture of explants on semi-solid MS medium without growth regulators for 2 days prior to inoculation with *Agrobacterium*, may stimulate the initiation of hairy roots.

[c]100 µmol/l acetosyringone in the culture medium may promote the development of hairy roots; acetosyringone may enhance the transfer of T-DNA from bacteria to plant cells.

Culture of hairy roots in liquid medium

Following the elimination of agrobacteria, hairy roots are cultured in liquid medium with shaking to aerate the cultures. Excise a single root tip (approx. 5–10 cm in length) and inoculate into MS-based liquid medium without growth regulators. Culture the roots at 25 °C in the dark, on a rotary shaker at 120 rpm.

The growth of hairy roots follows a sigmoid curve, the maximum biomass being reached at the end of the exponential phase, generally after about 20 days of culture. Hairy roots are fast growing and require no external supply of growth regulators [17].

16.2.2 Scale-up and regulation of secondary metabolite production

Hairy roots in a bioreactor

Hairy roots grow rapidly and branch extensively in culture medium lacking growth regulators. The stability of production of secondary metabolites is an interesting characteristic of hairy root cultures [18]. Recent progress in the scale-up of hairy root cultures has made this system an attractive tool for industrial processes.

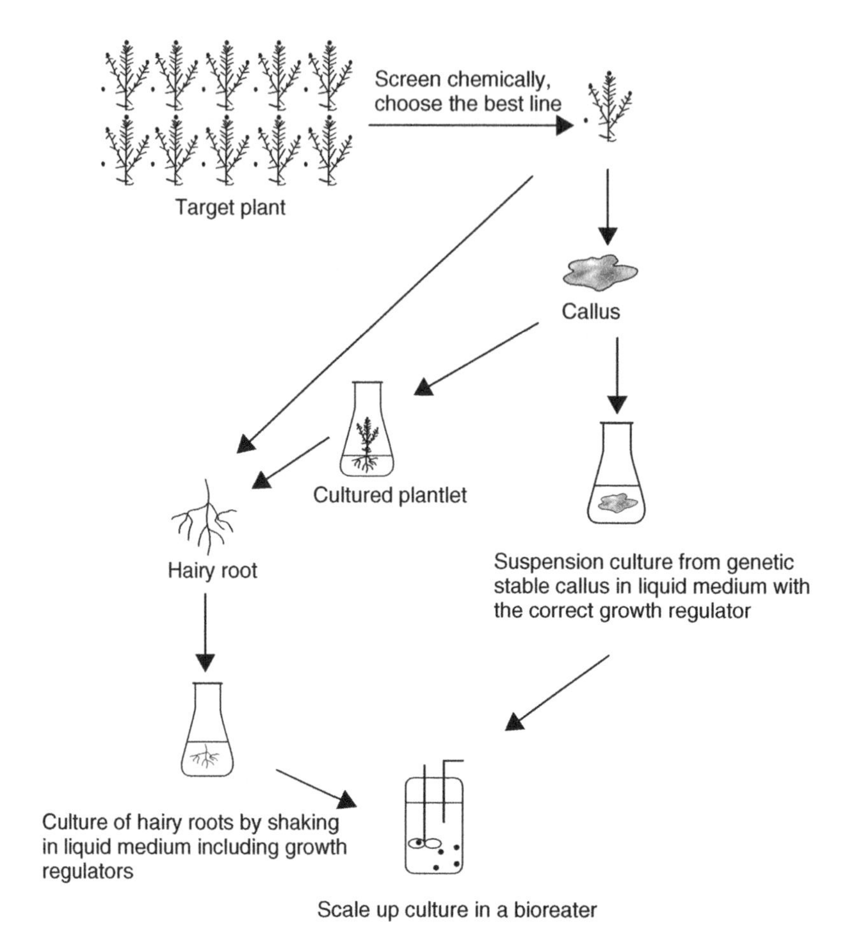

Figure 16.1 Production of secondary metabolites by plant cells. After choosing the target plant, the production of secondary metabolites commences with the induction of callus (Protocol 16.2) or hairy roots (Protocol 16.3). The culture conditions of callus and hairy roots are optimized, followed by scale up.

The scaling-up of hairy root cultures is of paramount importance for their biotechnological application (Figure 16.1). Two culture systems have been used with success. The first, based on the airlift concept, has been used to scale-up hairy roots of *Beta vulgaris* and *Artemisia annua* (bubble column bioreactor) [19, 20] and *Astragalus membranaceus* [21], and the coculture of *Genista tinctoria* shoots and hairy roots (basket bubble reactor) [22]. The airlift mesh-draught with wire-helices reactor, tested with *Solanum chrysotrichum* hairy roots, resulted in the production of homogeneous biomass in the culture medium [23]. The second culture system, the mist reactor, derives from the ability of hairy roots to grow hanging on a mesh support. This system offers advantages in that it reduces the volume of culture medium and obtains a concentrated form of the secreted metabolites. It has been used both in the laboratory and industry [24]. If the metabolite produced is stable and does not necessitate sterile conditions for production, a hydroponic plant culture system could be combined with the normal growth of photosynthetic aerial plant parts and a hairy root system. This assembly is likely to be the least expensive and the easiest to use.

PROTOCOL 16.4 Hairy Root Cultures of *Hyoscyamus Muticus* in an Air-Sparged Glass Bioreactor [25]

Equipment and Reagents

- Air-sparged glass bioreactor (Laborexin Oy.)[a]
- Orbital shaker
- Ammonium electrode (Oriola Co.)
- Filter papers (100 mm diam.)
- B5 basal salts mixture (Sigma) [26]
- Colorimetric szechrome-reagent (Polysciences Inc.)
- Sucrose, glucose, and fructose (Biopharm AG)
- Freeze dryer (Laborexin Oy.)
- Ammonium molybdate
- $NaH_2PO_4.H_2O$ solution (162.5 mg/l)
- KNO_3 solution (2500 mg/l)
- $(NH_4)_2SO_4$ solution (134 mg/l)

Method

1 Subculture hairy roots to B5 basal salts mixture (0.3 g root tissue/20 ml medium) twice for 2 weeks in 100 ml flasks on an orbital shaker at 70 rpm[b].

2 Transfer the inoculum into the bioreactor and culture in the light at 24 °C.

3 Remove the roots at the end of the culture period from the bioreactor and wash the roots with distilled water.

4 Dry the roots on filter paper, freeze dry and lyophilize the roots.

Notes

[a]The air-sparged bioreactor manufactured by Laborexin Oy has a working volume of 3.5 l, an internal diameter of the vessel of 15 cm and a total height of 35 cm. Sterile air (300 ml/min) is introduced through a glass tube and sparged via a sintered glass plate at the bottom of the bioreactor. A net of stainless steel prevents the roots from floating and drying.

[b]Incubate the cultures at 24 °C in the light (24/13 °C day/night, 16 h photoperiod of 200 μmol/m^2/s (Osram Cool White/Osram Fluora fluorescent tubes, 1 : 1 on a watt basis).

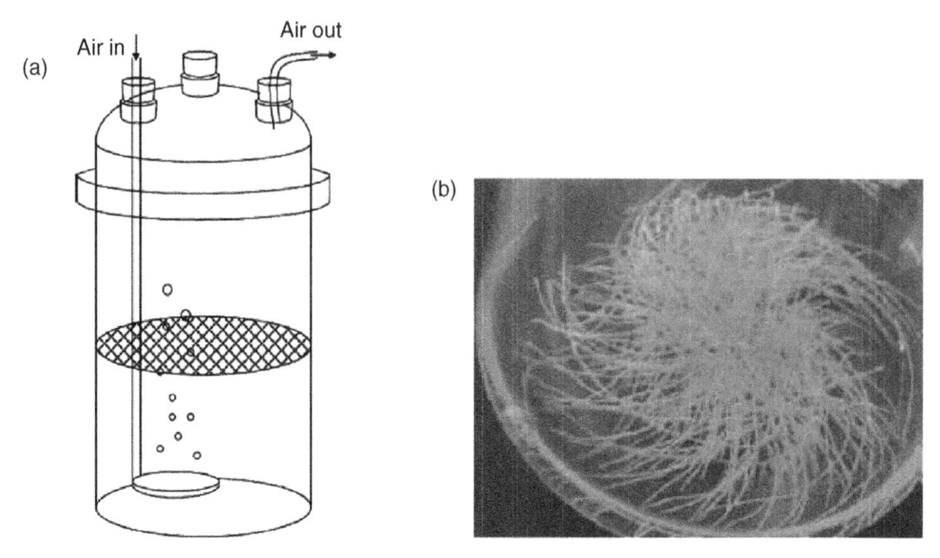

Figure 16.2 Scale-up of hairy root production. (a) Schematic diagram of an air-sparged bioreactor. (b) Hairy roots in agitated liquid medium in a 100 ml Erlenmeyer flask used to inoculate the bioreactor.

Scale-up of plant cell cultures for secondary metabolite production

Plant cell cultures have received much attention as a useful technology for the production of valuable plant-derived secondary metabolites such as paclitaxel, scopolamine and ginseng saponin [27, 28]. Several factors, including selection, design and optimization of bioreactor hardware, manipulation of environmental factors, such as medium components, illumination, shear stress and oxygen supply, need detailed investigations for optimization and scale-up [29]. Recent advances in plant cell processes, including high-density culture in suspension, continuous culture, process monitoring, modelling and controlled scale-up, are used extensively in the biosynthesis of products of commercial interest. Further developments in bioreactor culture processes and in metabolic engineering of plant cells for metabolite production are expected in the near future.

Metabolic engineering

Metabolic engineering is a new approach to understand and explain plant biochemical pathways. The availability of gene transfer techniques have led to increased interest in using this technique to redirect metabolic fluxes in plants for industrial purposes. Thorough mapping of a biosynthetic pathway is a prerequisite for any metabolic engineering programme. Engineering secondary metabolite pathways can be accomplished by several strategies [30], including: (1) elicitor treatment to stimulate plant defence responses system; (2) enhancing the expression or activity of a rate-limiting enzyme; (3) preventing feedback inhibition of a key enzyme;

(4) decreasing the flux through competitive pathways; (5) enhancing expression or activity of all genes involved in the pathway; (6) compartmentalization of the desired compound; (7) conversion of an existing product into a new product; and (8) decreasing the catabolism of the desired compound.

PROTOCOL 16.5 Elicitor-Induced Secondary Metabolite Accumulation from *Catharanthus Roseus* Suspension Cultures [31, 32]

Equipment and Reagents

- Orbital shaker, freeze dryer and refrigerator (Laborexin Oy.)
- Erlenmeyer flasks (100 ml, 200 ml)
- Ethanol
- Dimethyl sulfoxide (DMSO)
- Distilled water
- Paper towels
- Vacuum-driven filtration
- Storage devices (Millipore) such as 2 ml centrifuge tubes (Axygen)
- Yeast extract (Sigma)
- Silver thiosulfate ($Ag_2S_2O_3$; Sigma)
- Salicylic acid (SA, Sigma)
- Methyl jasmonate (MeJA; Sigma)
- MS basal salts mixture (Sigma)

Method

1 Inoculate 2 g fresh weight of cells into 100 ml Erlenmeyer flasks each containing 25 ml liquid medium, on an orbital shaker at 110–120 rpm and 25 °C, in the dark.

2 Prepare the elicitor by ethanol or DMSO precipitation[a].

3 Apply elicitor(s) treatments such as SA and MeJA over a range of concentrations to the suspension culture on day 6 after inoculation[b]. Use MeJA at a final concentration of 50 μM. Add an equivalent volume of ethanol as a control to the culture.

4 Harvest suspension cells from the culture medium by vacuum filtration after 0, 1, 4, 8 and 12 h of elicitation. Lyophilize the samples and store at −20 °C until extracted.

5 Determine both the biomass and metabolite content by HPLC (see Detection of Secondary Products, below).

Notes

[a]Dissolve 50 g yeast extract in 250 ml of distilled water. Add ethanol to 80% (v/v). Allow the precipitate to settle for 4 days at 6 °C. Decant and discard the supernatant solution. Dissolve the gummy precipitate in 250 ml of distilled water. Repeat the ethanol precipitation. Dissolve the second ethanol precipitate in 200 ml of distilled water, yielding the crude preparation without further purification.

[b]The elicitor dose is expressed by the total carbohydrate content determined by the phenol–sulfuric acid method [33] using sucrose as a standard. The final concentrations should be $Ag_2S_2O_3$ (30 μM), yeast extract (100 μg/ml), SA (0.1 mM) and MeJA (50 μM).

PROTOCOL 16.6 Engineering Tropane Alkaloid Biosynthetic Pathways [9, 27, 31, 32]

Equipment and Reagents

- Thermocycler (Takara)
- Hybrid heaters (Takara)
- Gel electrophoresis system (BioRad)
- Polymerase chain reaction (PCR) microtubes (1.5 ml capacity)
- PCR reaction reagents (Buffer with Mg^{2+}, dinucleotide triphosphate (dNTP), Taq DNA polymerase; TaKaRa)
- PCR DIG Probe Synthesis Kit (Roche)
- Plant expression vector (Invitrogen & Amersham Pharmacia)
- Restriction endonuclease and ligase (Amersham Biosciences)
- TRIzol reagent (Gibco/BRL)
- 100 mM potassium phosphate buffer, pH 7.5, containing 5 mM ethlenediaminetetra-acetic acid (EDTA), 10 mM mercaptoethanol, 0.5% (w/v) sodium ascorbate, and 2% (w/v) polyethyleneglycol 4000
- 50 mM potassium phosphate buffer, pH 8, containing 1 mM EDTA and 5 mM mercaptoethanol
- Hybond-N^+ nylon membrane (Amersham Pharmacia)
- Sephadex G25 prepacked PD-10 column (Amersham Pharmacia Biotech)
- Putrescine, *N*-methylputrescine, dansylchloride, acetonitrile and *S*-adenosylmethionine (Sigma)
- Sterile, purified water (Takara)
- Microcentrifuge (Takara)

Method

1 Construct a plant binary expression vector[a].

2 Introduce the binary vector into *Agrobacterium tumefaciens* strain LBA4404 by electroporation and generate transgenic plants by *Agrobacterium* or particle gun-mediated gene delivery[b].

3 Identify genetically modified plants by molecular detection procedures[c].

4 Extract, purify and assay the target enzyme[d].

5 Measure target compounds and relative concentration of metabolites[e].

6 See references [9] and [23] for detailed examples.

Notes

[a]Tobacco putrescine *N*-methyltransferase (PMT) cDNA with an introduced *Nco*I site at the first ATG, which had been cloned in pcDNAII (Invitrogen), is excised as an *Nco*I–*Bam*HI fragment and cloned into pRTL2 under the CaMV35S promoter with a duplicated enhancer. The PMT overexpression gene cassette is excised with *Hin*dIII and cloned in the binary vector pGA482 (Amersham Pharmacia).

[b]*Hyoscyamus niger* may be transformed by a leaf disc method that uses *A. tumefaciens*, essentially as described [26]. Transgenic plants are regenerated from the leaf discs, grown to maturity in a glasshouse, and selfed.

[c]In PCR analysis for detecting the presence of the *pmt* gene, genomic DNA is isolated from transformed *H. niger* samples using the acetyl trimethyl ammonium bromide (CTAB) method. PCR primers are FPMT (5′-GCCATTCCCATGAACGGCC-3′) and RPMT (5′-CCTCCGCCGATGATCAAAACC-3′). PCR is carried out in total volumes of 50 µl reaction mixtures, containing 1 µl of each primer (10 µmol/l), 1 µl of 10 mmol/l dNTPs, 5 µl of 10 × PCR buffer (Mg^{2+} plus) and 2.5 units of *Taq*r DNA polymerase (TaKaRa) with 200 ng of genomic DNA as template. The template is denatured at 94 °C for 5 min followed by 35 cycles of amplification (1 min at 94 °C, 1 min at 60 °C, 45 s at 72 °C) and then by 5 min at 72 °C. Total RNA is isolated from plant tissue by using TRIzol Reagent (Gibco/BRL) and subjected to Northern blot analysis for the expression of *pmt*. Aliquots of total RNA (10 µg/sample) are denatured and separated on a 1.1% (w/v) formaldehyde-denatured agarose gel. After electrophoresis, the RNA is transferred onto a positively charged Hybond-N^+ nylon membrane (Amersham Pharmacia) through capillary transfer. The probe is generated by PCR (PCR DIG Probe Synthesis Kit; Roche). PCR labelling of the probes with digoxigenin (DIG)-dUTP and hybridization (30 min prehybridization at 50 °C, followed by 16 h hybridization at 50 °C) are performed according to the manufacturer's instructions (Roche). Hybridizing bands are detected using the DIG Luminescent Detection Kit (Roche), and signals are visualized by exposure to Fuji X-ray film at 37 °C for 10 min.

[d]PMT enzymatic activity is evaluated as follows: tissues (0.5–1.0 g fresh weight.) are extracted on ice with 3 vol. of 100 mM potassium phosphate buffer (buffer A), followed by centrifugation at 27 000 g for 30 min. The supernatant is loaded onto a Sephadex G25 prepacked PD-10 column (Amersham Pharmacia Biotech) equilibrated and eluted with 50 mM potassium phosphate buffer (buffer B). The reactions are performed by incubating 100 µl of the purified supernatant with 20 µl of 25 mM putrescine (final concentration: 3.6 mM), 8 µl of 10 mM *S*-adenosylmethionine (final concentration: 0.6 mM) and 12 µl of

buffer B at 37 °C for 30 min. After terminating the reactions by heating in boiling water, 65 mM borate–KOH buffer and a solution of dansylchloride (5.4 mg/ml acetonitrile) are added to the incubation mixture. Following heating at 60 °C for 15 min, the dansylated amines are extracted by adding 0.5 ml of toluene, followed by vortex mixing for 30 s. After an aliquot (400 μl) of the toluene is removed, the residue is dried and resuspended in a fixed volume of acetonitrile, which is then injected into the HPLC. The retention time of *N*-methylputrescine is 24 min.

[e]Chemical extraction and analysis, see Protocols 16.6–16.8.

16.2.3 Detection of secondary products

The detection of secondary products is a crucial step in the analysis of plant materials, whether wild-type or genetically modified, with the aim to ultimately evaluate plant quality, effect of tissue culture or transgene manipulation. Secondary product detection includes the following steps:

1 Homogenization of freeze-dried plant material and preliminary sample preparation. The techniques commonly employed for sample preparation at this stage involve drying or distillation [34].

2 Extraction/leaching of soluble components of the material with suitable solvents, or their mixtures, or a supercritical fluid, including desorption, hydrolysis and saponification [34]. The selective extraction of compounds is based on differences in their chemical and physical properties.

3 Metabolite analysis by techniques such as HPLC, mass spectrometry (MS) or nuclear magnetic resonance (NMR) spectroscopy for qualification and quantitation of compounds.

Preliminary sample preparation

Preliminary sample preparation is carried out with the purpose of making materials homogeneous, facilitating subsequent extraction. The procedure commonly includes drying and homogenization, as in Protocol 16.6. However, if the metabolites of interest contain components that are volatile or that sublime, such as essential oils, some amines and organic acids, drying should be replaced with distillation.

PROTOCOL 16.7 Preliminary Sample Preparation for Compounds with Thermal Stability

Equipment and Reagents

- Drying oven (conventional or microwave)
- Thermometer capable of measuring to 110 °C (if using a conventional drying oven)

- Samples in containers suitable for use in the drying oven
- Balance (gram or milligram grade)
- Cutting instruments (FZ-102 miniature plant sample pulverizer; Qingdao Shengfang Apparatus Co.)
- Sieve (100 μm mesh)
- Homogenizers (ceramic; agate pestle and mortars) or mechanical homogenizers

Method

1 Weigh the samples and record their mass.

2 Dry the samples in the oven[a].

3 Record the dry weight and calculate the water loss.

4 Cut the dried materials mechanically or manually; sieve the cut materials to the same size.

5 Homogenize the materials.

Note

[a]Drying temperature and time should be selected according to the plant material. The drying of natural materials is frequently performed at 70 °C [34].

Extraction/leaching

This step is to isolate the components of interest from plant material, accompanied by enrichment of compounds and removal of unwanted materials. Plant metabolites often occur as complex mixtures of many compounds of a wide range of chemical and physical properties. For instance, with respect to polarity, the most important groups of substances in plant material are low polar (waxes, terpenoids), semipolar (lipids, phenolic compounds, low-polar alkaloids), and high polar compounds (polar glycosides, polar alkaloids, saccharides, peptides, proteins). Thus, the selective extraction of analytes is based on differences in their chemical and physical properties. Although different extraction methods are employed, there is also a general process for extraction of analytes, as in Protocol 16.8.

PROTOCOL 16.8 A General Process for the Extraction of Secondary Products

Equipment and Reagents

- Extraction apparatus[a] (e.g. Soxhlet extraction apparatus, separation funnel, microwave oven)

- Flasks (50 ml, 100 ml) for extracts and waste liquids
- Filter apparatus (Buchner funnel)
- Evaporation and concentration apparatus (rotary evaporator; Büchi; solid-phase extraction apparatus)
- Volumetric flasks (10 ml, 50 ml); transferpettor (Eppendorf)
- Solvents[b] (e.g. water, methanol, ethanol, chloroform)

Method

1. Dissolve the homogenized samples in a suitable solvent according to the chemical and physical properties of the secondary products; stir to ensure dissolution.
2. Add solvent for interference (unwanted materials) extraction with separating funnel; discard the extract.
3. Extract the analytes using a suitable method (e.g. Soxhlet extraction, accelerated solvent extraction, supercritical fluid extraction).
4. Filter the extract using the Buchner funnel; evaporate the filter to dryness or concentrate with solid-phase extraction for analyte enrichment.
5. Dissolve fully, in a volumetric flask, the enriched analytes in a suitable solvent; make up to volume in preparation for further chromatographic separation and final characterization.

Notes

[a]Extraction apparatus can be varied according to the different methods, e.g. Soxhlet extraction apparatus or microwave oven.

[b]The selection of the solvent is frequently based on the polarity of the components to be extracted or separated.

Metabolite analysis

The last step is to qualify and quantitate secondary products in test materials. Several analytical methods may be employed according to the physicochemical properties of the analytes, such as the presence of UV chromophores within their structures, their reactivities due to specific functional groups present in their structure, and their molecular weight. These methods include HPLC [35], NMR [36] and MS [37]. Of these methods, HPLC has developed as the most widely used technique for routine analysis in laboratories due to the advantage of simultaneous qualitative and quantitative analyses, as well as being a convenient and high performance procedure (see Protocol 16.8) However, sample determination with HPLC depends on the standards available. With further confirmation and identification of analytes required, more advanced and sophisticated methods such as LC-MS [38] may be employed, which enables simultaneous separation and identification of compounds.

PROTOCOL 16.9 Metabolite Analysis Using HPLC

Equipment and Reagents

- HPLC (Agilent 1100) equipment comprising an on-line degasser, auto-sampler, column temperature controller and diode array detector, coupled to an analytical workstation
- Balance (milligram grade)
- Volumetric flasks (50 ml), transferpettor (Eppendorf)
- Organic membrane (0.2 µm, Rf-Jet, Shanghai RephiLe Bioscience & Technology Co. Ltd.)
- Compound standards (Sigma)
- Mobile phase solvent; usually various mixtures of water and acetonitrile
- Organic solvent for preparation of standards and sample solutions

Method

1 Filter the sample solution through an organic membrane and inject into the HPLC[a].

2 Investigate the chromatographic conditions, including fixed phase, mobile phase, column temperature, detection wavelength and flow rate.

3 Weigh standards and prepare standard solutions using volumetric flasks. Dilute samples to an appropriate concentration range.

4 Run a blank sample first, e.g. injecting only methanol helps the system to settle down and ensures reproducible retention times.

5 Inject standard solutions into the HPLC, to establish a calibration curve. Each calibration curve should be analysed at least three times, usually with five to six different concentrations.

6 Determine sample components by matching retention time of the peak in the sample chromatograms with the peak of a standard.

7 Quantitate sample components using the calibration curve.

Note

[a]Make sure never to pump particles or air though the column, inject when the injector is dry, inject solution containing particles, or apply large (sudden) pressure drops over the column.

16.3 Troubleshooting

- It is essential to perform preliminary experiments with different strains of *A. rhizogenes* to optimize the induction of hairy roots. Based on the types of opines synthesized by transformed cells and hairy roots, *A. rhizogenes* can be classified into octopine, agropine (A4, ATCC15834, 16834), nopaline, mannopine (5196, TR101, TR7) and cucumopine (2635, 2657, 2659) strains.

- It is important to avoid any increase in temperature in the case of raw materials containing essential oils. The material should be processed in small batches to prevent the loss of essential oils.
- When working with HPLC it is recommended to ensure that the degasser is on and purge the pump thoroughly; keep solvent reservoir bottles and the solvent waste barrel closed as much as possible (organic solvents are volatile and can be toxic).

References

1. Harborne JB (2001) *Nat. Prod. Rep.* **18**, 361–379.

2. Dixon RA (2001) *Nature* **411**, 843–847.

3. Wink M (2000) *Trends Biotechnol.* **18, R321–322.

A general introduction to plant secondary metabolites.

4. Rachel DR, Yaron S, Yaakov T, *et al.* (2007) *Nature Biotechnol.* **25, 899–901.

The original publication describing the use of metabolic engineering for improvement of fruit quality.

5. Verpoorte R, Memelink J (2002) *Curr. Opin. Biotechnol.* **13, 181–187.

The significance of secondary metabolites for humans.

6. William RS (2001) *Metab. Eng.* **3**, 4–14.

7. Goleniowski ME (2000) *Biocell.* **24**, 139–144.

8. Chung I, Hong S, San KY, *et al.* (2007) *Biotechnol. Prog.* **23**, 327–332.

9. Zhang L, Ding R, Tang K, *et al.* (2004) *Proc. Natl Acad. Sci. USA* **101, 6786–6791.

Signficant production of secondary metabolites in engineered hairy roots.

10. Bourgaud F, Gravot A, Gontier E, *et al.* (2001) *Plant Sci.* **161, 839–851.

Detailed review of the production of plant secondary metabolites.

11. Murashige T, Skoog F (1962) *Physiol. Plant.* **15**, 473–479.

12. Van der heijden R, Verpoorte R, Hens-Hoopen JGT (1989) *Plant Cell, Tissue Organ Cult.* **1**, 231–280.

13. Moreno PRH, Van der heijden R, Verpoorte R (1995) *Plant Cell Tissue Organ Cult.* **42**, 1–25.

14. Butcher DN, Ingram DS (1976) *Plant Tissue Culture*. Edward Arnold, London.

15. Peebles CA, Gibson SI, San KY, *et al.* (2007) *Biotechnol. Prog.* **23**, 1517–1518.

16. Bhadra R, Vani S, Shanks JV (1993) *Biotechnol. Bioeng.* **41**, 581–592.

17. Giri A, Narasu ML (2000) *Biotechnol. Adv.* **18**, 1–22.

18. Hu ZB, Du M (2006) *J. Integr. Plant Biol.* **48**, 121–127.

19. Thimmaraju R, Bhagyalakshmi N, Ravishankar GA (2004) *Biotechnol. Prog.* **20**, 777–785.

20. Souret FF, Kim Y, Wyslouzil BE, Wobbe KK, Weathers PJ (2003) *Biotechnol. Bioeng.* **83**, 653–667.

21. Du M, Wu XJ, Ding J, Hu ZB, White KN, Branford-White CJ (2003) *Biotechnol. Lett.* **25**, 1853–1856.

22. Luczkiewicz M, Kokotkiewicz A (2005) *Plant Sci.* **169**, 862–871.

23. Caspeta L, Quintero R, Villarreal ML (2005) *Biotechnol. Prog.* **21**, 735–740.

*24. Stéphanie G, Jocelyne TG, Pratap KP, Marc R, Pascal G (2006) *Curr. Opin. Plant Biol.* **9**, 341–346.

Discussion of the advances in hairy root culture in large-scale bioreactors.

25. Wilhelmson A, Häkkinen ST, Pauli T. Kallio PT, Oksman-Caldentey KM, Nuutila AM (2006) *Biotechnol. Prog.* **22**, 350–358.

26. Gamborg OL, Miller RA, Ojiva K (1968) *Exp. Cell Res.* **50**, 151–158.

27. Sato F, Hashimoto T, Hachiya A, *et al.* (2001) *Proc. Natl Acad. Sci. USA* **98**, 367–372.

28. Tabata H (2006) *Curr. Drug Targets* **7**, 453–461.

29. Taticek RA, Lee CW, Shuler ML (1994) *Curr. Opin. Biotechnol.* **5**, 165–74.

30. Oksman-Caldentey KM, Inzé D (2004) *Trends Plant Sci.* **9**, 433–440.

31. Memelink J, Verpoorte R, Kijne JW (2001) *Trends Plant Sci.* **6**, R212–219.

*32. Ge XC, Wu JY (2006) *Plant Sci.* **170**, 853–858.

Elicitors applied to plant samples.

33. Fox JD, Robyt JF (1990) *Anal. Biochem.* **195**, 93–96.

34. Romanik G, Gilgenast E, Przyjazny A, Kamiński M (2007) *J. Biochem. Biophys. Methods.* **70, 253–261.

Discussion of the preparation techniques for samples of plant material for chromatographic analysis.

35. Meyer V (1999) *Practical High Performance Liquid Chromatography*, 3rd edn. John Wiley & Sons, Chichester, UK.

36. Krishnan P, Kruger NJ, Ratcliffe RG (2005) *J. Exp. Bot.* **56**, 255–265.

37. Fenn JB, Mann M, Meng CK, Wong SF, Whitehouse CM (1990) *Mass Spectr. Rev.* **9**, R37.

38. Niessen WMA (1999) *Liquid Chromatography-Mass Spectrometry*, 2nd edn. Marcel Dekker, New York.

17
Plant Cell Culture – Present and Future

Jim M. Dunwell
School of Biological Sciences, University of Reading, Whiteknights, Reading, UK

17.1 Introduction

Plant cell culture has a long history of development and exploitation that has been reviewed many times [1]. There are also several monographs and textbooks that provide detailed discussion on specific techniques and their application [2, 3]. By necessity therefore, the present review is limited in scope and, for the most part, will focus on recent publications. The various sections below each concentrate on a specific aspect of the technology; these are followed by a summary section relating to commercialization.

17.2 Micropropagation

Micropropagation *in vitro* is well established as a method to propagate [4], preserve and transport germplasm of many species including horticultural [5–7], medicinal [8, 9] and woody [10, 11] plants. It also reduces the risk of moving pathogens and insects with the germplasm owing to the inherent pathogen detection capabilities of aseptic cultures. Since this technology is usually limited to the multiplication of pre-existing meristems and does not involve the regeneration of plants from single cells or tissue, it will not be discussed in detail here.

Plant Cell Culture Edited by Michael R. Davey and Paul Anthony

17.3 Embryogenesis

17.3.1 Background

The process of embryogenesis *in vitro*, from either somatic or gametophytic cells, has been reviewed recently [12], and the section below should be read in combination with this review. In most cases, the production of an embryo *in vitro* occurs by induced division in a specific cell, either a constituent of a multicellular explant or in a specialized, often isolated, cell such as a guard cell [13], trichome [14] or a microspore [15].

17.3.2 Commercial exploitation of somatic embryos

Probably the most commercially valuable application of somatic embryogenesis is the propagation of conifers such as loblolly pine, where the method is well established on a large scale by several companies in North America and elsewhere [16, 17]. Multiplication techniques developed recently for other crop species include those for sugar cane [18], chicory [19], peach palm [20] and potato [21]. Other more specialist applications include the use of bioreactors for the growth of somatic embryos for multiplication [22] or for the production of high value compounds. Probably the most impressive example of this latter process is the use of transgenic somatic embryos of Siberian ginseng for production of the B subunit of *Escherichia coli* heat-labile toxin (LTB), a potent mucosal immunogen and immunoadjuvant for coadministered antigens [23].

17.3.3 Molecular aspects of somatic embryogenesis

For obvious reasons there have been many efforts made to combine molecular information from studies of zygotic and somatic embryogenesis. For example, a simple and efficient system has been developed to induce *Arabidopsis* somatic embryos at high frequency via ovule culture [24]. This method provides a useful system to create sufficient numbers of somatic embryos for use in biochemical, molecular and genetic studies. Amongst the family of genes studied in most detail in *Arabidopsis* is the *leafy cotyledon* (*lec*) genes that encode B3 domain proteins. Results from these studies provide evidence that, besides their key role in controlling many different aspects of zygotic embryogenesis, *lec* genes are also essential for the induction of somatic embryos *in vitro* [25–28]. In addition, ectopic expression studies have shown that expression of the BABY BOOM ERF/AP2 transcription factor is sufficient to induce spontaneous somatic embryo formation in *Arabidopsis, Brassica napus* and tobacco [29, 30].

In a similar context, the plant hormone auxin has been long recognized for its effects on post-embryonic plant growth. Recent genetic and biochemical studies have revealed that much of this regulation involves the Skp1/Cullin/F-box protein (SCF) (transport inhibitor response 1/auxin signalling F-box; TIR1/AFB)-mediated proteolysis of the Aux/IAA family of transcriptional regulators. With the finding that the TIR1/AFB proteins also function as auxin receptors, a potentially complete, and surprisingly simple, signalling pathway is suggested [31]. Related molecular studies suggest that localized surges in auxin within the embryo occur through a sophisticated transcellular transport pathway and cause the proteolysis of key transcriptional repressors [32]. As a result, downstream gene activation establishes much of the basic body plan of the embryo. The establishment of polarity at early stages of plant embryogenesis also depends on the role of programmed cell death (PCD). The emerging knowledge of PCD, and the role of metacaspases during plant embryogenesis, has been reviewed recently [33, 34].

Although the greatest emphasis in molecular studies of embryogenesis has been on *Arabidopsis*, some relevant information is available from crop species, including *Medicago* [35] and wheat [36]. In another example, expression patterns of about 12 000 genes were profiled during somatic embryogenesis in a regeneration-proficient maize hybrid line, in an effort to identify genes that might be used as developmental markers or targets to optimize the regeneration of maize plants from tissue culture [37].

17.3.4 Microspore derived embryos

The closely related topic of regeneration from gametophytic rather than somatic cells is discussed in Section 17.4.2 below.

17.4 Haploid methodology

17.4.1 Haploids and their exploitation

The ploidy level of a somatic cell is defined as the number of sets of the haploid number of chromosomes that the cell contains. Haploid organisms contain the same number of chromosomes (n) in their somatic cells as do the normal gametes of the species. The term haploid sporophyte is generally used to designate sporophytes having the gametic chromosome number. Apart from their intrinsic value because of their overall reduction in size compared with diploids, haploids also have value in allowing the isolation of mutants, which may be masked in a diploid. They also have value in transformation programmes [38, 39]; if haploids are transformed directly, then true breeding diploid transgenics can be produced in one step following doubling of chromosomes. However, the most commercially important use of haploids is based on the fact that significant improvements in the economics of plant breeding can be achieved via doubled haploid production, since selection and other procedural efficiencies can be markedly increased by using true-breeding (homozygous) progenies. With doubled haploid production systems, homozygosity is achieved in one generation and consequently, an efficient doubled haploid technology enables breeders to reduce the time and the cost of cultivar development relative to conventional breeding practices [40–43].

As well as having value in their own right as potential new varieties in inbreeding crops, homozygous plants are required in order to generate F_1 hybrid plants, involving crosses between selected homozygous males and females. These F_1 plants often exhibit so-called hybrid vigour (heterosis), a characteristic and often dramatic increase in yield compared with either parent.

17.4.2 Induction of haploid plants

In vitro methods for haploid production [44–46] can be classified into several categories.

Anther and microspore culture

During the 1960s, a major breakthrough in the production of haploids was achieved by the discovery that immature pollen grains (microspores), either in the form of isolated cells or still confined within the anther wall, could develop into haploid embryos if cultured under specific conditions *in vitro* [47]. This discovery, made in the non-crop plant *Datura*, stimulated much research activity in the succeeding years [48–52] and the process has since been extended not only to many other species, principally to members of the Solanaceae, Brassicaceae and Poaceae [40, 53], but also to species of herbaceous and woody crops [54].

Ovule culture

This technique is the female equivalent of the process described in the paragraph above, and has been applied to species including sugar beet [55, 56], onion [57, 58], squash [59], gerbera [60], rice [61], maize [62], niger [63] and tea [64]. Ovules have also been used as a transformation target [39].

Wide hybridization

Haploids can also be induced by a process of selective chromosome elimination that follows certain interspecific pollinations. This phenomenon was discovered first in barley [65] with crosses between *Hordeum vulgare* and *H. bulbosum*, and is now used routinely in wheat [66] and other cereal breeding programmes; haploids are induced in these species following pollination with maize pollen [67]. The process involves a phase of embryo rescue *in vitro*, usually followed by chromosome doubling with colchicine.

17.4.3 Molecular aspects of haploid induction from microspores

Although haploids were first isolated from cultured microspores more than 40 years ago [47], and there were many investigations in the following decade [48–52], it is only comparatively recently that progress has been made in understanding the molecular basis of this switch from a gametophytic to sporophytic pathway [68]. A variety of transcriptomic and metabolomic methods have now been applied

[15, 69–72]. These investigations have been complemented by a range of other molecular investigations [68, 73, 74]. Such approaches include the use of gene expression profiling in *Brassica napus*, a technique that has revealed the expression of several embryogenesis-related genes like the BABY BOOM ERF/AP2 transcription factor [29], LEC1 and LEC2 [25–28] as early as 48–72 h after the initiation of microspore culture [70]. Other related studies on *Nicotiana tabacum* have identified the important role of the *ntsml0* gene in the induction of embryos from microspores [69].

17.4.4 *Ab initio* zygotic-like embryogenesis from microspores

Probably the most interesting recent advance is the work on the direct induction of zygotic-like embryogenesis in microspores of *B. napus* [15]. Although regeneration from microspores of this species has been known for many years, recent modifications to culture conditions have provided a process most analogous to zygotic embryogenesis. Using the cultivar 'Lisandra', isolated microspores at the late unicellular stage are subject to a mild heat stress ($32\,^{\circ}C \pm 0.2\,^{\circ}C$) for 8 h. This treatment induces transverse divisions in each microspore to form a filamentous structure, of which the distal tip cell forms the embryo proper. The early division pattern of these embryos mimics exactly that observed during zygotic embryo production, with the lower end of the filamentous structure resembling the zygotic suspensor. These findings represent a major breakthrough and will facilitate the study of plant embryogenesis in an isolated system. Other recent improvements in methodology [68, 74] include the development of a simple and efficient isolated microspore culture system for producing doubled haploid wheat plants in a wide range of genotypes, in which embryogenic microspores and embryos are formed without any apparent stress treatment [75]. The regeneration frequency and percentage of green plants using this protocol are significantly greater than is found with the culture of shed microspores. However, despite this continuous range of improvements, there is still no method that can be universally recommended with a new species of interest. Much progress still depends on long and tedious comparisons of media and environmental conditions.

17.5 Somaclonal variation

The production and identification of valuable genetic variants among the regenerants from tissue culture was first proposed many years ago, but there are few successful examples of such so-called somaclonal variation. However, the molecular and genetic basis for this potentially beneficial as well possibly disadvantageous variation is now being investigated in many crops including barley [76], Bermudagrass [77], chrysanthemum [78], pear [79], oil palm [80], rice [81] and tea [82].

17.6 Transgenic methods

17.6.1 Background

The ability to regenerate whole plants from single cells *in vitro* was the starting point for the development of transgenic methods and the growth of the agricultural biotechnology industry during the 1980s. Recent advances in the production and exploitation of genetically modified (GM) crops have been reviewed extensively [83, 84]. In 2006, the global biotech crop area reached 102 million hectares, an increase from 90 million hectares planted in the previous year. The global biotech crop area has now increased more than 60-fold in the first 11 years of commercialization, making biotech crops the fastest adopted crop technology in recent history.

Information on the status of field trial applications for transgenic crops provides a means of estimating the time course of future commercial priorities and longer term trends. This information is available on-line for each of the main countries where such tests are undertaken. For the USA, access to the USDA APHIS data is most easily achieved through the Information Systems for Biotechnology (ISB) web site (http://www.nbiap.vt.edu/cfdocs/fieldtests1.cfm). Data for the EU are available from http://mbg.jrc.ec.europa.eu/deliberate/gmo.asp for trials conducted under Directive 90/220/EEC and http://gmoinfo.jrc.ec.europa.eu/for those conducted more recently under directive 2001/18/EC. Web sites are correct as of August 2009.

17.6.2 Regeneration and transformation techniques

The most commonly used transformation technologies [85, 86] are those involving either particle bombardment or *Agrobacterium* [87, 88] and these protocols will not be considered in detail. A noticeable recent trend is the development of efficient *Agrobacterium*-mediated methods for cereals and other crops previously considered recalcitrant [89]. Most of the recent studies have focused on wheat [90–94], with less emphasis on barley [95–97] and the model cereal *Brachypodium* [98–100]. In addition to improvements in the cell culture technologies for these species, there have been significant advances in the design of more efficient transformation constructs [101–103] for both biolistics [104, 105] and *Agrobacterium*-based methods [106, 107]. Similarly, recent improvements have been reported for banana [108], clementine [109], legumes [110], opium poppy [111] and strawberry [112]. There have also been some claimed improvements in the development of novel methods for direct gene transfer [113–115].

17.6.3 Chloroplast transformation

Chloroplast transformation [116, 117] has also become a method attracting both academic and commercial interest in recent years, partly because of the ability of this organelle to accumulate introduced proteins at very high yield but also because of the theoretical ecological advantages of reduced transfer of the transgene via pollen dispersal. Similar efforts are being made to develop methods of modifying mitochondrial-encoded traits in plants.

17.6.4 Biopharming

The so-called 'first generation' of commercial GM products comprised varieties either resistant to non-selective herbicides or to insect predation. These traits are now being combined, but in the near future a much greater variety of different transgenic lines are being developed [84]. These include ones with improved nitrogen use efficiency and tolerance to abiotic stresses. Also, there is likely to be a range of crops, including barley [118], wheat [119] and rice [120], that express higher value products such as bioactive proteins or other molecules. Some of these compounds can also be generated by plant cells in culture [121] and these systems are considered below (Section 17.8).

17.7 Protoplasts and somatic hybridization

The use of protoplasts is a well established plant cell culture technology, and has been exploited in numerous crops. First, it is useful for the testing of transgene constructs by transient expression [122, 123], but also as a means of producing somatic hybrids that might not have been possible via conventional sexual crossing. Specific recent examples of the latter process include the production of novel potato [124] and citrus lines [125].

17.8 Bioreactors

17.8.1 Production of plant products

Non-differentiated cells, tissues and whole organs have all been used for the production of valuable products *in vitro* [121, 126, 127], and selected examples will be described below. Some of these examples involve the use of isolated cells, but many utilise *Agrobacterium rhizogenes*-mediated hairy-root cultures [128]. The advantages of such cultures include a characteristic capacity for secondary metabolite production, an inherent genetic stability reflected in stable productivity and the possibility of genetic manipulation to increase biosynthetic capacity. One of the most important limitations for the commercial exploitation of hairy roots and other cells is the development of technologies for large-scale culture [129, 130].

17.8.2 Production of pharmaceuticals

The most well known and commercially significant example is the production of the polyoxygenated diterpene paclitaxel from cells of the Pacific yew, *Taxus brevifolia* [131–133]. DFB Pharmaceuticals (formerly the owner of Phyton Biotech) owns and operates the world's largest cGMP plant cell fermentation facility with bioreactors up to 75 000 l in size, in Ahrensburg, Germany. This facility produces paclitaxel for Bristol-Myers Squibb's Taxol oncology product. The production of alkaloids, terpenoids and indigo has been reviewed recently [134–136]. Protein products produced by cell culture include those of potential value in vaccine development [137].

17.8.3 Production of food ingredients

This subject has been included in a lengthy review of flavour biotechnology [126]. Recent specific examples include the production of betalains [138] and chichoric acid [139].

17.8.4 Production of cosmetics

One recent example of this is the successful establishment of an apple suspension culture producing a high yield of biomass, cultured in disposable, middle-scale bioreactors [140]. To obtain a suitable cosmetic product the authors used a high pressure homogenization technique to decompose the plant cells and release all the beneficial constituents while encapsulating these components at the same time in liquid nanoparticles.

17.8.5 Analytical methodology

One of the recent technical advances in this area has been the development of improved analytical methods for the qualitative and quantitative assessment of specific molecules produced by plant cells. Such methods have been extremely valuable in a range of proteomic and metabolomic investigations [82, 141, 142].

17.9 Cryopreservation

One of the consistent themes in the field of plant tissue culture is the value of cryopreservation [143] for the long-term maintenance of isolated cell, tissues or organs, particularly of vegetatively propagated crops such as apple or pear. The techniques required for this process have been improved gradually [144] so that protocols now exist for the preservation of many crop species [145, 146]. The balance is gradually shifting from the conservation of such crops in glasshouses, plantations or orchards to the use of flasks of liquid nitrogen. This trend is likely to continue.

17.10 Intellectual property and commercialization

17.10.1 Background

Much of the investment in novel tissue culture techniques over the last few decades have come from commercial companies aiming to exploit these methods. An integral part of this procedure is the need for such companies to protect their intellectual property by the award of patents or the use of other means. This subject has been discussed in a number of reviews focused on embryogenesis [12], transgenic methods [147, 148] and haploid induction technologies [149]. This information will not be repeated here. The consequence of this approach has been the consolidation of many companies in the agricultural biotechnology sector [150–152].

17.10.2 Sources of patent and other relevant information

Useful information on novel tissue culture methods is freely available from patent databases in the US (http://www.uspto.gov/patft/index.html), Europe (http://ep.espacenet.com/), World International Patent Organization (http://pctgazette.wipo.int/) and other international sites (eg http://www.surfip.gov.sg/; http://www.google.com/patents; http://www.freepatentsonline.com/; http://www.pat2pdf.org/) and the Patent Lens section of BiOS, Biological Innovation for Open Society, an initiative of CAMBIA (Center for the Application of Molecular Biology to International Agriculture) (http://www.bios.net/daisy/bios/patentlens.html). A very useful site with a summary of granted US ag-biotech patents from 1976–2000 is provided by the Economic Research Service (ERS) of the US Department of Agriculture (USDA) (http://www.ers.usda.gov/Data/AgBiotechIP/). It should be noted that the most detailed forms of patent analysis require commercial subscription from companies such as Derwent (http://www.derwent.com) or MicroPatent (http://www.micropat.com/static/index.htm). Much of the information is published in these patent sites prior to its appearance in the conventional research literature, and they should therefore be consulted on a regular basis in order to avoid wasteful repetition of research already conducted. All web sites are correct as of August 2009.

17.11 Conclusion

This brief review has described only a small proportion of the available information concerning the exploitation of plant cell culture techniques. The reader is referred to the literature for more detail. However, it is obvious that the diversity of technologies is still expanding and existing techniques are being improved with further research. Several of these techniques have been exploited commercially. These include micropropagation, somatic and gametophytic embryogenesis, somatic hybridization and transgenic technologies. It is hoped that the efficiency of these methodologies will all be improved by replacing mere empirical approaches with more focused methods based on improved analytical and molecular techniques.

References

**1. Loyola-Vargas VM, Vázquez-Flota F (2005) (Eds) *Plant Cell Culture Protocols*, 2nd edn. *Methods in Molecular Biology*. Vol. 318. Humana Press, Totowa, NJ, USA.

Good selection of protocols.

2. Gupta SD, Ibaraki Y (2006) (Eds) *Plant Tissue Culture Engineering. Focus on Biotechnology*, Vol. 6. Springer-Verlag, Berlin, Heidelberg.

***3. Xu Z, Li J, Xue Y, Yang W (2007) *Biotechnology and Sustainable Agriculture 2006 and Beyond*. Springer, Dordrecht, Netherlands.

The most comprehensive recent review.

4. George EF, Hall, MA, De Klerk G-J (2008) (eds) *Plant Propagation by Tissue Culture*, Vol. 1. Springer-Verlag, Berlin, Heidelberg.

5. Liu Z, Gao S (2007) *In Vitro* Cell. Dev. Biol. Plant **43**, 404–408.

6. Rout GR, Mohapatra A, Jain SM (2007) *Biotechnol. Adv.* **24**, 531–560.

7. Arditti J (2008) *Micropropagation of Orchids*. Blackwell Publishing, Oxford, UK.

8. Kayser O, Quax WJ (2007) (eds) *Medicinal Plant Biotechnology: From Basic Research to Industrial Applications*. Wiley-VCH, USA.

9. Chaturvedi HC, Jain M, Kidwai NR (2007) *Indian J. Exp. Biol.* **45**, 937–948.

10. Thakur RC, Karnosky DF (2007) *Plant Cell Rep.* **26**, 1171–1177.

11. Jain SM, Häggman H (2007) (Eds) *Protocols for Micropropagation of Woody Trees and Fruits*. Springer-Verlag, Berlin, Heidelberg.

12. Dunwell JM (2007) In: *Biotechnology and Sustainable Agriculture 2006 and Beyond*, Edited by Z Xu, J Li, Y Xue and W Yang. Springer, Dordrecht, The Netherlands, pp. 35–46.

13. Nobre J, Keith DJ, Dunwell JM (2001) *Plant Cell Rep.* **20**, 8–15.

14. Kim TD, Lee BS, Kim TS, Choi YE (2007) *Ann. Bot.* **100**, 177–183.

15. Joosen R, Cordewener J, Supena ED, *et al.* (2007) *Plant Physiol.* **144**, 155–72.

16. Nehra N, Becwar MR, Rottmann WH, *et al.* (2005) *In Vitro* Cell Dev. Biol.-Plant **41, 701–717.

Good review of the subject.

17. De Silva V, Bostwick D, Burns KL, *et al.* (2008) *Plant Cell Rep.* **27**, 633–646.

18. Lakshmanan P, Geijskes RJ, Wang L, *et al.* (2006) *Plant Cell Rep.* **25**, 1007–1015.

19. Legrand S, Hendriks T, Hilbert JL, Quillet MC (2007) *BMC Plant Biol.* **7**, 27.

20. Steinmacher DA, Krohn NG, Dantas AC, Stefenon VM, Clement CR, Guerra MP (2007) *Ann. Bot.* **100**, 699–709.

21. Sharma SK, Bryan GJ, Millam S (2007) *Plant Cell Rep.* **26**, 945–950.

22. Koskey RG, Barranco LA, Perez BC, Daniels D, Vega MR, Silva MD (2006) *Euphytica* **150**, 63–68.

23. Kang TJ, Lee WS, Choi EG, Kim JW, Kim BG, Yang MS (2006) *J. Biotechnol.* **121**, 124–133.

24. Wei J, Li XR, Sun MX (2006) *Plant Cell Rep.* **25**, 1275–1280.

25. Gaj MD, Zhang S, Harada JJ, Lemaux PG (2005) *Planta* **222**, 977–988.

26. Braybrook SA, Stone SL, Park S, *et al.* (2006) *Proc. Natl Acad. Sci. USA* **103**, 3468–3473.

27. Fambrini M, Durante C, Cionini G, *et al.* (2006) *Dev. Genes Evol.* **216**, 253–264.

28. Alemanno L, Devic M, Niemenak N, *et al.* (2008) *Planta* **227**, 853–866.

29. Boutilier K, Offringa R, Sharma VK, *et al.* (2002) *Plant Cell* **14**, 1737–1749.

30. Srinivasan C, Liu Z, Heidmann I, *et al.* (2007) *Planta* **225**, 341–351.

31. Quint M, Gray WM (2006) *Curr. Opin. Plant Biol.* **9**, 448–453.

32. Jenik PD, Barton MK (2005) *Development* **132**, 3577–3585.

33. Bozhkov PV, Filonova LH, Suarez MF (2005) *Curr. Top. Dev. Biol.* **67**, 135–179.

34. Bozhkov PV, Suarez MF, Filonova LH, *et al.* (2005) *Proc. Natl Acad. Sci. USA* **102**, 14463–14468.

35. Nolan KE, Saeed NA, Rose RJ (2006) *Plant Cell Rep.* **25**, 711–722.

36. Singla B, Tyagi AK, Khurana JP, Khurana P (2007) *Plant Mol. Biol.* **65**, 677–692.

37. Che P, Love TM, Frame BR, Wang K, Carriquiry AL, Howell SH (2006) *Plant Mol. Biol.* **62**, 1–14.

38. Kumlehn J, Serazetdinova L, Hensel G, Becker D, Lörz H (2006) *Plant Biotechnol J.* **4**, 251–261.

39. Holme IB, Brinch-Pedersen H, Lange M, Holm PB (2006) *Plant Cell Rep.* **25**, 1325–1335.

**40. Maluszynski M, Kasha KJ, Forster BP, Szarejko I (2003) (eds) *Doubled Haploid Production in Crop Plants: A Manual*. Kluwer Academic Publishers, Dordrecht, The Netherlands.

A comprehensive review.

41. Thomas WTB, Foster BP, Gertsson B (2003) In: *Doubled Haploid Production in Crop Plants: A Manual*. Edited by M Maluszynski, KJ Kasha, BP Forster and I Szarejko. Kluwer Academic Publishers, Dordrecht, The Netherlands, pp. 337–349.

42. Forster BP, Thomas WTB (2005) *Plant Breed. Rev.* **25**, 57–88.

43. Palmer CE, Keller WA (2005) In: *Haploids in Crop Improvement* II, Edited by CE Palmer, WA Keller and KJ Kasha. Springer-Verlag, Berlin, Heidelberg, Germany, Vol. 56. pp. 295–303.

44. Maluszynski M, Kasha KJ, Szarejko I (2003) In: *Doubled Haploid Production in Crop Plants: A Manual*. Edited by M Maluszynski, KJ Kasha, BP Forster and I Szarejko. Kluwer Academic Publishers, Dordrecht, The Netherlands, pp. 309–335.

45. Forster BP, Heberle-Bors E, Kasha KJ, Touraev A (2007) *Trends Plant Sci.* **12, 368–375.

Good recent review.

46. Xu L, Najeeb U, Tang GX, *et al.* (2007) *Adv. Bot. Res.* **45**, 181–216.

47. Guha S, Maheshwari SC (1964) *In Vitro* production of embryos from anthers of *Datura*. *Nature* **204**, 497.

48. Sunderland N, Dunwell JM (1974) In: *Tissue Culture and Plant Science 1974*. Edited by HE Street. Academic Press, London, UK, pp. 141–167.

49. Dunwell JM (1976) *Environ. Exp. Bot.* **16**, 109–118.

50. Dunwell JM (1978) In: *Frontiers of Plant Tissue Culture 1978*. Edited by TA Thorpe. Internatl. Assoc. Plant Tissue Cult. Calgary, Canada, pp. 103–112.

51. Dunwell JM (1985) In: *Biotechnology in Plant Science*. Edited by P Zaitlin, P Day and A Hollaender. Academic Press, Orlando, FL, USA, pp. 49–76.

52. Dunwell JM (1992) In: *Reproductive Biology and Plant Breeding*. Edited by Y Dattée, C Dumas and A Gallais. Springer-Verlag, Berlin, Germany, pp. 121–130.

53. Seguí-Simarro JM, Nuez F (2007) *J. Exp. Bot.* **58**, 1119–1132.

54. Pintos B, Manzanera JA, Bueno MA (2007) *J. Plant Physiol.* **164**, 1595–1604.

55. Van Geyt J, Speckmann GJ, Halluin KD, Jacobs M (1987) *Theor. Appl. Genet.* **73**, 920–925.

56. Doctrinal M, Sangwan RS, Sangwan-Norreel BS (1989) *Plant Cell Tissue Organ Cult.* **17**, 1–12.

57. Muren RC (1989). *Hortsci.* **24**, 833–834.

58. Geoffriau E, Kahane R, Martin-Tanguy J (2006). *Physiol. Plant.* **127**, 119–129.

59. Metwally EI, Moustafa SA, El-Sawy BI, Haroun SA, Shalaby TA (1998). *Plant Cell Tissue Organ Cult.* **52**, 117–121.

60. Tosca A, Arcara L, Frangi P (1999) *Plant Cell Tissue Organ Cult.* **59**, 77–80.

61. Rongbai L, Pandey MP, Pandey SK, Dwivedi DK (1999) *Euphytica* **106**, 197–203.

62. Tang F, Tao Y, Zhao T, Wang G (2006) *Plant Cell Tissue Organ Cult.* **84**, 100210–100214

63. Bhat JG, Murthy HN (2007) *Plant Growth Regul.* **52**, 241–248.

64. Gugsa L, Sarial AK, Lörz H, Kumlehn J (2007) *Plant Cell Rep.* **25**, 1287–1293.

65. Kasha KJ, Kao KN (1970) *Nature* **225**, 874–876.

66. Mochida K, Tsujimoto H, Sasakuma T (2004) *Genome* **47**, 199–205.

67. Sidhu PK, Howes NK, Aung T, Zwer PK, Davies PA (2006) *Plant Breed.* **125**, 243–247.

68. Segui-Simarro JM, Barany I, Suarez R, Fadon B, Testillano PS, Risueno MC (2006) *Eur. J. Histochem.* **50**, 35–44.

69. Hosp J, Tashpulatov A, Roessner U, *et al.* (2007) *Plant Mol. Biol.* **63**, 137–149.

70. Malik MR, Wang F, Dirpaul JM, *et al.* (2007) *Plant Physiol.* **144**, 134–54.

71. Tsuwamoto R, Fukuoka H, Takahata Y (2007) *Planta* **225**, 641–652.

72. Tsuwamoto R, Fukouka H, Takahata Y (2008) *Plant J.* **54**, 30–42.

73. Tang XC, He YQ, Wang Y, Sun MX (2006) *J. Exp. Bot.* **57**, 2639–2650.

74. Ribarits A, Mamun AN, Li S, *et al.* (2007) *Plant Biotechnol J.* **5**, 483–494.

75. Shariatpanahi ME, Belogradova K, Hessamvaziri L, Heberle-Bors E, Touraev A (2007) *Plant Cell Rep.* **25**, 1294–1299.

76. Bednarek PT, Orłowska R, Koebner RM, Zimny J (2007) *BMC Plant Biol.* **7**, 10.

77. Lu S, Peng X, Guo Z, *et al.* (2007) *Plant Cell Rep.* **26**, 1413–1420.

78. Zalewska M, Lema-Ruminska J, Miler N (2007) *Scientia Hort.* **113**, 70–73.

79. Palombi MA, Lombardo B, Caboni E (2007) *Plant Cell Rep.* **26**, 489–496.

80. Morcillo F, Gagneur C, Adam H, *et al.* (2006) *Tree Physiol.* **26**, 585–594.

81. Ngezahayo F, Dong Y, Liu B (2007) *J. Appl. Genet.* **48**, 329–336.

82. Thomas J, Raj Kumar R, Mandal AK (2006) *Phytochemistry* **67**, 1136–1142.

83. Dunwell JM (2002) *Phytochem. Rev.* **1**, 1–12.

84. Dunwell JM (2004) In: *Methods in Molecular Biology*, Vol. 286. *Transgenic Plants: Methods and Protocols*. Edited by L Peña. Humana Press, Totowa, NJ, USA, pp. 377–396.

85. Pua EC, Davey MR (2007) (Eds) *Transgenic Crops* IV. *Biotechnology in Agriculture and Forestry*. Vol. 59. Springer-Verlag, Berlin, Heidelberg, Germany.

86. Pua EC, Davey MR (2007) (Eds) *Transgenic Crops* V. *Biotechnology in Agriculture and Forestry*. Vol. **60**. Springer-Verlag, Berlin, Heidelberg, Germany.

87. Wang K (2006) (Ed) Agrobacterium *Protocols Vol. 1. Methods in Molecular Biology*. Vol. 343. Humana Press, Totowa, NJ, USA.

88. Wang K (2006) (Ed) Agrobacterium *Protocols Vol. 2. Methods in Molecular Biology*. Vol. 344. Humana Press, Totowa, NJ, USA.

89. Shrawat AK, Lörz H (2006) *Plant Biotechnol. J.* **4**, 575–603.

90. Ding L, Li S, Gao J, Wang Y, Yang G, He G (2009) *Mol. Biol. Rep.* **36**, 29–36.

91. Patnaik D, Vishnudasan D, Khurana P (2006) *Curr. Sci.* **91**, 307–317.

92. Supartana P, Shimizu T, Nogawa M, *et al.* (2006) *J. Biosci. Bioeng.* **102**, 162–170.

93. Wu HX, Sparks CA, Jones HD (2006) *Mol. Breeding* 18, 195–208.

94. Zhao, TJ, Zhao SY, Chen HM, *et al.* (2006) *Plant Cell Rep.* **25**, 1199–1204.

95. Hensel G, Valkov V, Middlefell-Williams J, Kumlehn J (2008) *J. Plant Physiol.* **165**, 71–82.

96. Lange M, Vincze E, Moller MG, Holm PB (2006) *Plant Cell Rep.* **25**, 815–820.

97. Shrawat AK, Becker D, Lörz H (2007) *Plant Sci.* **172**, 281–290.

98. Pacurar DI, Thordal-Christensen H, Nielsen KK, Lenk I (2008) *Transgenic Res.* **17**, 965–975.

99. Vain P, Worland B, Thole V, *et al.* (2008) *Plant Biotechnol J.* **6**, 236–245.

100. Vogel J, Hill T (2008) *Plant Cell Rep.* **27**, 471–478.

101. Lee LY, Kononov ME, Bassuner B, Frame BR, Wang K, Gelvin SB (2007) *Plant Physiol.* **145**, 1294–1300.

102. Qu S, Desai A, Wing R, Sundaresan V (2008) *Plant Physiol.* **146**, 189–199.

103. Verweire D, Verleyen K, De Buck S, Claeys M, Angenon G (2007) *Plant Physiol.* **145**, 1220–1231.

104. Yao Q, Cong L, Chang JL, Li KX, Yang GX, He GY (2006) *J. Exp. Bot.* **57**, 3737–3746.

105. Yao Q, Cong L, He G, Chang J, Li K, Yang G (2007) *Mol. Biol. Rep.* **34**, 61–67.

106. Himmelbach A, Zierold U, Hensel G, *et al.* (2007) *Plant Physiol.* **145**, 1192–1200.

107. Karimi M, Bleys A, Vanderhaeghen R, Hilson P (2007) *Plant Physiol.* **145**, 1183–1191.

108. Khanna HK, Paul JY, Harding RM, Dickman MB, Dale JL (2007) *Mol. Plant Microbe Interact.* **20**, 1048–1054.

109. Cervera M, Navarro A, Navarro L, Peña L (2008) *Tree Physiol.* **28**, 55–66.

110. Eapen S (2008) *Biotechnol Adv.* **26**, 162–168.

111. Facchini PJ, Loukanina N, Blanche V (2008) *Plant Cell Rep.* **27**, 719–727.

112. Hanhineva KJ, Kärenlampi SO (2007) *BMC Biotechnol.* **7**, 11.

113. Badr YA, Kereim MA, Yehia MA, Fouad OO, Bahieldin A (2005) *Photochem. Photobiol. Sci.* **4**, 803–807.

114. Chen CP, Chou JC, Liu BR, Chang M, Lee HJ (2007) *FEBS Lett.* **581**, 1891–1897.

115. Schinkel H, Jacobs P, Schillberg S, Wehner M (2008) *Biotechnol. Bioeng.* **99**, 244–248.

116. Karcher D, Kahlau S, Bock R (2008) *RNA* **14**, 217–224.

117. Lutz KA, Azhagiri AK, Tungsuchat-Huang T, Maliga P (2007) *Plant Physiol.* **145**, 1201–1120.

118. Joensuu JJ, Kotiaho M, Teeri TH, *et al.* (2006) *Transgenic Res.* **15**, 359–373.

119. Brereton HM, Chamberlain D, Yang R, *et al.* (2007) *J. Biotechnol.* **129**, 539–46.

120. Sardana R, Dudani AK, Tackaberry E, *et al.* (2007) *Transgenic Res.* **16**, 713–21.

121. Boehm R (2007) *Ann. NY. Acad. Sci.* **1102, 121–134.

Recent review.

122. Mazarei M, Al-Ahmad H, Rudis MR, Stewart CN Jr (2008) *Biotechnol J.* **3**, 354–359.

123. Miao Y, Jiang L (2007) *Nature Protocols* **2**, 2348–2353.

124. Borgato L, Conicella C, Pisani F, Furini A (2007) *Planta* **226**, 961–969.

125. Xu XY, Hu ZY, Li JF, Liu JH, Deng XX (2007) *Plant Cell Rep.* **26**, 1263–1273.

**126. Chin CK (2008) In: *Biotechnology in Flavor Production*. Edited by D Havkin-Frenkel and F Belanger. Wiley-Blackwell, Ames, IO, USA, pp. 104–117.

Recent review.

127. McCoy E, O'Connor SE (2008) *Prog. Drug Res.* **65**, 329, 331–370.

128. Mishra BN, Ranjan R (2008) *Biotechnol. Appl. Biochem.* **49**, 1–10.

129. Eibl R, Eibl D (2008) *Phytochem. Rev.* **7**, 593–598.

130. Perullini M, Rivero MM, Jobbágy M, Mentaberry A, Bilmes SA (2007) *J. Biotechnol.* **127**, 542–548.

131. Gao H, Gong YW, Yuan YJ (2007) *Biotechnol. Prog.* **23**, 673–679.

132. Vongpaseuth K, Nims E, St Amand M, Walker EL, Roberts SC (2007) *Biotechnol. Prog.* **23**, 1180–1185.

133. Vongpaseuth K, Roberts SC (2007) *Curr. Pharm. Biotechnol.* **8**, 219–236.

134. Roberts SC (2007) *Nat. Chem. Biol.* **3**, 387–395.

135. Sato F, Inui T, Takemura T (2007) *Curr. Pharm. Biotechnol.* **8**, 211–218.

136. Warzecha H, Frank A, Peer M, Gillam EM, Guengerich FP, Unger M (2007) *Plant Biotechnol. J.* **5**, 185–191.

137. Lienard D, Tran Dinh O, van Oort E, *et al.* (2007) *Plant Biotechnol. J.* **5**, 93–108.

138. Pavlov A, Georgiev M, Bley T (2007) *Z. Naturforsch.* **62**, 439–446.

139. Wu CH, Murthy HN, Hahn EJ, Paek KY (2007) *Biotechnol. Lett.* **29**, 1179–1182.

140. Schürch C, Blum P, Zülli F (2008) *Phytochem. Rev.* **7**, 599–605.

141. Engelsberger WR, Erban A, Kopka J, Schulze WX (2006) *Plant Methods* **2**, 14.

142. Williams BJ, Cameron CJ, Workman R, Broeckling CD, Sumner LW, Smith JT (2007) *Electrophoresis* **28**, 1371–1379.

143. Grout BW (2007) *Methods Mol. Biol.* **368**, 153–61.

144. Sakai A, Engelmann F (2007) *CryoLetters* **28**, 151–172.

145. Fang JY, Wetten A, Johnston J (2008) *Plant Cell Rep.* **27**, 453–461.

146. Mandal BB, Ahuja-Ghosh S (2007) *CryoLetters* **28**, 329–336.

147. Dunwell JM (2005) *Plant Biotech. J.* **3**, 371–384.

148. Pray CE, Naseem A (2005) *AgBioForum* **8**, 108–117.

149. Dunwell JM (2008) In: *Advances in Haploid Production in Higher Plants*. Edited by A Touraev, BP Forster and SM Jain. Springer-Verlag, Berlin, Heidelberg, Germany, pp. 97–125.

150. Chan HP (2006) *AgBioForum* **9**, 59–68.

151. Kukier KN (2006) *Nat. Biotech.* **24**, 249–251.

152. Dunwell JM (2009) *Acta Hort*. (in press).

Index

Numbers in *italics* refer to Figures and Protocols

Plant Cell Culture Edited by Michael R. Davey and Paul Anthony

Printed and bound by CPI Group (UK) Ltd, Croydon, CR0 4YY

16/04/2025

14658558-0002